PROBLEMS OF RELATIVE GROWTH

HETEROCHELY AND REGENERATION IN THE COMMON LOBSTER

1. MARINE LOBSTER, *Homarus* (*Gammarus*) *vulgaris*, Normal Left-handed Specimen.
2. The Same Regenerating the Left 'Crusher' Claw after Autotomy.
3. Shed Skin of the Same at the Next Moult.
4. The Same after this Moult, showing Direct Regeneration of Left Crusher.

Note the slenderer propus and absolutely longer dactylus in the smaller ('nipper') claw.

PROBLEMS OF RELATIVE GROWTH

BY

SIR JULIAN S. HUXLEY

WITH A NEW INTRODUCTION BY THE AUTHOR
AND A SUPPLEMENTARY ESSAY BY
E. C. R. REEVE AND JULIAN HUXLEY

WITH 105 ILLUSTRATIONS

Second Edition

DOVER PUBLICATIONS, INC.
NEW YORK

This Dover edition, first published in 1972, is an unabridged republication of the work originally published by Methuen & Co., London, in 1932. A new Introduction has been written by the author specially for the present edition.

This edition also contains as a new Appendix an essay by E. C. R. Reeve and Julian S. Huxley entitled "Some Problems in the Study of Allometric Growth," reprinted with permission from *Essays on "Growth and Form" Presented to D'Arcy Wentworth Thompson,* edited by W. E. Le Gros Clark and P. B. Medawar (Oxford: Clarendon Press, 1945).

International Standard Book Number: 0-486-61114-0
Library of Congress Catalog Card Number: 75-175422

Manufactured in the United States of America
Dover Publications, Inc.
180 Varick Street
New York, N. Y. 10014

TO

D'ARCY WENTWORTH THOMPSON

'The morphologist, when comparing one organism with another, describes the differences between them point by point, and " character " by " character." If he is from time to time constrained to admit the existence of " correlation " between characters (as a hundred years ago Cuvier first showed the way), yet all the while he recognizes this fact of correlation somewhat vaguely, as a phenomenon due to causes which, except in rare instances, he cannot hope to trace; and he falls readily into the habit of thinking and talking of evolution as though it had proceeded on the lines of his own descriptions, point by point, and character by character. But if, on the other hand, diverse and dissimilar fishes can be referred as a whole to identical functions of very different co-ordinate systems, this fact will of itself constitute a proof that a comprehensive " law of growth " has pervaded the whole structure in its integrity, and that some more or less simple and recognizable system of forces has been at work.'—D'ARCY THOMPSON (*Growth and Form*, p. 727).

INTRODUCTION TO THE DOVER EDITION

THE concept of different rates of growth of parts of the body relative to that of the body as a whole, or *allometry* as it has been called since 1936 (J. S. Huxley and G. Teissier, *Nature,* Vol. 137, p. 780), has now been shown to have many more effects on animal (and plant) form than I mentioned in my book on relative growth, here reprinted.[1]

However, I was the first to generalize the simple formula for allometry: $y = b.x^a$, where y represents the size (linear or volumetric) of a particular organ or part of the body; b the relative size of the organ at its first appearance; x that of the total body, less the size of the organ at the same stage of development; and a the percentage rate of growth of the organ y.

Clearly when $a = 1$, the relative size of the organ remains constant during life, as with the small feeding claw of both male and female fiddler-crabs; when it is less than 1, the organ becomes *relatively* smaller during post-natal development, as with the legs of ungulates, which must be large at birth to enable the animal to walk almost as soon as born. The same is true of the human brain, which must be large at birth, in order to cope with the multifarious problems of life in a complex environment. Furthermore, man's brain became larger during his evolution, as his group life became more elaborate and necessitated greater intelligence and improved methods of communication.

In addition, the relative growth of parts of the brain and head have changed during evolution. In early species of man like the Neanderthalers, the frontal lobes, which deal with more

[1] See *Essays on Growth and Form* (Oxford, 1945); "Discussion on the Measurement of Growth and Form," with thirteen participants *(Proc. Roy. Soc.* [London], Series B, **137** (1950), 433); and discussions of the evolutionary aspects of allometry in general books like B. Reusch, *Evolution above the Species Level* (Methuen, London, 1959) and my own *Evolution: the Modern Synthesis* (Allen & Unwin, 1942; 2nd revised ed., 1961).

complex situations, were smaller than in modern *Homo sapiens,* while the receding chin gave less space for the tongue, thus limiting their capacity for elaborate speech.

Equally adaptive is the converse case, of small limbs at birth in creatures like mice, rabbits and marsupials, where the young are born helpless and small, and are protected in a burrow or pouch until sufficiently developed to leave their sheltered existence, after which their limbs grow at a high rate to keep up with body-size.

On the other hand, when a is greater than 1, the organ becomes relatively larger so long as growth continues—as in the large claws of male fiddler-crabs, which, like the small claws of both sexes, start at about 2% of the weight of the rest of the body, but may eventually reach 70% of this figure.

Since 1932, when my book was published, a number of new consequences of this differential growth of parts have been discovered, and many further facts of development have been shown to follow the simple allometric formula. Even before that date, the relative brain-weight of mammals was shown to decrease allometrically with increasing size, both in individuals and between species, and the same is true for relative heart-weight in birds (see references in the new appendix to this book, the essay by Reeves and myself entitled "Some Problems of Allometric Growth," quoted from *Essays on Growth and Form,* Oxford, 1945).

Various complications exist. Thus, many secondary sexual organs show a sharply increased relative growth-rate at sexual maturity. However, all organs with very high relative growth-rates grow more slowly in old age, with the result that they do not become disproportionately enormous. This happens with the large claws of male fiddler-crabs, which otherwise would become too big to wave in sexual display, and might indeed exceed the weight of the rest of the body.

When high relative growth of an organ occurs, and external conditions later become less favorable, the species may fail to survive. This seems to have taken place in the so-called Irish Elk, a giant type of fallow deer, whose antlers (like that of all deer) show positive allometry, but higher than that of other species. Thus they eventually become so large that they constituted a grave handicap during the ice-age, when there was less food available. In consequence, the species became extinct.

During geological time, the growth-rate and relative size of horns and antlers and other allometrically growing organs may vary in different lines of a group, doubtless in relation to the animal's final size and to its feeding habits. This seems to have happened in the large-horned reptiles known as Titanotheres, where Hersh (1934, *Amer. Nat. 58,* 537) showed that there were 'six allometric tribes,' each with its own set of growth-coefficients for limbs and horns.

The same sort of thing has happened in the domestic dog. Man's unconscious or deliberate selection has produced five or six types of growth-mechanism in different breeds, resulting in forms as different as pekinese and foxhound, bulldog and saluki, pug and greyhound, each with its own system of growth, involving total size, length of muzzle relative to cranium, of leg and tail-length relative to body-size, etc.

Another factor affecting final shape is the existence of what may be called *growth-gradients.* Thus, as I showed in my original studies, the paired limbs of many Crustacea show a gradually decreasing growth-rate from front to rear, interrupted by local regions of high growth-rate when an especially large claw or other appendage is present.

The adaptive significance of the general growth-gradient probably lies in the biological need to develop the brain and its associated sense-organs as early as possible, leaving the full formation of the trunk and posterior regions to the time when the whole organism becomes free-living.

There seems also to be a certain quantity of what we may call growth-potential available for appendages (limbs). For when one of them is enlarged, e.g. for sexual display as in male fiddler-crabs, growth (and final size) of the appendage immediately posterior to it is slightly accelerated, and those in front slightly reduced.

There may also be growth-gradients within a single limb. Thus, in the male fiddler-crab's large claw, its terminal segment has the highest growth-rate, while the growth of the remaining segments falls as they approach their point of origin, the first segment having a growth-rate equal to that of the rest of the body.

Furthermore, any organ, at a given stage of development, appears to have a specific growth-capacity. Thus in axolotls (and other amphibians), when an eye or limb is grafted onto an older and therefore larger animal, its growth-rate is accelerated until it becomes of the 'right' dimensions for the older host. Conversely, the transplant's growth-rate is decreased when grafted from an older to a younger or smaller host.

Similarly, in regeneration of a limb, the regenerated bud at first grows very rapidly; but its growth-rate gradually slows down until the limb is of the 'right' size for the animal at its present stage, after which its growth-rate becomes normal.

Furthermore, organ-growth in such cases is similar to that of antlers in deer, except that their weight shows high positive allometry relative to that of the body during life; they are shed every autumn, and when they re-grow next spring, they reach the size deducible from their allometric formula. Thus it is not the organ itself (the pair of antlers) which shows allometry, but its *growth-potential*.

Another fact relevant to the growth and final stage of various organs, and in some cases of the whole body, is the existence of what Ford and I called *rate-genes,* genetic units which determine the rate at which various processes occur, and which may begin acting at different stages of the developmental process. Thus in the little amphipod crustacean *Gammarus chevreuxi,* the eye is red in the youngest free-living stage, but then gradually becomes black through an increasing rate of melanin deposition. The actual *rate* of this process is also relevant. If rapid, the eye becomes completely black when still small, but appears even darker than when it has reached adult size, because the black ommatidia are crowded into a smaller surface.

The *stage* at which rate-genes begin acting is also relevant. Thus when an organ ceases to have biological value, and therefore tends to be reduced in size by natural selection (since smaller size means less waste of tissue-forming material), it will be advantageous for the process of reduction to start as early as possible, and to proceed rapidly to the limit. This has happened to the human tail, which starts small in the embyro, and disappears entirely before birth. In other cases, as in the hind limbs of whales, anatomical reduction may not be complete, but the organ remains functionless, a tiny vestige from a much earlier stage in cetacean evolution, using up a negligible amount of material.

Such changes in rate of development of an organ may lead to its recapitulating its evolutionary history during individual development, as with man's tail, the appearance of gill-slits in the embryos of terrestrial vertebrates, the size and form of horns and antlers, etc.

When the growth-rate of the whole organism is slowed down, the effect may be to do away with various characters of earlier evolutionary stages. In some cases, this produces *neoteny*—

the capacity of a larval form to reproduce itself sexually. This has happened with the Mexican salamander, the axolotl—but only in extremely arid regions, where a moist-skinned amphibian would be at a disadvantage in the hot, dry air. In moister areas, the gilled and finned larva metamorphoses into a typical land salamander, as do our common newts.

In this case (and in frogs) it is known that thyroid extract will cause metamorphosis from aquatic larva to terrestrial adult, so that presumably the rate of thyroid growth is decisive. This view is confirmed by the state of affairs in the blind, albino and permanently neotenous cave salamander, *Proteus*. Not only is its own thyroid a tiny vestige, but its tissues have lost the power to respond to the secretion of active thyroid glands by metamorphizing into a terrestrial form. Here again the climate outside the caves is very arid, and the environment extremely barren: natural selection has promoted permanent and irreversible neoteny.

A process akin to neoteny also occurs in man, though, as de Beer and others have pointed out, a better term would be foetalization. The process of development *in utero* has been so much slowed down that human beings when born resemble a late foetal ape, in having negligible hair, except on the cranium, and in the large size of the skull (and therefore brain) relative to that of the body and limbs. Development and ageing is also much slower after birth than in apes.

The adaptive significance of this last fact is clear. Though the new-born infant is physically helpless, it can be looked after by its mother, and protected by the family and social group, more effectively than in apes or monkeys. Furthermore, man's slower rate of development and longer life-span gives more time for the young human being to learn, to acquire more varied mental and physical skills, and also more time for the older men and women to pass on their learning and their skills to the rest of the social group, including the new generation.

I must briefly refer to a few other important factors affecting the form and body-build of higher animals. The first is temperature in relation to size, both of the whole body and of the organs promoting heat-loss. Total size in closely related forms of warm-blooded animals tends to be higher in colder regions, since heat is conserved by larger bulk. This applies even to some invertebrates: thus the hive-bee cannot survive in the low summer temperature of Spitsbergen, whereas bumble-bees can and do. In this and many mammalian cases, degree of hairiness is also concerned. The hairy body of the bumble-bee and the thick pelt of large northern mammals like polar bears, or small inhabitants of cold high altitudes like chinchillas, are also adaptations to prevent excessive heat-loss.

Conversely, exposed parts of the mammalian body like limbs, and especially ears, are much reduced in cold climates, as in the Arctic as against the common European fox; and are enlarged as organs for losing heat in tropical species, like the desert Fennec fox and the elephants, especially the African species, whose habitat is hotter than that of the smaller-eared Indian species. In elephants and other large tropical mammals, the cooling function of the large ears is increased by their constant flapping.

The endocrine glands may also affect both absolute size and proportions. Thus an enlarged pituitary gland promotes both increased general growth, and also the exaggerated development of face and limbs, in many terrestrial vertebrates. This condition is known as acromegaly. Acromegaly is also found in various extinct lines of very large reptiles in which an enlarged pituitary may be inferred from skull-shape and bodily proportions as well as from the over-large recess which held the pituitary gland in life.

Excess of pituitary secretion, with resultant large size and acromegaly (as shown by enlarged jowl and feet), has been produced unwittingly by human selection in some strains of domestic dog, e.g., the St. Bernard. Large size alone, however, need not be associated with hyperpituitary growth: thus Great Danes are very large but show no trace of acromegaly.

On the other hand, large size itself has various consequential effects on form—notably on relative size of parts of the body. Thus extremely tall human races like Dinka and Watusi have relatively (as well as absolutely) longer legs than pygmies.

Size and activity of the thyroid gland may also exert both individual and evolutionary effects on form. Thus the shorter face and jowls of bulldogs and pugs are associated with abnormal thyroid secretion, and the same is probably true of the numerous fish species with abbreviated faces. Short faces in fish seems to be an adaptive character, permitting close browsing, whereas in bulldogs it has been brought about by the whim of fanciers, and the whole metabolism of the breed has been thrown out of gear, as shown by the frequent abnormality of their development.

As already mentioned, level of thyroid activity influences final form in various amphibia; if too low, metamorphosis fails to occur, and the animal remains permanently in aquatic larval form, though it can still develop reproductive organs. If too high, it causes precocious metamorphosis, producing adult-type forms too small to cope with their environment, while in man it produces exophthalmia, high blood-pressure and often serious emaciation.

I must also mention the effect on human body-build of growth in different dimensions, as revealed by the work of Kretchmer and W. H. Sheldon. Though variability of body-build in man is extremely high, three types are particularly common—the slender, poorly muscled and often small hypomorph, the fat robust mesomorph, and the tall, well-muscled and athletic hypermorph.

Furthermore, there appears to be some correlation between body-build and psychology. The fat mesomorph tends to be contented and placid; the hypomorph to have an active inner life, sometimes of a manic-depressive type, alternating between over-enthusiasm and depression; and the hypermorph to be more concerned with physical than mental activity—doing rather than thinking.

These differences in body-build and temperament *may* be related to differences in endocrine balance, but we are not yet sure, and much research remains to be done on this interesting problem.

Finally I must just touch on the question of growth-limitation. In warm-blooded vertebrates, there exists an upper limit to absolute size, which, however, varies slightly with environmental temperature, subspecies living at lower temperature being of slightly greater bulk, so as to conserve internal heat better. The imagos of higher insects also have a genetically limited size, owing to the fact that they never moult after emerging from the pupal stage. The only variations found in total size in adult insects depend on the greater or lesser growth achieved by the larva, and in warm-blooded vertebrate species on differences in nutrition during their development, both pre- and post-natal.

In other organisms, notably certain fish, growth is potentially unlimited, and is cut short only by the wearing-out process of old age, or by extraneously caused death. On the other hand, the *rate* of growth in such forms gradually falls, so that there is a brake on the achievement of excessive size.

Unlimited growth is also the central feature of all types of cancer. The problem here, as yet not fully solved, is *how* a cell or group of cells escapes the growth-limitation of ordinary tissues and how and why some benign tumors become malignant—though we know a good deal about the various agencies which initiate or promote unlimited growth and the cancer-proneness of different tissues in different genera or families of animals and plants (see my book *Biological Aspects of Cancer,* Allen & Unwin, 1957).

It was D'Arcy Thompson's remarkable book on *Growth and Form* (Cambridge, 1917), together with my first sight, in 1913, of male fiddler-crabs, with their display claw steadily increasing in relation to body-size, which first put me on to the study of relative growth. Since then, myself and many other biologists have quantified the process of allometry, have measured growth-gradients (which were explicit in D'Arcy Thompson's studies), have shown the relation of animals' final form to the differential growth of their parts, have studied the effect of endocrine activity and environmental temperature on size and shape, have demonstrated that allometry is not a function of time but of size, and that it is of almost universal occurrence in animals, and in some aspects of plant growth, besides drawing attention to the fact of unlimited growth in some aquatic animals, and the bearings of these various facts on cancerous disease.

JULIAN HUXLEY

London
October, 1971

PREFACE

IN this book I have attempted to give some account of the chief results emerging from a study of the relative growth of parts in animals which I have undertaken during the last ten years. I have tried to correlate my own findings and conclusions with those of other workers in the same and related fields, but am well aware of the many gaps that remain. However, it has not been my main intention to produce an exhaustive survey of the subject, but rather to set forth certain new facts and ideas and some of their chief implications.

There are, I think, four chief points in the book which are more or less new. One is the quantitative formulation of heterogonic growth (Chapters I and II) ; a second is the discovery of the widespread existence of growth-gradients, and their quantitative analysis (Chapters III and IV) ; a third is the recognition that growth of logarithmic spiral type as seen in Molluscan shells, etc., operates with the same growth-mechanisms (growth-centres and growth-gradients) as does growth of ordinary type as seen in a Crustacean antenna or a sheep's leg (Chapter V) ; and the fourth is the application of these results to certain evolutionary problems, as set forth in the final chapter.

I owe a great deal to previous work in this field : first and foremost to D'Arcy Thompson's *Growth and Form*, but also to the books and papers of Champy, Teissier, Schmalhausen, and others too numerous to mention.

I have to thank Professor L. T. Hogben and Dr. R. A. Fisher, F.R.S., for reading the book in typescript and making various useful suggestions ; and I owe a great deal to Professor H. Levy for helping me with some of the mathematical problems involved. My thanks are also due to Dr. C. F. A. Pantin and Professor Selig Hecht, whose discussions with me of various problems raised in this book have helped greatly in clarifying my ideas. And especially I would like to thank my pupils and co-workers, Mr. E. B. Ford, Mr. F. N. Ratcliffe, Miss M. Shaw (Mrs. White), Miss M. A. Tazelaar, Mr.

S. F. Bush, Professor F. W. Kunkel, Mr. J. A. Robertson, Miss I. Dean, Mr. A. S. Edwards and Mr. F. S. Callow, without whose collaboration I should never have been able to collect and analyse the data on which this treatment of the subject is founded. Finally, I must not forget my secretary, Miss P. Coombs, whose aid has been invaluable in preparing the book for the press.

Many of the figures have been drawn for this book. As regards the others, I would like to express my thanks for the willingness of the authors and publishers concerned for allowing me to reproduce them. Acknowledgements have been made in the list of illustrations: the citations there made refer to the literature list for fuller details.

JULIAN S. HUXLEY

KING'S COLLEGE, LONDON
December, 1931

CONTENTS

PAGE

CHAPTER IV

GROWTH-GRADIENTS AND THE GENERAL DISTRIBUTION OF GROWTH-POTENTIAL IN THE ANIMAL BODY

CHAPTER V

GROWTH-CENTRES AND GROWTH-GRADIENTS IN ACCRETIONARY GROWTH

CHAPTER VI

HETEROGONY, GROWTH-GRADIENTS AND PHYSIOLOGY

CHAPTER VII

BEARINGS OF THE STUDY OF RELATIVE GROWTH ON OTHER BRANCHES OF BIOLOGY

LIST OF ILLUSTRATIONS

PROBLEMS OF RELATIVE GROWTH

CHAPTER I

CONSTANT DIFFERENTIAL GROWTH-RATIOS

§ 1. Introductory

THE problem of differential growth is a fundamental one for biology, since, as D'Arcy Thompson especially has stressed (1917), all organic forms, save the simplest such as the spherical or the amoeboid, are the result of differential growth,—whether general growth which is quantitatively different in the three planes of space, or growth localized at certain circumscribed spots. But the subject has received little consideration. D'Arcy Thompson's own treatment, though exhaustive on certain points (e.g. the logarithmic spiral), profoundly original and important in others (e.g. his use of Cartesian transformations to illuminate the evolution of one form from another), and interesting throughout, is admittedly incomplete. Certain large bodies of data, such as those included in Donaldson's *The Rat* (1924) and in various treatises on physical anthropology, e.g. R. Martin (1928) exist on differential growth in mammals, but have so far not been analysed save by the use of purely empirical formulae ; Champy (1924) has written a very stimulating book on differential growth of such extreme type as to warrant the term 'dysharmonic', and has later given further examples (1929) ; Przibram has recently (1930) collated some of his interesting results and ideas. But, apart from this, little that is connected or general has been written on the subject ; and even the individual papers dealing with the topic are few and on the whole disconnected.

Since 1920 I have been studying certain phases of the problem : the purpose of the present review is to bring together the various aspects which have presented themselves, to demonstrate the existence of certain broad empirical laws which appear to govern most cases of differential growth so far studied, to discuss their bearing on other branches of biology, and to point the way to further attack on the subject by those trained in other methods.

The first step, it appeared to me, was to study a number of clear-cut cases of differential growth and to see whether they were capable of quantitative expression. My own mathematics are regrettably deficient, but I was able (see Section 2) to obtain a simple formula which appears to be at any rate a first approximation to a general law for differential growth. Among many morphologists and systematists there appears still to linger a distrust of the application of even such elementary mathematics to biological problems. The usual criticism is that the formulae arrived at may have a certain convenience, but can tell us nothing new, and nothing worth knowing of the biology of the phenomenon. This appears to me to be very ill-founded. In the first place, to have a quantitative expression in place of a vague idea of a general tendency is not merely a mild convenience. It may even be a very great convenience, and it may even be indispensable in making certain systematic and biological deductions. But further, it may suggest important ideas as to the underlying processes involved; and this is precisely what the quantitative analysis of relative growth is doing. As will be seen in this and the subsequent chapters, there are certain hypotheses which square with the formula, others which do not: without the quantitative expression, we should be largely theorizing in the air. I would not trouble to spend my time on this point if it had not been urged on several occasions in my hearing; otherwise, one would expect that the interaction of quantitative theory with observation and experiment devoted to testing the theory, so fruitful not only in other sciences but in genetics within the field of biology, would automatically be welcomed.

Furthermore, the establishment of one quantitative rule leads on to the discovery of others. Chapters I and II will be devoted to showing that, when we consider the growth of whole organs relative to the rest of the body, the results can be understood if we postulate that the ratio between the intensity (or relative rate) of growth of the organ and that of the body remains constant over long periods of the animal's life. To borrow a term from another branch of science, there is a constant partition-coefficient of growth-intensity between organ and body. It was next found that in many organs, especially those growing at markedly different rates from the body as a whole, growth-intensity was not distributed uniformly, but in a more or less regular pattern. This led on to the notion, already suggested on different grounds by D'Arcy Thompson, that the growth-intensity of the body as a whole (or, if you prefer it, the relative growth-rates of its various parts) is distributed according to an orderly system of 'growth-gradients'. These conclusions will be discussed in Chapters III to V.

The physiological mechanism underlying these general rules still remains very obscure, in the absence of experiment specifically directed to the point: but there are some interesting hints and possibilities, and these will be discussed in Chapter VI.

Finally, the facts derived from the study of relative growth have a number of important bearings upon other branches of biology; and the concluding chapter will be devoted to these. I hope to convince the systematist that by a knowledge of the laws of relative growth we are put in possession of new criteria bearing on the validity of species, sub-species, and 'forms'; the nature of certain dwarf forms; and the importance (or the reverse) of size-differences in general for systematics. In regard to that special branch of systematics usually called physical anthropology, it will be found that these laws have a bearing on the important question as to whether true evolutionary change has taken place in civilized populations during historical time. As regards evolution, it will be found that the subject throws light upon the question of adaptation, on the general theory of orthogenesis, and on the selection problem. Furthermore, the existence of growth-gradients, as D'Arcy Thompson has already pointed out, makes it much easier for us to understand how certain types of evolutionary transformation can have been brought about, since a single genetic change affecting a growth-gradient will automatically express itself in a changed relation in the size of a large number of organs or regions.

Then comparative physiologists will find it necessary to know precisely how to discount the effects of differences in total absolute size when they wish to estimate the comparative development of an organ in a series of related species or groups; and will further find interesting hints as to the nature of factors which tend to limit the size of an organ at high absolute sizes.

Nor can genetics be left out. A constant partition of growth-intensity between different regions implies constant differences in their *rates* of growth. Thus any genes controlling relative size of parts will have to exert their action by influencing the rates of processes, and so fall into line with the numerous other rate-factors whose importance has been summarized by Goldschmidt (1927) and by Ford and Huxley (1929). The fact, however, that the *ratios* between growth-rates, and not their absolute values, are the determining factors introduces certain complications, whose discussion will be found to have an interesting bearing upon the analysis of other genetic 'characters'.

Finally, the ancient problem of embryological recapitulation will be found to be illuminated from a new angle; and many undoubted cases of recapitulation will be found to owe their origin not to any mysterious phyletic law, but to embryological convenience, adjusting evolutionary changes in the size of an organ to the general rules of relative growth during individual development.

This brief introductory sketch will, I hope, have shown some of the chief points of interest in the study of relative growth. We must now come to grips with the subject, and for the reasons above stated propose to do so by considering what at first sight seems a rather arid point—the quantitative expression of the relation between the body as a whole and an organ whose proportionate size changes during life.

§ 2. Constant Differential Growth-ratios

Champy (l. c.) and others have pointed out that certain organs increase in relative size with the absolute size of the body which bears them; but so far as I am aware, I (Huxley, 1924B) was the first to demonstrate the simple and significant relation between the magnitudes of the two variables. In typical cases, if x be the magnitude of the animal (as measured by some standard linear measurement, or by its weight *minus* the weight of the organ) and y be the magnitude of the differentially-growing organ, then the relation between them is $y = bx^k$, where b and k are constants.[1] The constant b is

[1] This can also be written $\log y = \log b + k \log x$, which means that any magnitudes obeying this formula will fall along straight lines if plotted on a double logarithmic grid.

here of no particular biological significance, since it merely denotes the value of y when $x = 1$—i.e. the fraction of x which y occupies when x equals unity. We may call it the

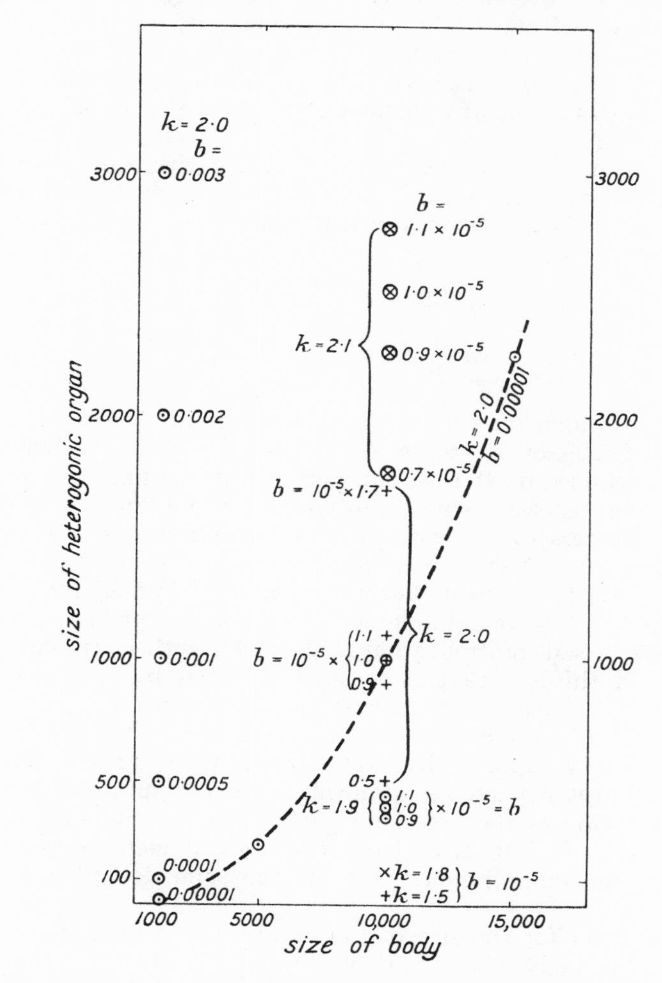

FIG. 1.—Diagram to show the quantitative effect of varying the constants in the simple heterogony formula, $y = bx^k$, assuming that the origin of growth in x and y begins at the same time.

The dotted line gives the growth of the organ (y) when $k = 2{\cdot}0$ and $b = {\cdot}00001$. The points to the left show values of y for different values of b when the rest-of-body is of size 1000. Those to the right show the effects, at body-size 10,000, of varying both k and b.

fractional coefficient. But the value of k has an important meaning.

It implies that, for the range over which the formula holds the ratio of the relative growth-rate of the organ to the relative growth-rate of the body remains constant, the ratio itself being denoted by the value of k. By relative growth-rate is meant the *rate of growth per unit weight,* i.e. the actual absolute growth-rate at any instant divided by the actual size at that instant.

This is at once seen by plotting the logarithm of y against the logarithm of x. In unit time the increase in the logarithm of y is k times the increase in the logarithm of x, which may be written:

$$\frac{d}{dt} . \log y = k\frac{d}{dt} \log x$$

or

$$\frac{dy}{dt}/y = k\frac{dx}{dt}/x$$

This formula, on which I have had the advantage of consulting Professor Levy, of the Imperial College of Science, can be deduced on the basis of very simple assumptions about growth in general. One essential fact about growth is that it is a process of self-multiplication of living substance—i.e. that the rate of growth of an organism growing equally in all its parts is at any moment proportional to the size of the organism. A second fundamental fact about growth is that the rate of self-multiplication slows down with increasing age (size); a third is that it is much affected by the external environment, e.g. by temperature and nutrition. The two latter considerations affect all parts of the body equally, so that we may suppose that the growth-rate of any particular organ is proportional simultaneously (*a*) to a specific constant characteristic of the organ in question, (*b*) to the size of the organ at any instant, and (*c*) to a general factor dependent on age and environment which is the same for all parts of the body.

If y stand for the size of the organ, and x for that of the rest of the body, we shall then have

$$\frac{dx}{dt} = \alpha x \mathrm{G} \text{ and } \frac{dy}{dt} = \beta y \mathrm{G},$$

where α and β are the specific constants for the rest of the body and for the organ in question, and G measures the

general conditions of growth as affected by age and environment,[1] then
$$\frac{dy}{dx} = \frac{\beta y}{\alpha x}.$$

Thus $\log y = \frac{\beta}{\alpha} \log x + \log b$, where b is a constant: i.e.
$$y = bx^{\beta/\alpha}.$$

And β/α, which can also be written k, is a constant, and is also the ratio of the specific components of the growth-rates of y and x respectively.[2]

I am, of course, aware that the existence of growth-cycles and other facts make it impossible to suppose that the expression for change of growth can be so simple as here set forth. We must suppose that each cycle may have its own general and specific components of the growth-rate—i.e. that α, β and γ may change comparatively abruptly during the life-cycle, and also it is quite possible that other inherent alterations, such as the gradual increase of viscosity of protoplasm with age (Ružicků, 1921), will cause gradual and progressive diminution of the specific constants which would account for the various distortions of the **S**-shaped curve of growth from the form expected on the simplest assumptions. But I am convinced that some such general method of envisaging growth is sound; and it is interesting to find our empirical formula for constant differential growth-ratios deducible from it. (See also Schmalhausen, 1927B, 1930.)

Exactly the same formula would apply to two sums of money put out at different rates of compound interest, provided that they were not accumulating discontinuously by quarterly or annual interest payments, as in financial fact, but continuously, as in the Compound Interest Law, and as in biological growth. k would here denote the ratio of the

[1] One might expect, from certain experimental data, that the factor G would be a simple function of the defect of the size of the organism at any given time from its final size; but this would not interfere with the validity of our more general formula.

[2] It may well be that the 'general factor' is not capable of such a simple formulation. But provided that such a general factor does exist—i.e. that the growth both of organ and of rest-of-body is related to some general law of growth affecting the organism as a whole, the deduction of constant differential growth-ratios remains valid. And that such a relation does exist is shown by the work of Przibram, Harrison and others discussed in Chapter VI.

two rates of interest. (In our biological parallels, we know nothing of the actual *rates* of growth, for since the organ and the body have both existed for the same length of time when we measure them, the time-factor cancels out, in point of fact.)[1] The actual rates, unlike those for the two sums of money, will obviously be altering continuously; they will be high in youth, low in age; increased by high temperature, decreased by low; and so forth. What concerns us is that if our formula holds, the *ratio* of the relative rates of growth remain constant.

In such cases, therefore, we have a *constant differential growth-ratio*, denoted by the value of k. If we prefer to concentrate upon the growth of the organ relative to the growth of the body considered as a standard, then we may speak of k as denoting the *growth-coefficient* of the organ. An organ which is thus growing at a different rate from the body as a whole may be called *heterogonic*, to use the convenient term coined by Pézard (1918). If it is growing at the same rate as the body it must be styled *isogonic*; as will be apparent, isogony is merely a special case of heterogony, as the circle is a special case of the ellipse.

It is clear that comparatively small variations in the value of k will have large results provided that growth continues over a considerable range of size. An attempt to show this graphically has been made in Fig. 1.

The best worked-out example of this law so far concerns the large chela of male fiddler-crabs, *Uca pugnax* (Huxley, 1927A). This obeys the law of constant growth-ratio from crabs of only about 60 milligrams total weight to the largest found, weighing sixty times as much; the value of k, however, changes quite abruptly at about 1·1 g. total weight, a point which probably denotes the onset of sexual maturity, decreasing here to less than 80 per cent. of its value for the earlier growth-phase. (It is a noteworthy and unexpected fact that the growth-coefficient of this secondary sexual character should be reduced instead of increased when the gonad begins to function.) See Figs. 2 and 3.

In our examples of the fiddler-crab, the weights of chela and rest-of-body behave, over the earlier and longer growth-phase, like £2 and £100 put out at 8 per cent. and 5 per cent. (continuous) compound interest respectively [2], and a calculation on this basis

[1] For cases where an organ is laid down considerably later than the body as a whole, see Chapter IV.

[2] Strictly speaking, of course, 8·1 and 5 per cent. (see p. 10).

will reproduce the actual figures for weight.[1] But we can be perfectly sure that the actual growth-rate of the crab and of its claw slows off with age, that it differs in summer and winter, and is further subjected to all kinds of irregular fluctuations due to temperature, food and other factors. The actual rates may be as 8,000 : 5,000 in early life, as 160 : 100 later, as 4 : 2·5 in maturity, and as 0·08 : 0·05 in extreme old age ; yet so long as the ratio 8 : 5 is preserved, claw-size will always be the same function of body-size—a body of given

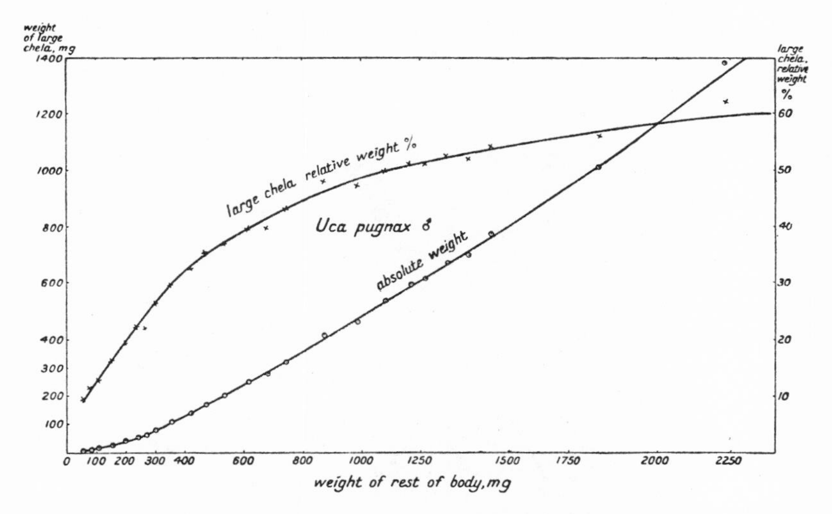

FIG. 2.—Increase of absolute and relative chela-weight in the large chela of the male fiddler-crab, *Uca pugnax*.

(*Constructed from the data of Huxley*, 1927A.)

weight will have attached to it a claw whose weight would be the same whether the body had taken three weeks or three years to reach its present size.[2] The differential growth-ratio remains the biologically and morphogenetically important factor.

The expression $y = bx^k$ can be written $\log y = \log b + k \log x$.

[1] $y_0 = 2$, $x_0 = 100$. At time t, $y_t = y_0 e^{0\cdot08t}$, $x_t = x_0 e^{0\cdot05t}$. After ten years, $y_{10} = 2e^{0\cdot8} =$ £4·45, and $x_{10} = 100e^{0\cdot5t} =$ £164·92. After twenty years, $y_{20} = 2e^{1\cdot6} =$ £9·91, and $x_{20} = 100e^{1\cdot0} =$ £271·9, and so on. Double logarithmic plotting of these figures gives a straight line.

[2] In all probability this is only true as an approximation. It is *a priori* unlikely that there is no differential effect of environmental agencies on the growth-rates of body and chela respectively.

In other words, if the logarithms of the magnitudes are plotted, we should expect a straight line, from the slope of which the value of k can be read off; (if α be the angle the line makes with the x axis, then $\tan \alpha = k$). Fig. 3 shows the excellent

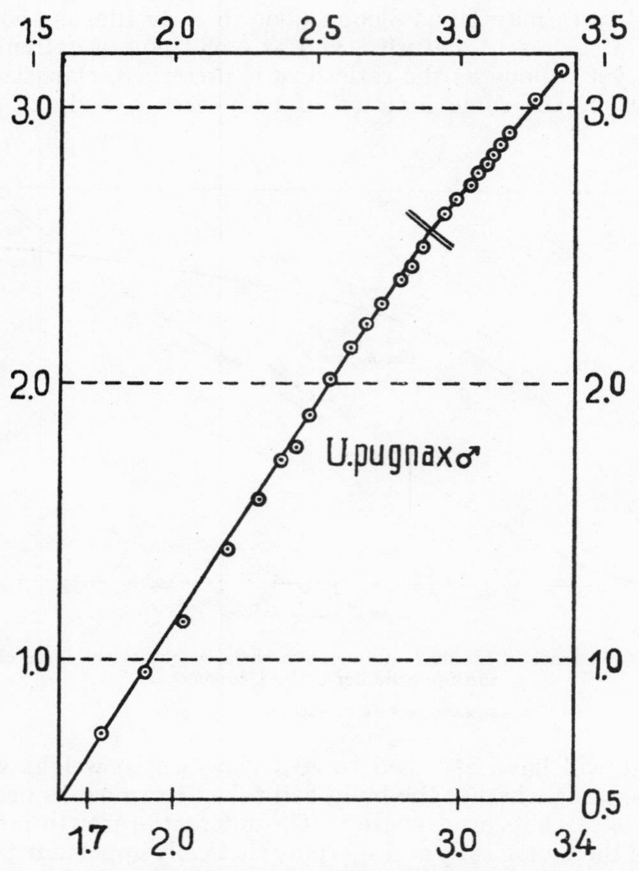

FIG. 3.—Increase of the logarithm of absolute chela-weight with the logarithm of body-weight in male fiddler-crabs.

approximation of the actual points to a straight line when so plotted. In this particular case, the constants are as follows —first phase (to total weight 1·1 g. or just over; rest-of-body weight about 0·75 g.): $b = 0{\cdot}0073$, $k = 1{\cdot}62$; second phase (from this point to maximum size, in this sample maximum total weight rather over 3·5 g.): $b = 0·083$, $k = 1·255$. As purely graphic methods, especially with logarithmic plotting, are not sufficient to establish the accuracy of an empirical formula of this sort (see, e.g., Gray, 1929), I have calculated the values to be expected from the formula. As will be seen from Table I, the deviations from expectation are slight—only in four cases over 5 per cent., and these all in the first phase, where errors in weighing are liable to be relatively greater; the mean deviation for the second phase is only + 0·19 per cent., for the second phase it is + 0·35 per cent., and would be smaller but for the large deviation of the last class, which consists of only a few individuals. Further, there is no trend of the deviations from predominantly positive to predominantly negative or vice versa. We may thus take the formula as a rather surprisingly close approximation to reality. It is possible that the delimitation of the beginning of the second phase after the 14th instead of after the 15th class would have improved matters; and also that small alterations in the values of k would have given an even better fit,[1] but I am only concerned to show that the data conform to this type of mathematical expression, not to obtain accuracy in an extra decimal place in the formula itself.

We are accordingly justified in saying that the large chela of the male Uca grows in close approximation to the formula of constant differential growth-ratio, namely: $y = bx^k$.

In passing, it is worth noting that the logarithmic method of plotting brings into true relief an important point which is entirely obscured by the usual method of plotting on the absolute scale—namely the fact that growth is concerned essentially with the *multiplication* of living substance. On the logarithmic scale, equal spaces on the graph denote equal amounts of multiplication, whereas on the ordinary absolute scale they denote equal additions. From the point of view of growth, the increase of weight of our fiddler-crabs from 5 mg. to 25 mg. is equivalent to that from 1 g. to 5 g.; but on the absolute scale the former interval cannot even be represented on the same graph as the latter. Thus when I speak of a fraction of the growth-period, I shall invariably be thinking in terms of multiplicative growth, in which an n-fold

[1] The last two columns of the table give the expectation if 0·0074 be substituted for 0·0073 as the value of b, and show that this gives a greater deviation, but one of opposite sign.

TABLE I

Uca pugnax (401 SPECIMENS) GROWTH-RATIO OF LARGE CHELA (y) AND REST OF BODY (x)

(a) First phase: to total weight $(x + y) = 1{\cdot}1$ g. $y = 0{\cdot}0073\ x^{1{\cdot}62}$ (mg.).

Class	x = mean weight of rest of body after removal of large chela	y = mean weight of large chela (actual)	y calculated	Per cent. deviation actual from calculated	y calculated on formula $y = 0{\cdot}0074x^{1{\cdot}62}$	Per cent. deviation actual from calculated.
	mg.	mg.	mg.		mg.	
1	57·6	5·3	5·16	+ 2·7	5·24	+ 1·1
2	80·3	9·0	8·89	+ 1·2	9·02	− 0·2
3	109·2	13·7	14·59	− 6·1	14·79	− 7·4
4	156·1	25·1	25·88	− 3·0	26·24	− 4·0
5	199·7	38·3	38·90	− 1·5	39·45	− 2·9
6	238·3	52·5	51·76	+ 1·4	52·48	+ 0·4
7	270·0	59·0	63·53	− 7·1	64·42	− 8·4
8	300·2	78·1	75·34	+ 3·7	76·38	+ 2·3
9	355·2	104·5	98·63	+ 5·9	100·0	+ 4·5
10	420·1	135·0	129·4	+ 4·3	131·2	+ 2·9
11	470·1	164·9	155·2	+ 6·2	157·4	+ 4·8
12	535·7	195·6	191·9	+ 1·9	194·5	+ 0·6
13	617·9	243·0	242·7	+ 0·1	246·0	− 1·2
14	680·6	271·6	283·8	− 4·3	287·7	− 5·6
15	743·3	319·2	327·3	− 2·5	331·9	− 3·8
	Algebraic sum of deviations			+ 2·9		− 16·9
	Mean deviation			+ 0·19		− 1·13

(b) Second phase: from total weight 1·2 g. onwards. $y = 0$.

	x	y	y calculated	Per cent. deviation	y calculated on formula $y = 0{\cdot}084\ x^{1{\cdot}255}$*	Per cent. deviation
16	872·4	417·6	406·8	+ 2·6	411·7	+ 1·4
17	983·1	460·8	472·4	− 2·5	478·0	− 3·6
18	1,079·9	537·0	531·9	+ 1·1	538·3	− 0·2
19	1,165·5	593·8	585·6	+ 1·4	592·7	+ 0·2
20	1,211·7	616·8	628·7	− 1·9	636·2	− 3·1
21	1,291·3	670·0	665·6	+ 0·7	673·6	− 0·5
22	1,363·2	699·3	720·6	− 3·0	729·3	− 4·1
23	1,449·1	777·8	769·1	+ 1·7	778·4	− 0·1
24	1,807·9	1,009·1	1,015·0	− 0·6	1,028·0	− 1·8
25	2,235·0	1,380·0	1,327·0	+ 4·0	1,344·0	+ 2·6
	Algebraic sum of deviations			+ 3·5		− 9·2
	Mean deviation			+ 0·35		− 0·92

* In Huxley, 1927A (p. 152), this was stated as 1·33 owing to an error.

increase from one absolute size is regarded as equivalent to an n-fold increase from another absolute size.

The same total increase could be subdivided into fractions of equal *absolute* size; but this method of subdivision in terms of additive growth has not the same biological value, and will not be adopted. Thus for an increase from 1 g. to 256 g., equal fractions of the growth-period are best represented by the equal multiplicative increases from 1 to 4, 4 to 16, 16 to 64 and 64 to 256 g.; and not by the equal additive increases to 64, 128, 192 and 256 g.

§ 3. Examples of Constant Differential Growth-ratios

The expression $y = bx_k$ I shall refer to as the simple heterogony formula. This formula, or an approximation to it, has been found to hold for a number of other organs, e.g. the

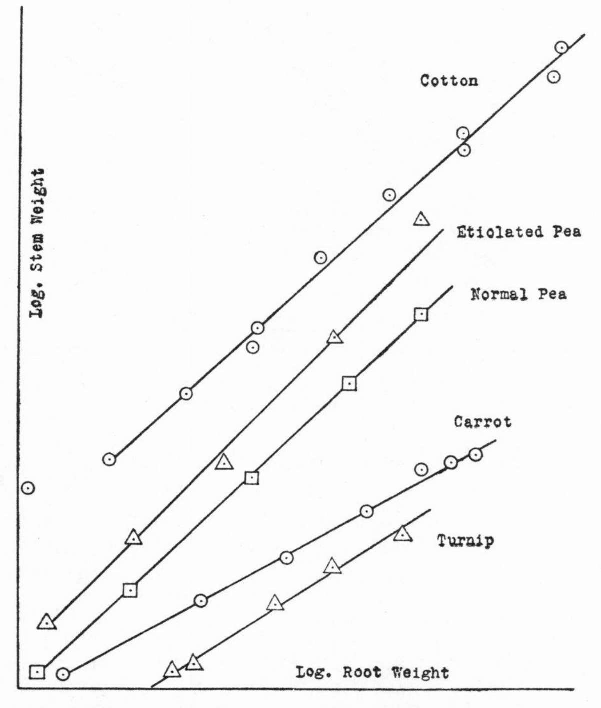

Fig. 4.—Increase of the logarithm of stem-weight against the logarithm of root-weight in various plants.

abdomen of some female crabs (Shaw, 1928; Sasaki, 1928); the chelae of many male and some female Decapoda (Huxley, 1927; Shaw, 1928; Tazelaar, 1930; Tucker, 1930 [1]); other appendages of various crustacea; the trunk of Planarians as

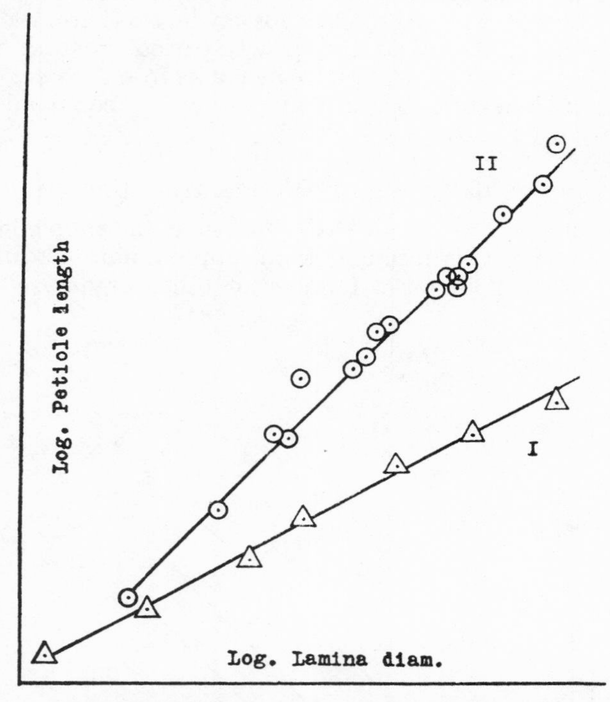

FIG. 5.—Increase of the logarithm of petiole-length against the logarithm of lamina diameter in nasturtium leaves, *Tropaeolum*.

against the head (Abeloos, 1928); the face as against the cranium of dog and baboon (Huxley, 1927, analysing Becher, 1923; Huxley, unpublished, analysing Zuckerman, 1926); the shoot as against the root of certain plants (Pearsall, 1927); the size (facet-number) of the two lobes of the eye in the

[1] Tucker states that his data indicate a linear relation between male chela length and breadth, and carapace length. However, his graphs and his percentage measurements indicate that this does not give an accurate fit, whereas logarithmic plotting (Fig. 55) gives an excellent approximation to a combination of two straight-line curves, as in male Uca.

bar-eye mutant of *Drosophila melanogaster* (Hersh, 1928); the linear dimensions of certain Molluscs (Nomura, 1928;

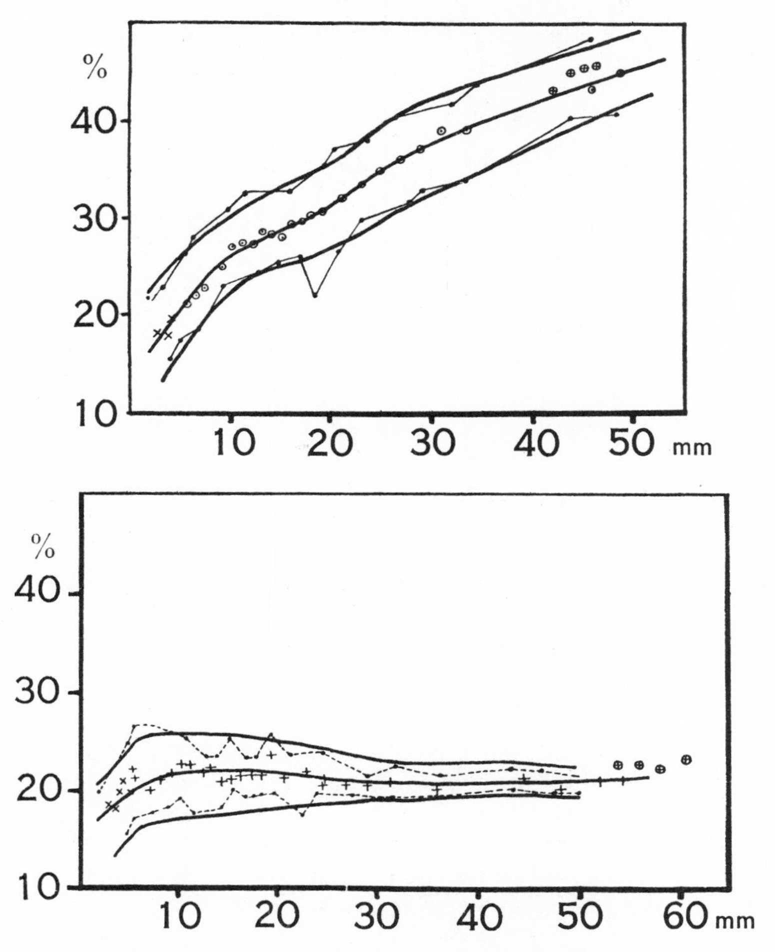

FIG. 6.—Increase in relative width of abdomen with increase of carapace length in the shore-crab, *Carcinus maenas*: above, female; below, male. The ordinates represent $\frac{\text{abdomen width}}{\text{carapace length}}$ %; the abscissae represent carapace length in mm.

Nomura and Sasaki, 1928); the tail of the mouse Phenacomys (Taylor, 1915); the dimensions of the casques of Hornbills (Banks, unpublished); the weights of various organs of the rat (analysis of the data of Donaldson, Hatai, Jackson, etc., —see Donaldson, l. c.); the length of the head of whalebone whales during post-natal life (Mackintosh and Wheeler, 1929); the lamina diameter and petiole length in Tropaeolum leaves

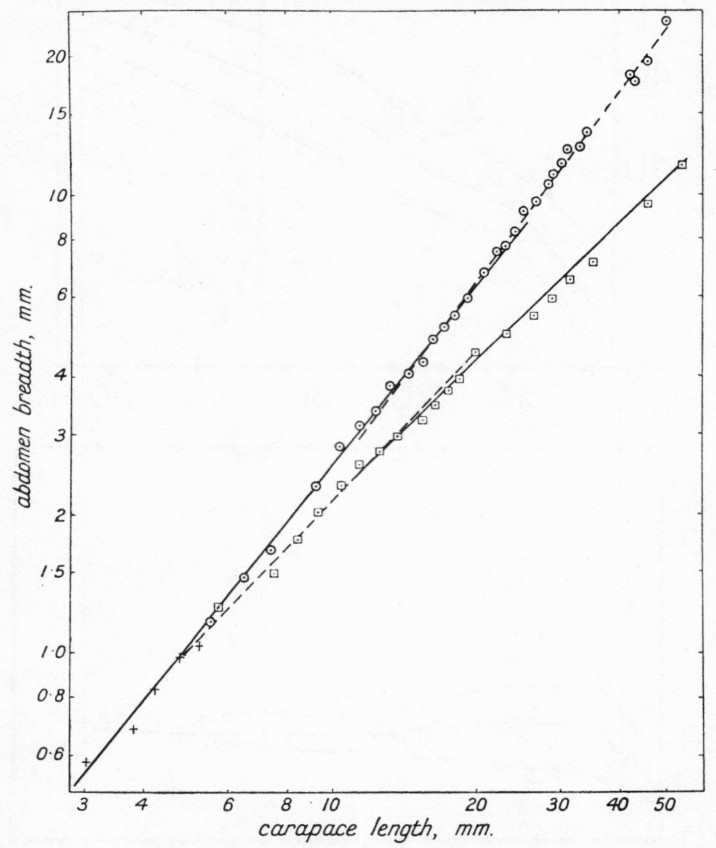

FIG. 7.—Increase of width of abdomen with increase of carapace length in the shore crab, *Carcinus maenas*: logarithmic plotting.

+, unsexables; ⊙, females; ⊡ males. The growth coefficient for unsexables and young females is 1·26, for older females 1·42, that for young males 1·07, for older males 0·94.

(Pearsall, 1927), and even for the amounts of various chemical substances in the growing larvae of the wax-moth Galleria and the meal-worm beetle Tenebrio (Teissier, 1929, 1931), as well as for organs which, physiologically speaking, represent special cases, as the antlers of deer which are shed every year

(Huxley, 1926, 1927, 1931), and the horns, mandibles, etc., of various holometabolous insects (Huxley, 1927—see Chapter II). It is probable that any other changes of proportion which have not yet been analysed from this point of view will turn out to obey the same law, e.g. the progressive increase of relative tail-length in the salamander Eurycea (Wilder, 1924), or that of wing-rudiments in dragonflies (Balfour-Browne, 1909). See also p. 258 (Daphnia), p. 263 (rat).

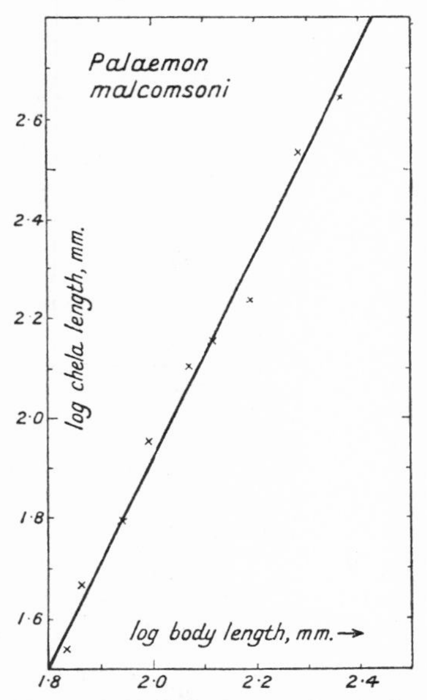

FIG. 8.—Relative growth of the chela in the prawn, *Palaemon malcomsoni*: logarithmic plotting. (*From the data of Kemp, and Henderson and Mathai.*)

Most of the organs thus far cited are *positively heterogonic*, increasing in relative size with growth; others, however, decrease in relative size with increase in absolute size, and are accordingly *negatively heterogonic*. Such of these cases as have been analysed also appear to obey the rule of constant differential growth-ratios: e.g. nucleus in oocytes of Hydractina (Teissier, 1927); brain in various mammals (Lapicque, 1907; Dubois, 1914, 1918); the number of nerve-fibres and/or neurones in mammals (Lapicque and Giroud, 1923; Dubois, 1918); heart in many vertebrates (Klatt, 1919; Clark, 1927, for references); limbs in post-natal sheep (Hammond, 1927, 1929); certain limbs in Hermit-crabs (Bush, 1930); pereiopods in the racing-crab Ocypoda (Cott, 1929); legs in Orthoptera (Przibram, 1930); Gammarus eyes (p. 260).

Some of the facts are graphically illustrated in Figs. 4, 5 (plant organs); 6, 7, 22, 23 (crab abdomen); 8, 21, 32, 55, 77 (Crustacean chelae); 9, 10 (mammalian cranium); 11 (mouse tail-length); 12 (facet-number, Drosophila eye); 13–15 (various organs of crabs); 16–20 (chemical substances in insect-larvae); 71 (head-length, whales); 25, 27 (deer antlers); 33–35, 40, 91, 92 (organs of holometabolous insects).

We may give some tables and figures in support of these statements.

SHEEP DOG (Fig. 9)
Data from Becher (1923)
Analysed in Huxley, 1927

x	y
42·0	22·0
65·3	48·3
74·5	58·0
85·5	73·5
99·3	89·1
112·6	102·0
120·0	112·0

$\Sigma = 30$.
x = cranial region (mm.).
y = facial region (mm.).
k = 1·5 (except for highest values of x).

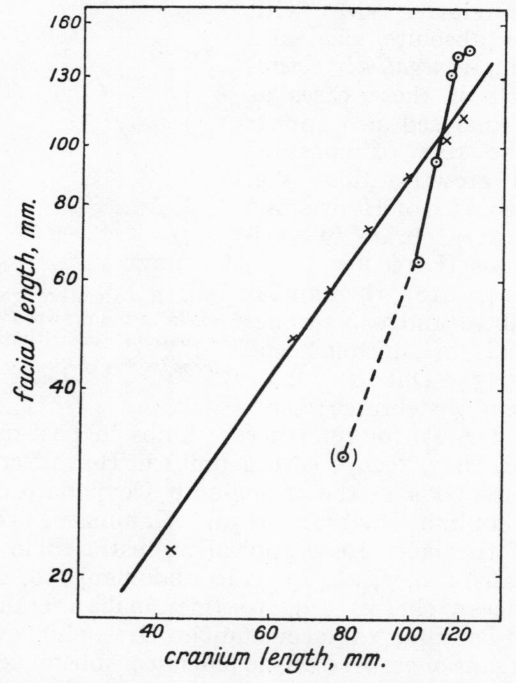

FIG. 9.—Growth of the facial region relative to the cranial region in the skulls of sheep-dogs (×) and baboons (⊙).

k for sheep-dog about 1·49; for baboon, points 2–5, about 4·25.

(*Constructed from the data of Becher*, 1923, *and Zuckerman*, 1926.)

VALUES OF CRANIUM-LENGTH AND FACE-LENGTH IN THE BABOON *Papio porcarius* AT DIFFERENT SIZES (from Zuckerman, 1926) (Figs. 9, 10)

No. of Cases	Mean Cranium-length mm.	Mean Face-length (Naso-prosthion) mm.
1	78·5	31·0
4	100·25	64·6
7	108·9	94·8
3	114·7	131·0
6	118·25	140·8
4	122·0	144·25

The growth-coefficient of face-length on cranium-length for Classes 2–5 is about 4·25, a very high figure. The curve shows irregularities at both ends.

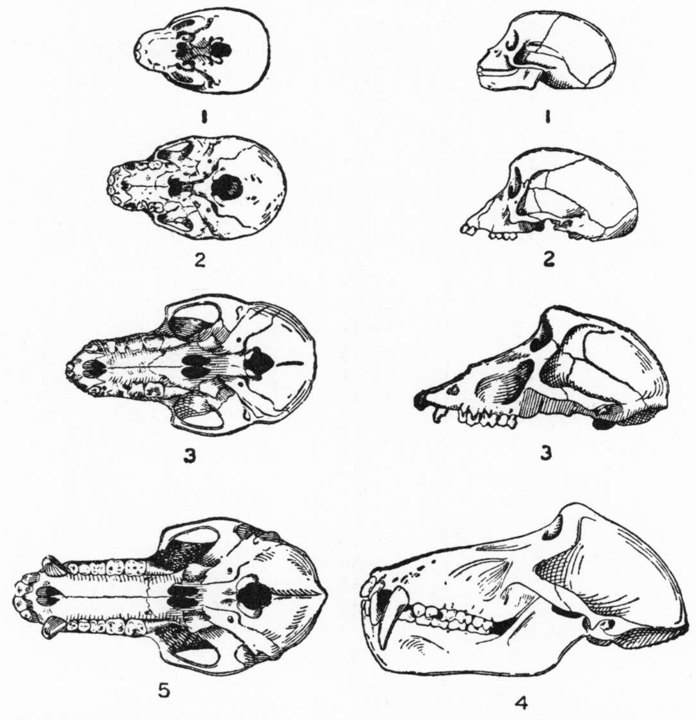

FIG. 10.—Baboon skulls of various sizes, to show the increase in relative size of facial region with absolute size of skull.

1, new-born; 2, juvenile (with milk dentition); 3, adult female; 4, adult male.

TABLE IA

ABDOMEN-BREADTH AGAINST CARAPACE-LENGTH IN CARCINUS MAENAS FROM PLYMOUTH (Figs. 6, 7)

(Data of Huxley and Richards, 1931)

Σ	Mean car.-l. (mm.)	Mean abd.-br. (mm.)
(*a*) Unsexable (74 specimens)		
13	3·09	0·578
22	3·80	0·680
20	4·22	0·823
11	4·76	0·979
8	5·19	1·019
(*b*) Females (281 specimens)		
12	5·56	1·16
16	6·52	1·45
14	7·41	1·67
12	9·32	2·30
15	10·37	2·80
16	11·35	3·11
17	12·33	3·37
23	13·29	3·81
19	14·35	4·06
12	15·31	4·29
19	16·35	4·82
15	17·36	5·15
15	18·16	5·48
7	19·34	5·93
6	20·33	6·72
10	21·51	6·82
12	22·40	7·54
8	23·34	7·76
6	24·33	8·33
5	25·52	9·16
10	26·45	9·59
6	27·65	9·79
2	28·45	10·40
3	29·30	11·03
4	30·55	11·62
4	31·49	12·46
3	32·37	12·75
5	33·32	12·70
4	34·33	13·56
1	43·50	17·50
2	45·30	19·20
1	46·30	19·90
1	50·20	23·70

VALUES OF CONSTANT DIFFERENTIAL GROWTH-RATIOS FOR SHOOT-WEIGHT AGAINST ROOT-WEIGHT IN VARIOUS PLANTS (from Pearsall, 1927) (Fig. 4)

Plant	Value of k
Daucus carota (carrot)	0·55
Brassica rapa (turnip)	0·65
Gossypium roseum (cotton) . .	0·90
Impatiens sp.	1·00
Pisum sativum (pea)	0·90–1·15 (3 expts.).
,, ,, (etiolated) . . .	1·75–2·65 ,,
Triticum vulgare	1·05
Hordeum distichum (low N) . .	1·20
,, ,, (high N) . .	1·55
Linum usitatissimum	1·30

VALUES OF CELL-DIAMETER AND NUCLEAR DIAMETER IN THE OOCYTES OF HYDRACTINIA ECHINATA, IN μ. (FROM TEISSIER, 1927)

$k = 0{\cdot}69$; $b = 1{\cdot}5$

x = oocyte diameter	6·8	10·0	13·6	17·5	22·2	25·3	34·0
y = nuclear ,,	5·6	7·3	9·0	11·4	12·5	14·4	16·9
Deviations from calculated values .	0·0	0·0	−0·1	+0·1	−0·2	+0·5	−0·2
x = oocyte diameter	43·0	53·5	70·0	88·0	118·0	136·0	168·0
y = nuclear ,,	22·5	24·8	29·6	30·3	36·5	42·7	52·0
Deviations from calculated values . .	+2·4	+1·5	+1·5	−2·6	−3·8	−2·5	+0·2

MEAN VALUES OF PRE-OCULAR LENGTH AND TOTAL LENGTH IN *Planaria gonocephala*, CALCULATED FROM ABELOOS (1928)

$k = 0{\cdot}63$ approx.

Total length . .	1·5	3·0	5·0	18·0 mm.
Pre-ocular length .	0·231	0·43	0·50	1·125 mm.

MEAN VALUES FOR FACET-NUMBER IN THE TWO LOBES OF BAR-EYED DROSOPHILA AT DIFFERENT TEMPERATURES (from Hersh, 1928) (Fig. 12)

Temperature	32·0°	29·5°	27·5°	25·5°	21·5°	18·0°	15·0°	
No. facets dorsal lobe	17·63	22·15	23·30	28·63	43·85	62·08	66·04	Homozygous bar ♀♀
No. facets ventral lobe	12·73	18·86	19·24	19·11	36·24	57·62	67·28	
No. facets dorsal lobe	81·22	152·15	150·29	193·54	196·21	222·45	256·12	Heterozygous (Bar × normal) ♀♀
No. facets ventral lobe	41·93	90·20	103·66	114·29	151·51	159·01	185·89	

In both cases k for ventral lobe on dorsal lobe is about 1·5, but the value of b is considerably higher for the homozygotes.

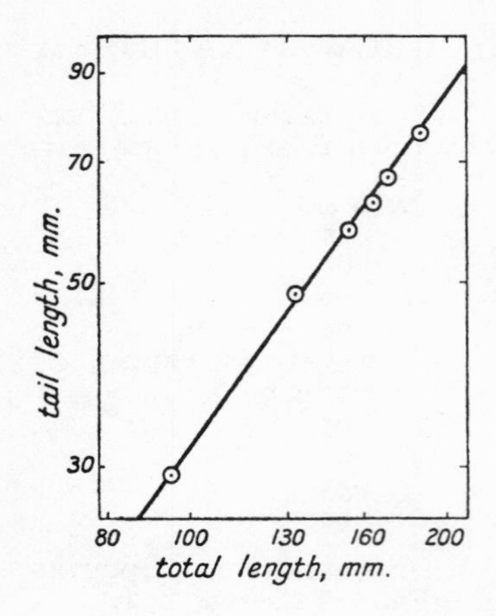

FIG. 11.—Tail-length against total length during growth in the mouse, *Phenacomys longicaudus*; logarithmic plotting.

k = about 1·41.

(*Recalculated from the data of Taylor*, 1915, *p.* 129.)

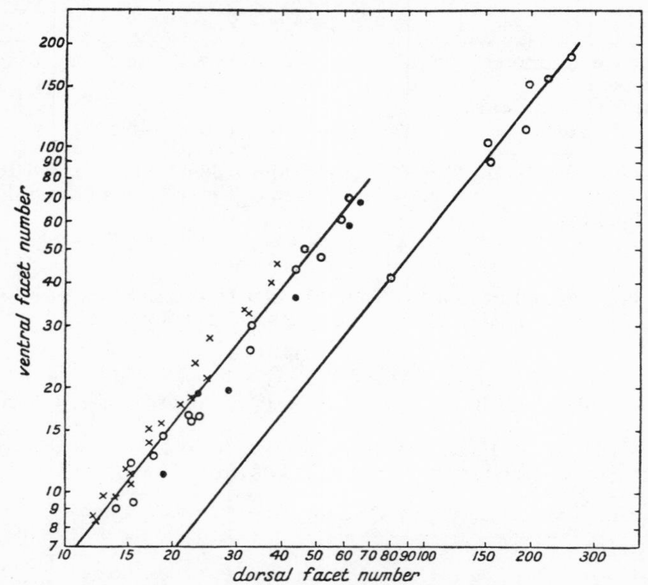

FIG. 12.—Relation of facet-number in dorsal and ventral lobes of mutant female fruit-flies of the bar-eye series; logarithmic plotting. With decreasing temperature, the total number of facets in the eye increases; but the number in the ventral lobe of the eye increases heterogonically relative to the number in the dorsal lobe.

The curve on the right denotes heterozygotes between bar and wild-type (full eye). In the curve on the left, × denotes homozygous ultra-bar *and* heterozygotes between ultra-bar and bar; o, heterozygotes between ultra-bar and wild-type; •, homozygous bar.

k for all is close to 1·5; b is lowered by admixture of the wild-type gene, and is at its maximum in pure ultra-bar (i.e. rises with decreasing absolute size of the eye). It is clear that facet-formation must begin ontogenetically in the dorsal lobe.

DIMENSIONS (IN INCHES) OF PARTS OF THE BILL AND CASQUE OF THE HORNBILL *Antheracoceros malaganus* (from E. Banks, unpublished)

	♀♀							♂
Length of gape along curve	3·5	3·8	4·25	4·25	5·0	5·3	5·6	5·8
Length of casque (straight) . . .	—*	—*	2·6	3·2	3·55	4·6	5·2	6·7
Height of casque above bill . . .	0·5	0·9	1·0	0·9	1·4	1·6	1·5	1·85
Height of bill proper	1·2	1·2	1·5	1·25	1·5	1·75	1·6	2·05

* Indistinguishable from bill.

Both in length and height the casque is highly heterogonic relative to the length and/or the height of the bill proper.

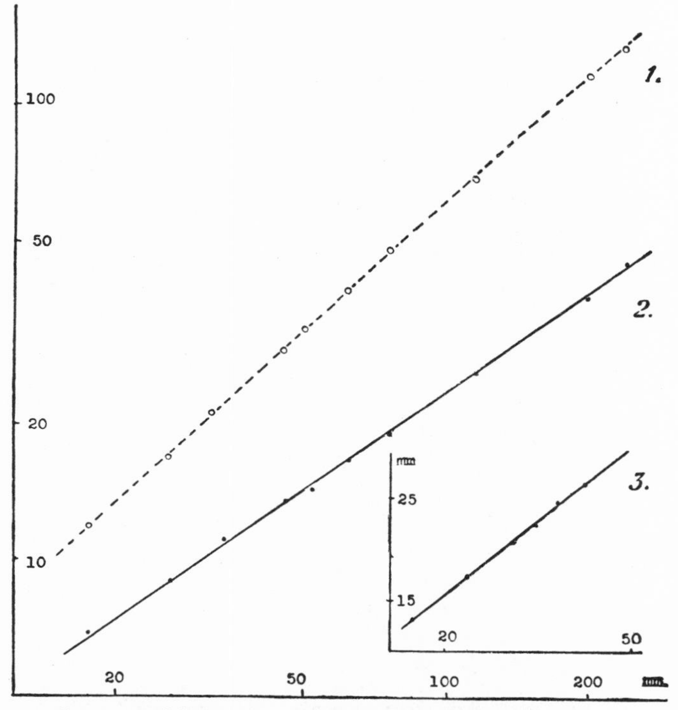

FIG. 13.—Increase of interocular distance with carapace width in the crabs Cancer (1) and Eriphia (3), and of carapace length with carapace width in Cancer (2); logarithmic plotting.

k for interocular distance, in Cancer 0·70, in Eriphia 0·76.

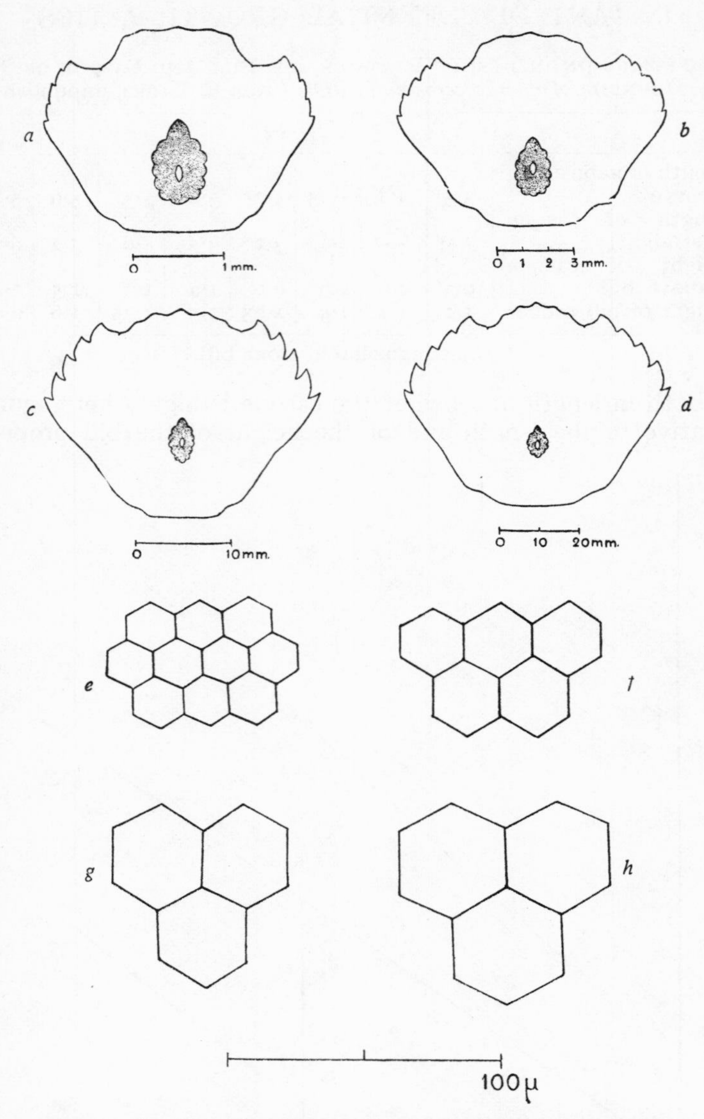

FIG. 14.—Change of proportions in the crab, *Carcinus maenas.*

A—D, outline of carapace and of thoracic ganglion in four specimens of different absolute size, to show negative heterogony of interocular breadth ($k = 0{\cdot}85$), and of thoracic ganglion ($k = 0{\cdot}6$). The carapace lengths have been made the same for all: actually they were 3·1, 10·9, 30·0 and 71·0 mm. respectively.

E—H, ommatidia of the same four specimens, all drawn to the same absolute scale. k for ommatidial diameter $= 0{\cdot}32$.

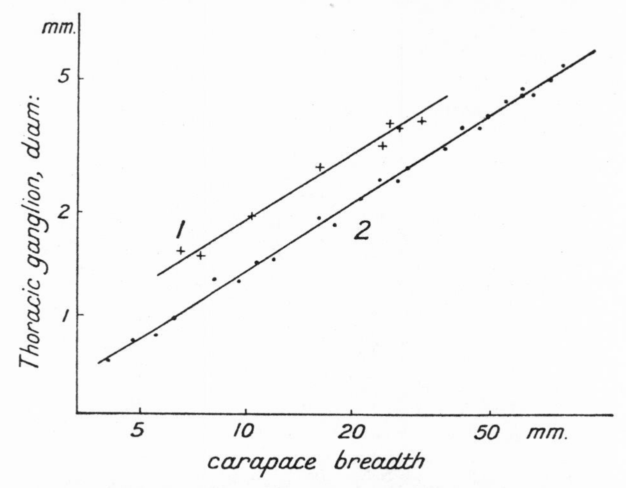

FIG. 15.—Diameter of the thoracic ganglion against carapace breadth in crabs: (1) Pachygrapsus; (2) Carcinus; logarithmic plotting.
k in both cases about 0·6. (After Teissier, 1931: in the original paper, the Carcinus curve is erroneously ascribed to Pachygrapsus and vice versa.)

Since the above was written, the important paper of Teissier (1931) has provided numerous fresh examples. These we may summarize in tabular form (and see Figs. 13–20).

Animal	y = organ	x = standard to which organ is compared	Value of k
Mealworm, Tenebrio molitor larva	weight of moulted skin or carapace	total fresh weight	0·8
Water-beetle, Dytiscus marginalis, larva	,, ,,	,, ,,	0·8
Water-boatman, Notonecta glauca	,, ,,	,, ,,	1·0
Shore-crab, Carcinus maenas	,, ,,	,, ,,	1·0
Crab, Pachygrapsus marmoratus	,, ,,	,, ,,	1·0
Stagbeetle, Lucanus cervus ♂	weight of desiccated mandibles	weight of desiccated elytra	2·0
,, ,, ,,	weight of desiccated head	,, ,,	1·5
,, ,, ,,	weight of desiccated legs	,, ,,	1·0
,, ,, ,,	mean diameter eye	length of elytron	1·0
,, ,, ,,	maximum breadth head	,, ,,	2·0

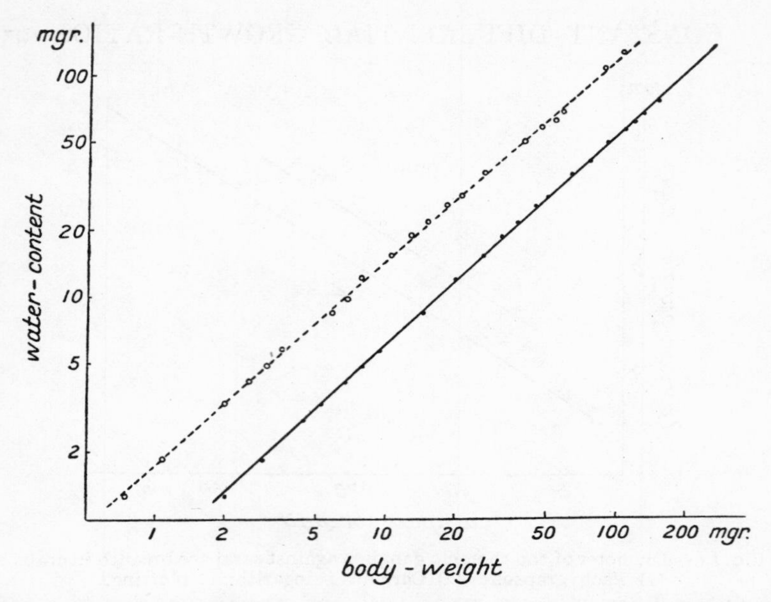

Fig. 16.—Water-content against body-weight in the larval mealworm, Tenebrio. Solid line, fresh weight ; dotted line, dry weight ; logarithmic plotting. k (fresh weight), 0·975 ; (dry weight), 0·92.

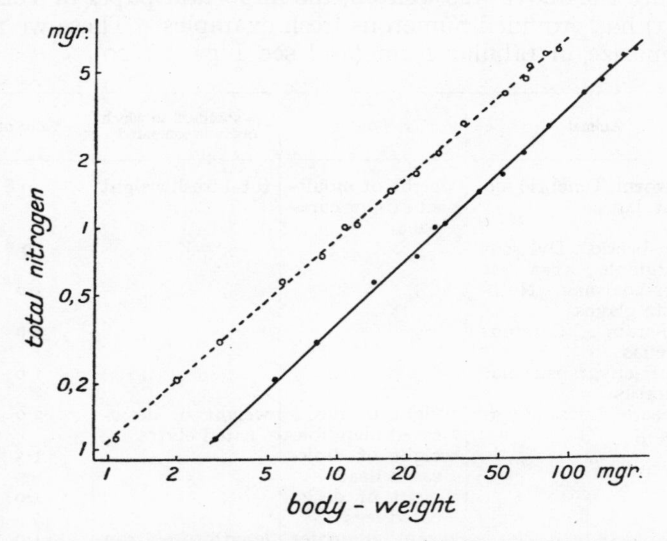

Fig. 17.—Total nitrogen against body-weight (solid line, fresh weight ; dotted line, dry weight) in the larval mealworm, Tenebrio ; logarithmic plotting. k for fresh weight, 0·965 ; for dry weight, 0·91.

Animal	y = organ	x = standard to which organ is compared	Value of k
Stick insect, Dixippus morosus	length prothorax	total length	1·0
,, ,, ,,	length head	,, ,,	0·71
,, ,, ,,	breadth head	,, ,,	0·71
,, ,, ,,	diameter eye	,, ,,	0·48
Mealworm, Tenebrio molitor, larva	breadth head	$\sqrt[3]{\text{total weight}}$	0·95
Crab, Carcinus maenas	interocular breadth	carapace breath	0·85
,, ,, ,,	carapace length	,, ,,	about 1·0
,, Pachygrapsus marmoratus	carapace length	,, ,,	about 1·0
,, ,, ,,	interocular breadth	,, ,,	0·85
,, Eriphia spinifrons	,, ,,	,, ,,	0·76
,, Cancer pagurus .	,, ,,	,, ,,	0·70
Water-boatman, Notonecta glauca	diameter of a single ommatidium	total length	0·46
Cockroach, Blatta orientalis	,, ,,	,, ,,	0·36
Stick insect, Dixippus morosus	,, ,,	,, ,,	0·37
Crab, Carcinus maenas	,, ,,	carapace breadth	0·32
,, Pachygrapsus marmoratus	,, ,,	,, ,,	0·47
Crayfish, Potamobius astacus	,, ,,	,, length	0·40
Crab, Carcinus maenas	diameter thoracic ganglion	carapace breadth	about 0·6
,, Pachygrapsus marmoratus	,, ,,	,, ,,	,, 0·6
Insect larva, Chaoborus crystallensis	diameter cerebral ganglion	body length	,, 0·6
,, ,, ,,	mean diameter abdominal ganglia	,, ,,	,, 0·6
Stick insect, Dixippus morosus	diameter cerebral ganglion	total length	,, 0·6
,, ,, ,,	mean diameter thoracic ganglia	,, ,,	,, 0·6
,, ,, ,,	mean diameter abdominal ganglia	,, ,,	,, 0·6
Mealworm, Tenebrio molitor larva	mean diameter abdominal ganglia	total length	,, 0·6
Water-boatman, Notonecta glauca	diameter 2nd thor. ganglion	,, ,,	,, 0·6
Insect larva, Chaoborus crystallensis	mean diameter nucleus, nerve-cells	body length	,, 0·4

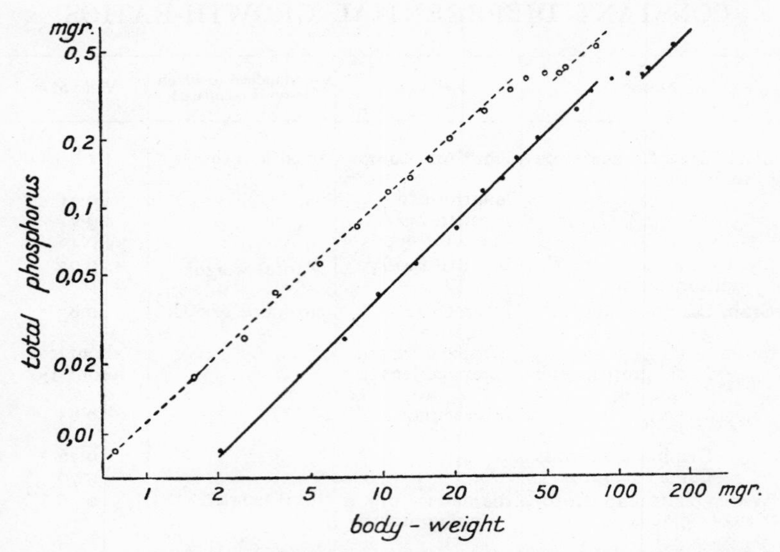

FIG. 18.—Total phosphorus against body-weight in the larval mealworm, Tenebrio; logarithmic plotting.

k for fresh weight, 1·03, later 1·08; for dry weight, 0·975, later 1·02.

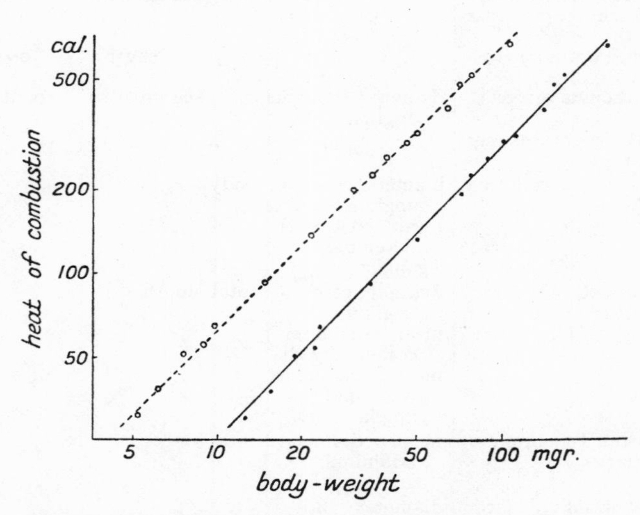

FIG. 19.—Heat of combustion against body-weight in the larval mealworm, Tenebrio; logarithmic plotting.

k for fresh weight, 1·07; for dry weight, 1·02.

Animal	y = organ	x = standard to which organ is compared		Value of k	
		(a)	(b)	(a)	(b)
Mealworm, Tenebrio molitor larva	total fat	fresh weight	dry weight	1·13	1·06
	total carbon	,,	,,	1·07	1·01
	dry weight	,,	,,	1·06	1·00
	dry weight, fat removed	,,	,,	1·04	0·98
	total phosphorus, until late in larval life	,,	,,	1·03	0·975
	(total phosphorus, end of larval life)	,,	,,	(1·08	1·02)
	protein nitrogen	,,	,,	1·03	0·975
	fresh weight	,,	,,	1·00	0·945
	total water	,,	,,	0·975	0·92
	total nitrogen	,,	,,	0·965	0·91
	extractives	,,	,,	0·87	0·82
	lipidic phosphorus	,,	,,	0·83	0·80
	nucleic phosphorus	,,	,,	0·82	0·78
	ash	,,	,,	0·81	0·76
	total heat of combustion	,,	,,	1·07	1·02
	Oxygen consumption at rest and fasting	,,	,,	about 0·8	
	Oxygen consumption under normal conditions, but on a ration permitting only slow growth (flour)	,,	,,	about 0·9	
	ditto, but fed a ration permitting normal growth	,,	,,	about 0·95	
Wax-moth, Galleria mellonella, larva	total water, early phase*	,,	,,	1·0	1·0
	total water, late phase	,,	,,	0·96	0·91
	dry weight, early phase	,,	,,	1·0	1·0
	dry weight, late phase	,,	,,	1·08	1·0
	total heat of combustion, early phase	,,	,,	1·13	1·09

* 'Early phase' and 'late phase' refer to the fact that after a period of regular differential growth lasting for most of the larval period, there occurs a short phase of irregularity, denoting rapid change in character of metabolism, followed by a second regular period resembling the first but with different quantitative relations.

Animal	y = organ	x = standard to which organ is compared	Value of k
		(*a*) (*b*)	(*a*) (*b*)
Wax-moth, Galleria mellonella, larva	total heat of combustion, late phase	fresh weight, dry weight	1·16 1·09
	total phosphorus, early phase; total phosphorus, late phase	total water	0·9
	total fat, early phase; total fat, late phase	dry weight	1·32
	dry weight, fat removed, early phase; dry weight, fat removed, late phase	,,	0·82

Now it is to be observed that a constant differential growth-ratio, during some at least of the period of growth, is not merely an empirical rule found over a large range of organs and groups, but is what one would expect on *a priori* grounds. For it is the biologically simplest method we can conceive of obtaining the enlargement (or diminution) of an organ.

If an organ is to begin its career small and end it large, the obvious method is to make it grow at a higher rate than the rest of the body; the difference once initiated (by whatever physiological means) there is no *a priori* reason why it should not be maintained at approximately the same relative level throughout, since changes which affect the growth of the body will be expected to have, within narrow limits, a proportionate effect on the growth of the organ (and see p. 6).

We shall later consider certain special cases where growth of organ and body appear not to be equally affected in certain circumstances (pp. 200, 259); others which show that the formula for constant differential growth-ratio can only be an approximation when we are dealing with an organ as a whole (p. 81); and still further facts which rule out some of the early stages of an organ's development from the operation of the law (p. 139, *seq.*). However, both empirical facts and theoretical considerations warrant us in regarding a constant differential growth-ratio, operating over a longer or shorter period of time, as the primary law of the relative growth of parts, once they have reached the stage of full histological differentiation. It may (like Boyle's law) prove only to be an approximation, and to be capable of modification in certain circumstances ; yet (again like Boyle's law) it may remain fundamental.

We may now proceed to consider a little more in detail some of the cases. In the first place, we can utilize our formula to deduce the moment of onset, in male *Uca pugnax*, of the large chela's heterogony. To do this we must first know the weight of the small claw. This in males is identical in form

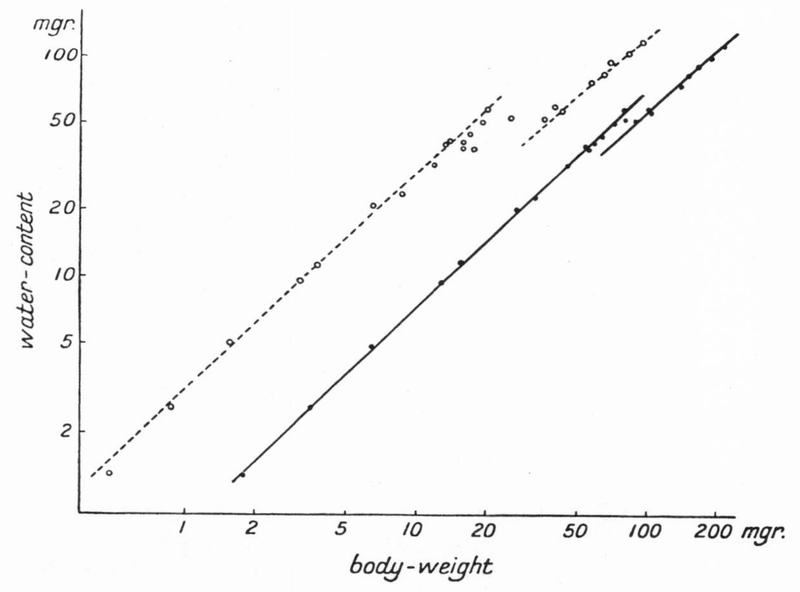

FIG. 20.—Increase of water-content with total weight in the larva of the wax-moth, Galleria ; logarithmic plotting.

k for fresh weight, early phase 1·0, late phase 1·0 ; for dry weight, early phase 0·96, late phase 0·91.

with both claws of females, and does not change its relative weight with increasing body-size : at all stages it weighs, as does a single female chela, almost precisely 0·02 of the rest-of-body weight. If we make the assumption that our formula for the first phase holds from the first moment of increase of the large chela, we have simply to extrapolate from our formula and find the point on the curve at which rest-of-body weight is fifty times chela weight. This is found to be close to 5 mg. body-weight, when the chela should weigh 0·1 mg.

We can now proceed to check this deduction. Morgan (1923A, 1924) has found that the very youngest post-larval *U. pugnax* he could obtain have both claws alike, of female or small type, in both sexes. In this stage, the chelae are autotomized very readily; and only when one is thus thrown off does the other proceed to transform into a large chela. After this has once happened, the fates of the two chelae are irreversibly determined, though initially either may become a large chela through the accident of the other's autotomy. The moment of determination of the large chela appears normally to take place very early, during the first or second instar of post-larval life.

Accordingly, I collected and weighed a number of the smallest fiddler-crabs to be found on the beach, in which the sexes could not be determined by casual inspection of either chela-size or abdomen-shape. Their mean weight was about 6·7 mg.—an excellent approximation to the 5 mg. prophesied on theoretical considerations.

Then again, we can compare relative growth in different species of the same genus. *Uca minax* is much scarcer near Wood's Hole, and the comparatively few specimens available were all of a size to be in the second phase of *U. pugnax*. However, they yielded one or two interesting results. The double logarithmic plot of chela against rest-of-body clearly approximated to a straight line; but owing to the smaller number available it was impossible to determine the growth-coefficient of the chela so accurately. It was, however, certainly between 1·58 and 1·66—in other words, almost exactly the same as that of *U. pugnax* for the first phase. Either *U. minax* has no change in the growth-coefficient of the chela at or near maturity, or at all periods its chelar growth-coefficient is higher than in *pugnax*. *U. minax* also differs from its relative in the greater size which it attains; the biggest specimens found weighed 17·8 g. as against 3·6 g. for *U. pugnax*. Correlated with this, as was to be expected, was the greater relative weight of the large chela to be found in *minax*. This in one specimen amounted to no less than 77 per cent. of rest-of-body weight, as against a maximum of 65 per cent. in *U. pugnax*.

It is clear that the large chelae of big specimens of *U. minax* must be getting close to their maximum limit of relative size. A claw as big as the rest of the body would not be very practicable, and these are already over three-quarters this relative size. In *U. pugnax*, where our figures are more accurate, the large chela has attained half the rest-of-body size when the total weight is about 1·65 g. If the animal could grow to 24 g. (less than 30 per cent. bigger than the biggest *U. minax*), its large chela would be the same weight as all the rest of it together. Twenty-four grams is a very small weight for many crabs, including forms of similar semi-terrestrial and burrowing habits to Uca, such as Ocypoda, yet no species of fiddler-crab has grown to a size much over that of *U. minax*. It may thus be plausibly suggested that the existence of this continuously high growth-ratio in the large claws of the male fiddler-crabs may have acted as a limiting factor in their size-evolution, any possible advantage to be obtained by increase in size not countervailing to cause selection to alter the chela's growth-mechanism.

Analysis of the

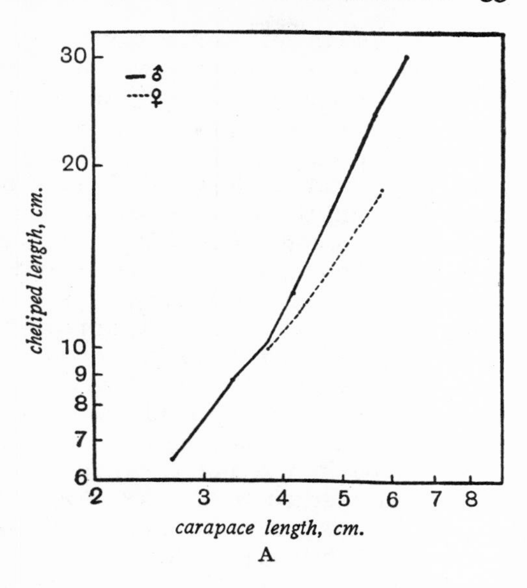

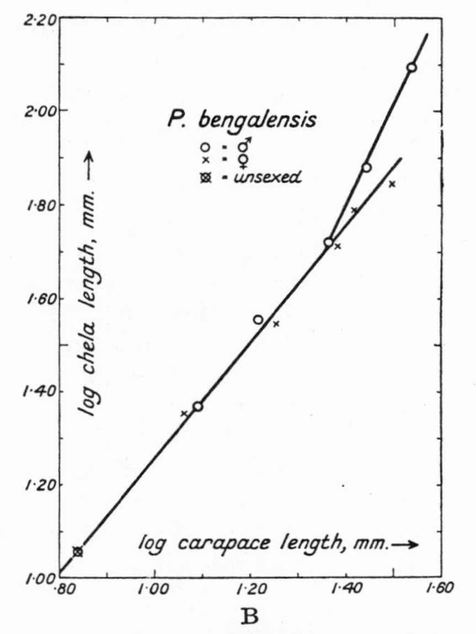

FIG. 21.—Relative growth of male and female chelae. (A) in *Palaemon carcinus*, (B) *Palaemon bengalensis*; logarithmic plotting.

(*From data of Kemp*, 1913–15.)

relative growth of homologous organs next shows that the same organ may behave very differently, as regards its growth-behaviour, in different forms. For instance, the chela of male *Uca pugnax* shows heterogony on one side of the body only, but shows it throughout all but the first instar of post-larval life, with a decrease in its growth-coefficient apparently at the time of sexual maturity. In the spider-crab *Maia squinado* (Huxley, 1927; and unpublished), both chelae are heterogonic, but heterogony does not set in until quite late in life, presumably at sexual maturity, and then continues till death. The same appears to be the case with the large prawn *Palaemon carcinus* (Tazelaar, 1930), though here the chelipeds are the second and not the first pereiopods; and with the spider-crabs of the genus Inachus. But in the latter the male chelae revert more or less completely to the female type in the non-breeding season; this reversion is much less marked in *I. dorsettensis* (Shaw, 1928) than in *I. mauritanicus* (Smith, 1906A). In various other crabs, and in lobsters, crayfish and pistol-crabs (Alpheus), both chelae are heterogonic, but with different growth-coefficients, leading to the condition of heterochely. Furthermore, the sex-difference as regards the growth-coefficient of the chela may vary, some forms having equal positive heterogony in both sexes, others showing positive heterogony in both sexes, but with a lower growth-coefficient in the female, and still others showing male heterogony but female isogony. This variability is particularly well shown in prawns (Palaemonidae); in these, further, the large chela is the second, not the first pereiopod. Finally in *Gammarus chevreuxi*, Kunkel and Robertson (1928) have shown that the marked heterogony of the male gnathopod begins shortly before sexual maturity and ends shortly after, its growth being roughly isogonic for the much longer previous and subsequent periods. (Cf. also birds' wings, p. 263.)

A similar state of affairs is seen in regard to the abdomen of female Brachyura. This must always be heterogonic for some part of its development, since it is always broad in the adult, always narrow and of male type in the young juvenile. A state of affairs similar to that of the large male chela of Uca is found in the female abdomen of *Carcinus maenas*—it is heterogonic (with a late increase in intensity) from the earliest stages until the end of life. The details for both sexes at all ages are shown and described in Fig. 7. In female Uca, on the other hand, while heterogony is initiated at the beginning of post-

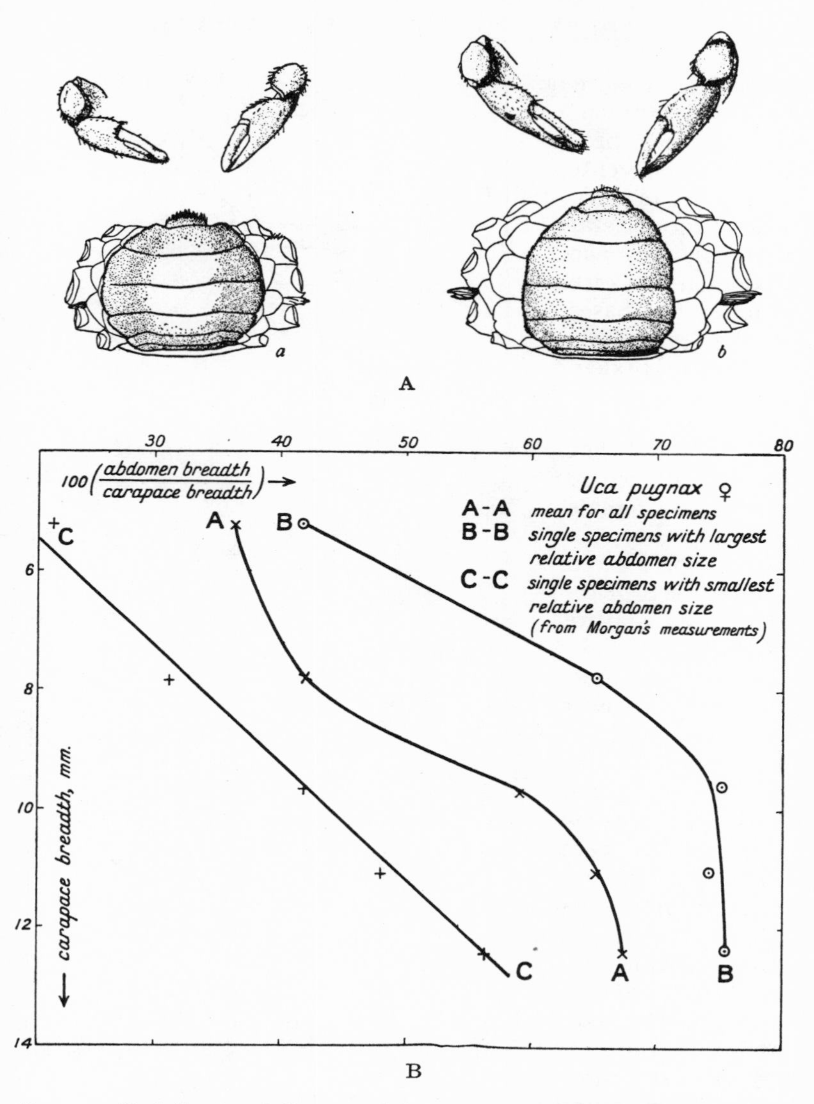

FIG. 22.—Variation in relative growth in the female abdomen of fiddler-crabs (*Uca pugnax*).

(A) Left, medium-sized crab (carapace breadth, 10 mm.; abdomen, 6 mm., broad and reaching the bases of the legs). Right, medium-sized crab (carapace breadth, 11 mm.; abdomen, 5½ mm., broad, not reaching the bases of the legs).

(B) Means (A—A) and extremes (B—B and C—C) of relative abdomen breadth in female crabs between 4 and 14 mm. carapace breadth. There is a clear bimodality of mean abdomen size, with marked heterogony between 8 and 11 mm. carapace breadth.

larval life, a state of equilibrium (adult female proportions) is eventually attained, when the lateral margins of the abdomen have reached the bases of the legs (Morgan, 1924; Huxley, 1924A) (Fig. 22).

(In passing, it may be noted that Morgan was led to postulate female intersexuality in this species on finding certain apparently mature females with abdomens proportionately narrower than the full female type. Huxley, however (l. c.), was able to show that these were merely the extreme minus variants for the normal variation curve of female abdomen-growth, and that the supposition of intersexuality was uncalled for—an interesting application of the study of heterogony.)

That the equilibrium-position is

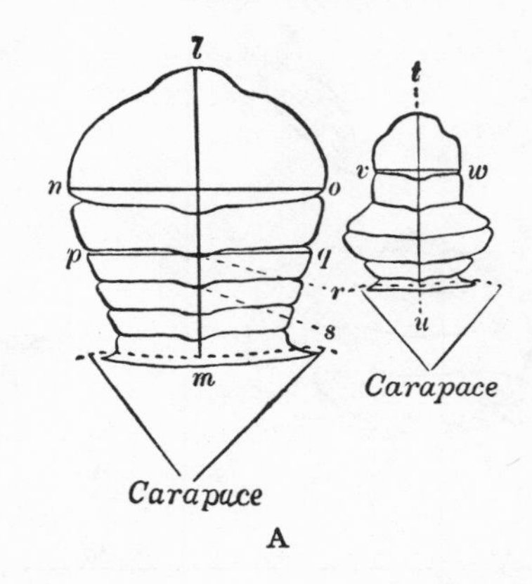

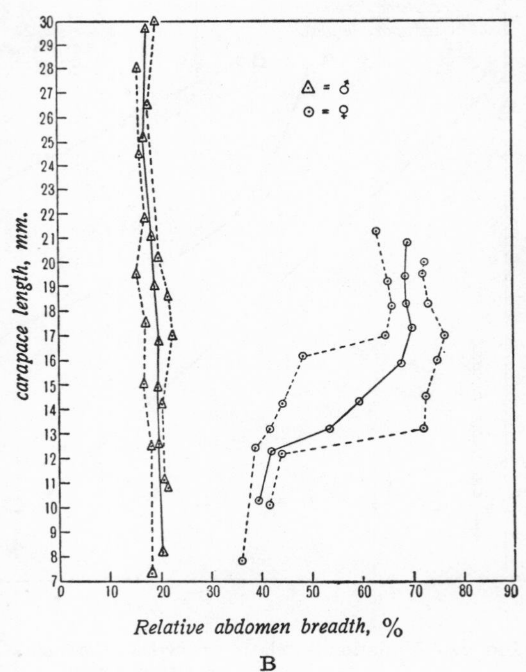

FIG. 23.—Relative growth in the male and female abdomen in the spider-crab, *Inachus dorsettensis*.

(A) Left, mature female abdomen; right, mature male abdomen. Abdominal breadth was taken on the 6th segment ($n - o$; $v - w$).

(B) Change of means and extremes of relative abdomen breadth with increase of carapace length in males and females.

not in any way automatically, and still less mechanically, brought about when the sides of the abdomen reach the legs is shown by Pinnotheres, the pea-crab (Atkins, 1926)

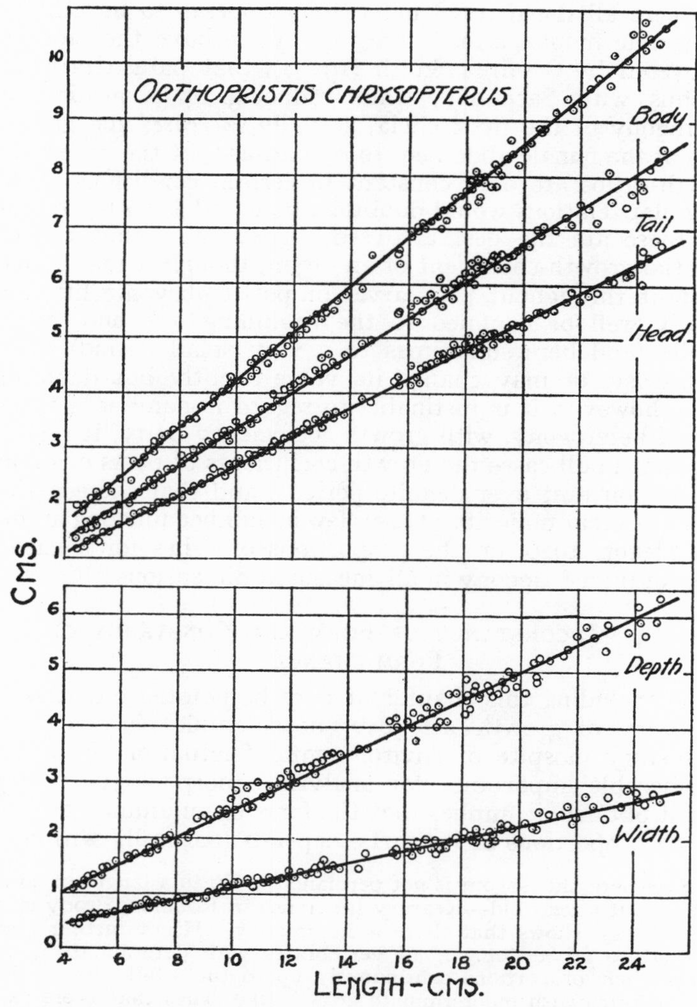

FIG. 24.—Relative growth in the teleost fish, *Orthopristis.*

The abscissae represent total length. The ordinates represent (above) the length of the head, trunk (body), and tail; the division between head and trunk is taken at the hind end of the opercular bone, between trunk and tail immediately above, or below, the end of the hindmost median fin; (below) maximum depth and width of body.

in the adult female of which the abdominal margins far overlap the leg-bases.

Then we have the quite different case of Inachus (Fig. 23) which resembles that of the male gnathopod in Gammarus, only here all the marked heterogony appears to be achieved in a single moult-period. And finally we have the fact first discovered by Geoffrey Smith (1906A), that parasitization of Inachus with Sacculina, while reducing or abolishing the heterogony of the male chela, actually *increases* the growth-ratio of the female abdomen, to remind us that the differential growth-ratios are only constant in certain conditions.

Similar relations would doubtless be found for other organs, but these are the best analysed. They show that the differential growth-coefficient of an organ, though it may remain constant throughout post-larval or post-embryonic life, may equally well be confined to the beginning, the end, or the middle (and here sometimes to a very small period) of the life-history, or may change its value slightly but definitely. Since, however, it is justifiable to regard isogony as a special case of heterogony, with growth-coefficiency unity, it remains true that in all cases the growth-coefficients of parts or organs remain constant over definite periods, and that these periods are in the great majority of cases few in number and long in time. (See Hecht, 1916, on the proportions of fishes for a case of long-continued isogony in all measured dimensions: Fig. 24.) [1]

§ 4. Inconstancy of Form and Constancy of Form-change

In concluding this chapter, it may be pointed out that the constancy of growth-ratio over considerable periods of the life-history in spite of environmental fluctuation, is of very considerable importance for analytical morphology. Whereever it obtains, it implies that the form of an animal, as given by the proportions of its parts, depends (naturally within the

[1] Even here, the isogony is not permanent. Up to a length of 30 cm.—i.e. about a year old—Kearney (reference in Robbins, Brody et al., 1928, p. 123) shows that there is heterogony. Hecht further points out that so far as known, all vertebrates with determinate growth change their proportions continuously up to the adult phase. It is only in forms with indeterminate growth like fishes that there exists a long-continued period with no change in external proportions (though even here the relative size of the viscera changes). The work of Keys (p. 259), however, indicates that Hecht's conclusions are not strictly accurate. Compare also Olmsted and Baumberger (p. 261).

limits of normal variation, of which more later) solely upon its absolute size, not upon the length of time it has taken to reach that size, nor upon changes in any other external variable. The modifications of and exceptions to this statement we shall consider later; here we can accept it as our first general rule.

As immediate corollary of this we have the fact that only animals in which all organs are growing at the same rates will preserve their form unchanged with increase of size; and this is as much as to say that no animal will keep its form identical while increasing in size, for it appears highly improbable that any animal will be found in which some organs do not grow at a different rate from the body as a whole. And even if for the moment we stick to external form, and further if we only consider quite large differences in growth-activity, we shall find many animals in which, as in the male fiddler-crab, the only constancy of form is the constancy of its mode of change. This is less obvious and in some ways less important in the higher animals (notably almost all mammals and birds among vertebrates and almost all insects and spiders among arthropods), in which growth ceases at a definite size, and there supervenes an adult stage of constant size and often of long duration. For here we can often afford to consider only the adult forms, in which the proportions of form have been fixed by the cessation of growth. It is the limitation of form at a fixed absolute size which confers this convenience upon the systematist and the morphologist.

Even here, however, as we shall see in detail later, the rule has many applications. To take the most obvious case, the absolute size at which growth ceases may be altered by treatment such as feeding; in such case, the permanently stunted individual will approximate in proportions to a normal juvenile stage, the well-fed, abnormally large specimen will have proportions not met with at all among the normal population of adults.[1] It is, in other words, a mere biological accident that adult proportions, even in species with limited growth, are relatively fixed; and to neglect the fundamental fact of change

[1] As a matter of fact, the degree of development (including growth) of a higher vertebrate appears to be simultaneously dependent upon at least two variables, size and age. This is well shown by Appleton (1925, see his chart 3) as regards the degree of ossification in new-born rabbits; and by Jackson (1925) for young rodents stunted by underfeeding. Jackson's results are discussed further in Chapter VI. See also the work of Adolph (p. 258).

of proportions with absolute size, and to proceed as if certain arithmetic (percentage) proportions were immutable 'characters' of the species, may lead to serious error.

But we must remember that the limitation of growth and the consequent establishment of a small range of stable adult size is a late and specialized feature in evolution. The majority of animals show unlimited growth: they continue growing, though usually at a constantly diminishing rate, until they die, or in asexually-reproducing forms, until they divide. A lobster or a plaice may increase its linear dimensions several fold after the attainment of sexual maturity. In such types, there is *no fixed or adult form*; the change of proportions continues unabated throughout life, and may be as obvious during post-maturity as during pre-maturity. An excellent example of this is provided by the detailed studies of Mrs. Sexton (1924) on the successive instars of *Gammarus chevreuxi*, supplemented by the work of Kunkel and Robertson (1928) on the same species.[1]

Even among mammals a change of proportions may continue throughout life. In the voles (Microtinae) Hinton (1926, Chap. II, 8–14, Pls. III, IV, IX) finds that slow growth occurs long after the adult state has been arrived at, the epiphyses of the long bones never uniting. This continuous growth is accompanied by continuous change of proportions. With increasing size of the skull, for instance, the rostrum becomes relatively narrower and slightly longer, the interorbital region narrower and the molars relatively smaller. Unfortunately the measurements given do not permit of any accurate statement as to the changes involved, or as to the distribution of growth-potential in different regions. Here is an interesting field for the student of relative growth. It would be particularly interesting to discover whether the relative growth-rates of tail and parts of skull, limbs, etc., remained the same after the attainment of sexual maturity as they did before. It would be easier to investigate this on the limb-segments than on the skull, which undergoes complex distortions and curvatures. It would also, of course, be necessary to keep

[1] Sexton states that sexual maturity occurs at the seventh instar, that proportions continue to change for two further instars in the male, one in the female, but after this no further proportion-changes occur (though the males at least may increase about 40 per cent. in length). That this statement is not accurate is shown by Kunkel and Robertson, whose graphs demonstrate a change in the proportions of several organs up to the largest sizes found.

the animals under standard conditions, since the work of Sumner and of Przibram has shown that increased temperature causes an increased relative size of appendages in rodents.

It is true the change will not usually be of the same extent after sexual maturity, for although the changes in absolute size may be greater between maturity and death than in the period from the post-embryonic or post-larval phase to maturity, yet the fraction of total growth which takes place after maturity is always a good deal less, if measured by the true criterion, namely the amount of multiplication of initial size. For the fiddler-crab, for instance, the pre-maturity multiplicative increase in weight is about 250-fold, the post-maturity increase about three- to four-fold, though the absolute (additive) increases are roughly as 1 to 2·5. None the less, the post-mature alterations may be very considerable. In the male fiddler-crab, after his attainment of sexual maturity, the proportion of the weight of the large chela increases from 43 per cent. of rest-of-body weight to nearly 62 per cent.—an increase of nearly 45 per cent. in relative size.

No two male *Uca pugnax* have the same proportions unless they happen to be of the same absolute size: any diagnosis made on the basis of percentage measurements of chelae (and also, though much less markedly so, for other organs such as the pereiopods) would be valueless. But in spite of the fact that the form of the animal is continually changing, it does so in an orderly way; and though percentage values for the limbs have no diagnostic significance, the constants in the growth-ratio formula are true specific characters. In a word, the systematist needs algebra as well as arithmetic in making any diagnoses based upon the size of parts of the body.

Note.—S. A. Allen (1894), *Amer. Mus. Nat. Hist. Bull.*, **6**, 233, also finds a progressive change of proportions in a rodent (see p. 40). Neotoma shows a steady increase of dolichopy and dolichocephaly with increase of absolute size.

CHAPTER II

THE COEFFICIENT OF CONSTANT GROWTH-PARTITION; AND SOME SPECIAL CASES

§ 1. The Heterogony of Deer Antlers

THERE are certain special cases so important to a study of relative growth that they deserve a chapter to themselves.

The first is that of the antlers of deer. As is well known, these are shed each year, and replaced the year after by a totally new growth. Usually, each new growth is larger than the preceding growths, but in old age, illness, or other especially unfavourable conditions, the weight (and number of 'points') may decrease. An analysis of the normal growth of the antlers of a number of individual red deer (*Cervus elaphus*) and of the factors affecting that growth, is given in Huxley (1926).

Further, casual inspection is sufficient to indicate that relative antler-weight increases with absolute body-weight, as is stressed by Champy (l. c.). To obtain quantitative data, however, was not easy. After much search, I hit on the papers of Dombrowski (1889–1892) published many years ago in an obscure periodical—the only papers to my knowledge to contain the body-weights and antler-weights of large numbers of Red deer and Roe deer. These data, supplemented by those of Rörig (1901), by scattered cases in the literature, and by information privately supplied to me by sportsmen, have now been analysed by me (Huxley, 1927, and 1931). It appears quite definitely that although there may be much individual variation even in one locality, and though extraneous agencies such as the amount of lime in the soil affect relative antler-weight considerably, yet when the mean of considerable

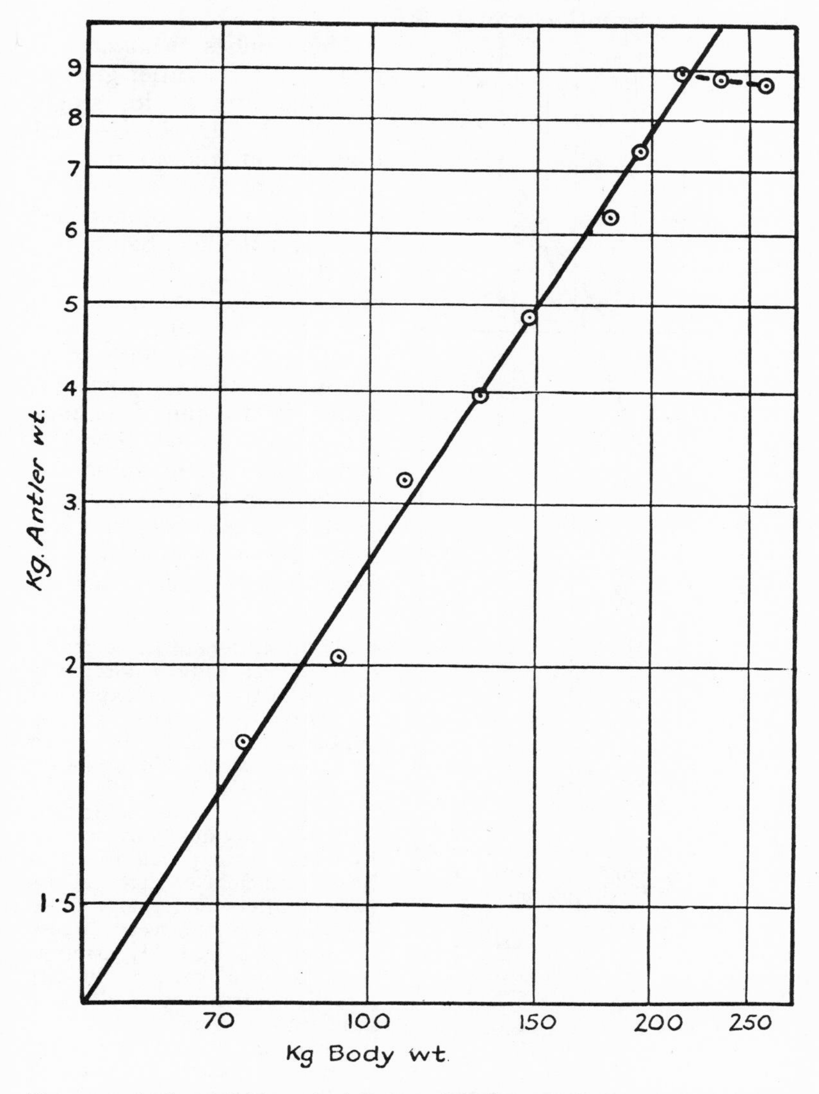

FIG. 25.—Antler-weight against body-weight in 527 adult red deer (*Cervus elaphus*); logarithmic plotting. See Table II.

k, except for the last two points, is close to 1·6.

numbers is taken, the results approximate to the formula for constant differential growth-ratio.[1]

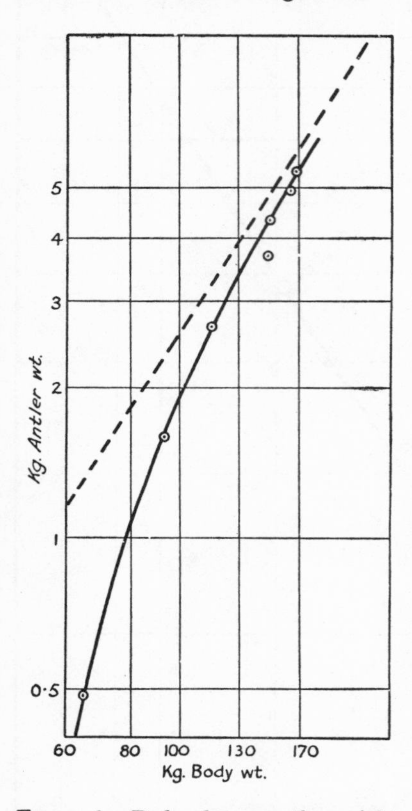

FIG. 26.—Red deer, antler-weight against body-weight at various ages; logarithmic plotting.

The antler- and body-weights for the first 7 years of life (in stags from Warnham Park, Sussex) are plotted. The curve (solid line) bends over and approximates to the straight line curve (dotted line) for antler-weight against body-weight in adults (see Fig. 25).

k begins with a value of 3·0 or over, and declines to 1·6 or under.

This applies to *adult* animals. When antler-growth is taken by age for single individuals, it will be seen that the differential growth-ratio of the antler-weight is not constant, but declines steadily with age, being first about 3·0, and declining to close to 1·0. (When regression of body-weight occurs, it appears certain that regression in antler-weight accompanies it, though it cannot yet be stated whether the regression is heterogonic.)

It is doubtless affected also by numerous subsidiary factors such as abundance of food and specific dietary

[1] A discrepancy occurs as regards the antlers of those beasts with highest body-weight; the weights of these when expressed as relative (percentage) weights, fall below those of the body-size class next below. This appears to be merely a classificatory phenomenon. There being considerable individual variation as to what we may call the partition-coefficient of material between antlers and body, those animals with the very largest body-weights are likely to represent extreme variants in the direction of heavy body but light antlers. Further, and possibly more important, since body-weight is extremely variable owing to fluctuations in amount of fat, most very heavy beasts are likely to owe their exceptional weight to exceptional nutritive conditions; and therefore their *relative* antler-weight will go down relatively to this excess of fat, which is presumably without immediate significance in determining antler-size.

TABLE II

BODY-WEIGHT, ANTLER-WEIGHT, POINT-NUMBER AND RELATIVE ANTLER-WEIGHT OF 527 RED DEER SHOT IN VARIOUS PARTS OF EUROPE (392 COLLECTED BY DOMBROWSKI; 10 BY BAILLIE GROHMAN; 125 BY HUXLEY. ANALYSED BY HUXLEY, 1931A)

Arranged by body-weight classes, all of 20-kg. interval (except the last class, of 40-kg. interval). Note that for Classes 2 to 8 (comprising over 90 per cent. of the animals) the relative antler-weight rises steadily with increasing body-weight. k for antler-weight (except for the last two classes) is about 1·6; b = ·00162.

Class	kg. body-weight	No. of specimens	Mean body-weight kg.	Mean antler-weight kg.	Mean point-number	Relative antler-weight per cent. of body	Relative pt. no. $\left(\frac{\text{pt. no.}}{\text{body-wt.}}\right)$
1	60– 80	19	74·4	1·64	7·50	2·20	0·101
2	80–100	119	93·4	2·03	8·20	2·17	0·088
3	100–120	106	110·4	3·16	9·81	2·86	0·089
4	120–140	113	130·6	3·96	11·64	3·03	0·089
5	140–160	65	148·9	4·78	11·74	3·21	0·079
6	160–180	29	170·7	6·21	13·10	3·64	0·077
7	180–200	33	191·1	7·28	14·77	3·81	0·077
8	200–220	18	211·8	8·91	15·41	4·21	0·073
9	220–240	14	231·7	8·79	13·62	3·79	0·059
10	240–280	11	259·1	8·63	13·78	3·33	0·053

TABLE III

BODY-WEIGHT, ANTLER-WEIGHT, POINT-NUMBER AND RELATIVE ANTLER-WEIGHT OF 405 ROE DEER SHOT IN VARIOUS PARTS OF EUROPE (DATA FROM DOMBROWSKI; ANALYSED BY HUXLEY, 1931A)

k for antler-weight = about 0·57; b = ·0455.

Class, Body-weight	No. of specimens	Mean body-weight kg.	Mean antler-weight g.	Mean point-number	Relative antler-weight per cent. of body
1–5 (13–18 kg.)	127	16·6	225·7	5·92	1·36
6–12 (19–25 kg.)	254	20·9	257·0	6·05	1·23
13–15 (26–39 kg.)	24	28·3	306·5	6·21	1·08

factors. The reason that the weights for adults fall upon the line corresponding to a constant growth-ratio with k = about 1·6 appears simply to be that during the decline of the antler's growth-ratio, a rather narrow range of values for the growth-coefficient is attained during adult life (see Fig. 26), the great majority falling between say 1·8 and 1·4.

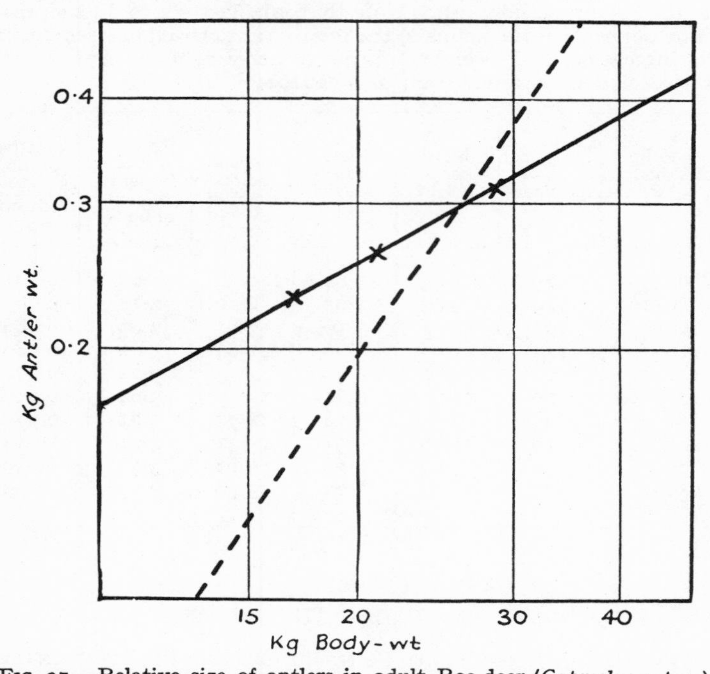

FIG. 27.—Relative size of antlers in adult Roe-deer (*Capreolus caprea*).

Solid line, antler-weight against body-weight in 405 Roe-deer ; k = 0·57. Dotted line, prolongation of corresponding curve for adult Red Deer (see Fig. 25). Logarithmic plotting. See Table III.

Corresponding data for the Roe deer (*Capreolus capreolus*) gave what at first sight appeared a quite paradoxical result—namely a *decrease* of relative antler-weight with increase of absolute body-weight among adult males (Fig. 27). There is thus negative heterogony of the antlers, with a growth-coefficient of about 0·57. Reflection suggests the probable explanation. There is no reason why the decline in the antler's growth-coefficient with age should not in another species proceed much faster than in the Red deer, and reach a stage where by the attainment of maturity it was normally below 1·0. This purely quantitative difference in the rate of change with age would suffice to explain the apparently contradictory results (Fig. 28). As to the biological causes underlying this

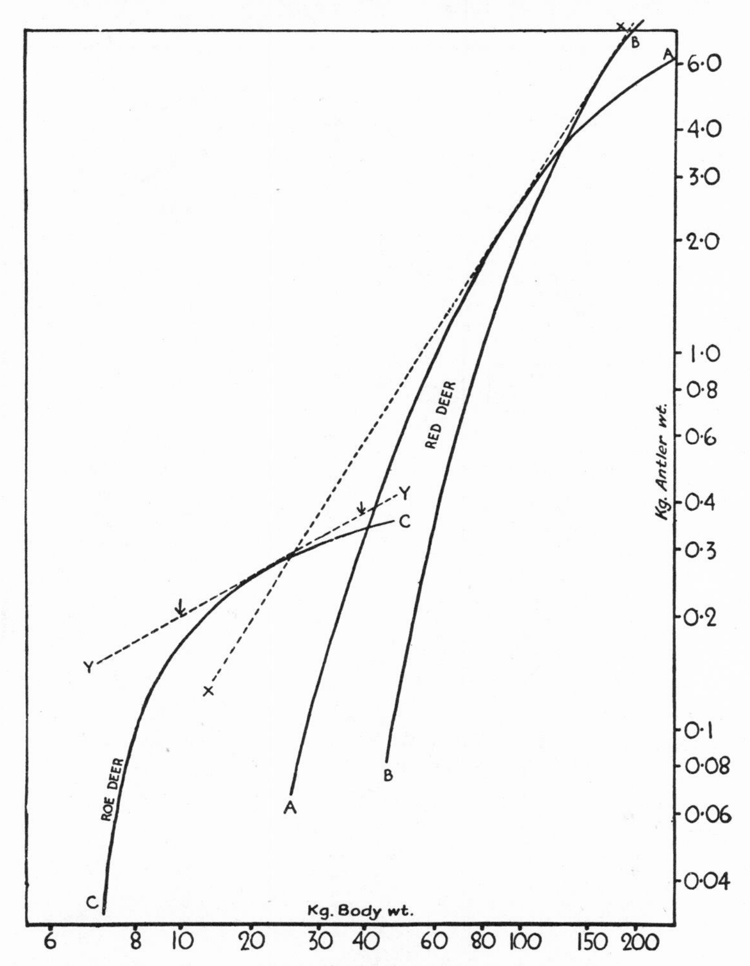

FIG. 28.—Diagram to compare probable method of antler-growth in Red and Roe-deer; logarithmic plotting.

X—X, the curve for adult red deer (see Fig. 25). A—A and B—B, probable curves for individual antler-growth with age in a small and a large specimen respectively. Y—Y, the curve for adult roe-door (see Fig. 27). C—C, probable curve for individual antler-growth in a typical roe-deer specimen, rising at first more rapidly, but then sinking much lower than the corresponding curve for red deer.

quantitative difference we can only speculate: it would seem probable that the Red deer type of slow decrease, with positive heterogony throughout, is the normal course of events in Cervidae, but that it was for some reason biologically desirable for the Roe deer to have small antlers.

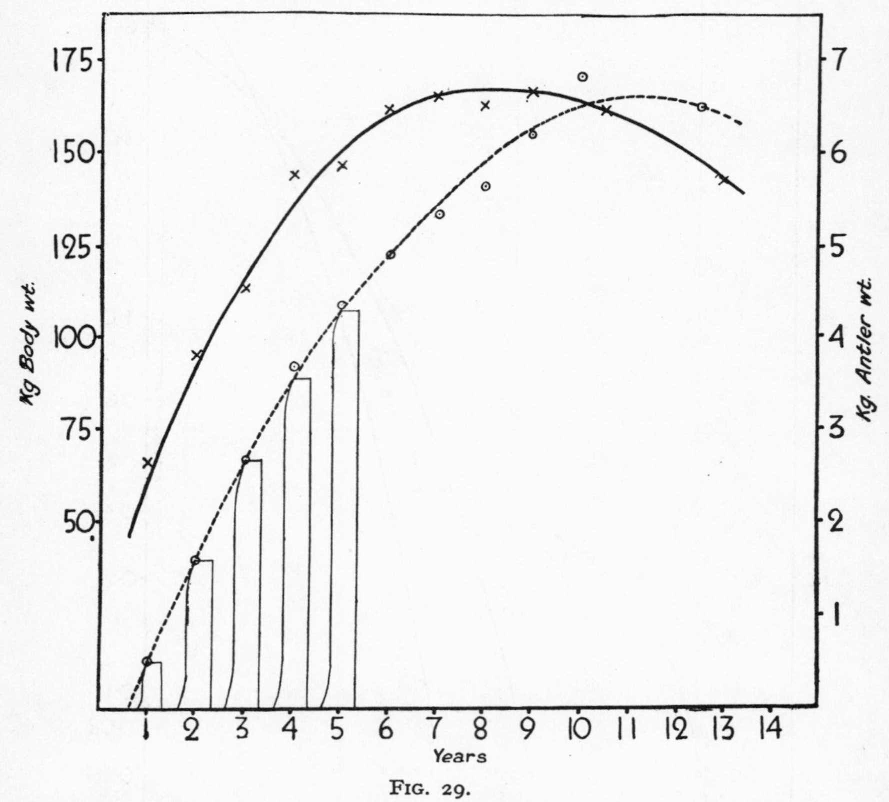

FIG. 29.

Solid line, body-weight against age in 212 male red-deer from Warnham Park. Dotted line, antler-weight against age in a smaller and selected group of stags from the same locality. The thin continuous lines below the curve for antler-weight represent diagrammatically the actual growth and shedding of the antlers year by year. The fact that the antler curve inflects later than that for body-weight is probably due to the antlers being from a selected group of beasts, of size above the average.

After this digression, we will return to the general problem involved in the growth-ratio of the antlers. We have seen that the apparent constancy of their growth-ratio, obtained by plotting antler-weight against body-weight in adults, is shown to be a particular consequence of the steady decline of individual growth-ratio with age. But even this does not exhaust the complexity of the phenomenon. The curve for age-change of growth-ratio is obtained by plotting the weights of fully-formed antlers of known age against body-weight for the same age. It will be at once clear that the actual growth-ratio of the antlers can never be the same as that thus obtained, but must always be higher (Fig. 29). For the points on the curve are those which would be obtained if the antler grew with a constant differential growth-coefficient from its inception; whereas actually, it has to begin its growth anew each year from zero.

Now this is of considerable importance, since it indicates that it is not necessarily the actual rate of growth which is regulated in accordance with our formula, but the *limitation of the total amount of growth achieved.* What our results tell us is that at any given body-size the total amount of material which can be incorporated in the organ is proportional to the body-size raised to a power (the exact value of the power also varying with age). The mechanism of this relation is at present obscure. We do not know whether the total bulk of material in the body imposes the relation directly, which is unlikely; whether some substance is formed in the body in this particular quantitative relation, as an exponential function of body-weight, and the final size of the organ is then directly proportional to the amount of this substance; or whether there be after all a true constant differential growth-ratio between organ and body, determined by some peculiarity of the organ, but that this growth-ratio represents a limiting value, higher values being possible and indeed necessary whenever the relative size of the organ is below its limiting amount. The last supposition is perhaps the most probable, on the close analogy with regeneration (see below), but experiment alone can decide the point.

§ 2. The Coefficient of Constant Growth-partition

In any case, to speak simply of *growth-coefficients* in such a case is misleading; yet we require a term for the exponent of body-size according to which relative organ-size is limited. Two terms are possible—either *coefficient of growth-limitation*, or else *growth-partition coefficient*; I shall adopt the latter.[1]

[1] Since writing this passage, I find that Robb (1929) had previously suggested the same idea of growth-partition, which has later been adopted by Twitty and Schwind (1931).

In general, it would appear that the existence of a growth-partition coefficient is the most fundamental fact in considering relative growth of parts, and that when true constant growth-coefficients or constant differential growth-ratios are found, they represent special limiting cases of this more general conception.

Let us now consider three further examples which support this conclusion. I have mentioned regeneration. I shall deal with this more fully in a later chapter. Here it suffices to recall the fact that in an animal capable of full regeneration, any organ, heterogonic or not, will, after amputation, be restored in favourable conditions to its normal proportionate

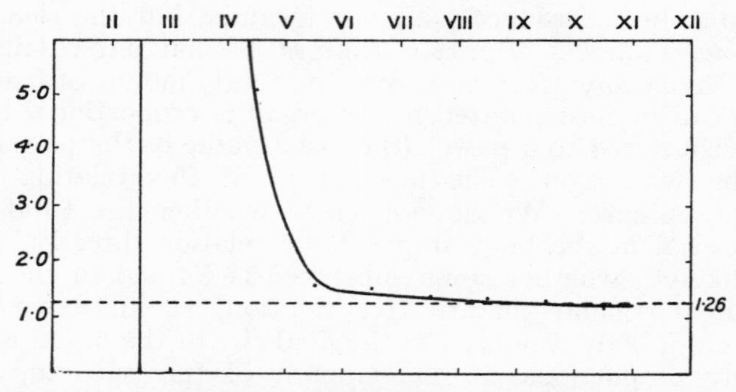

FIG. 30.—Decrease of growth-coefficient during regeneration in the legs of *Sphodromantis bioculata.*

The abscissae represent moult-stages. The ordinates are growth-quotients: i.e. the ratio of the length of the leg at a given moult to its length at the preceding moult. The dotted line represents the mean growth-quotient ($1\cdot26 = \sqrt[3]{2}$) for normally-growing limbs. The solid line is the curve for a middle-leg amputated before the IIIrd moult (mean of 3 specimens).

size. In other words, during the process of regeneration its growth-ratio will be much higher than normal, and will gradually sink until it reaches the normal value, at which it will then continue. This emphasizes the generally accepted idea that regeneration is simply a special case of growth, and furthermore makes it clear that here at least the normal growth-ratio of an organ merely represents a limiting value. Thus, as was suggested above with regard to deer-antlers, relative size of organs appears to be determined in the first instance as an equilibrium between amount of material in the organ and amount of material in the body, the equilibrium being determined according to our general formula $y = bx^k$. If the equilibrium be upset, regulation towards the equilibrium position will occur during later growth. The particular mechanism by which the equilibrium is attained does concern growth-ratio; the more the organ is below equilibrium-size, the higher will be its growth-ratio.[1] (See Fig. 30.)

These conclusions are supported by various lines of evidence. In the first place, in cases of grafting of organs we should expect the organ of a young animal grafted on to an older and larger animal to be accelerated in its growth until it reached a size prescribed by its growth-partition coefficient, and the organ of an older animal grafted on to a younger and smaller animal to be correspondingly retarded in its growth. For the first, we may turn to the results of Wachs (1914). When he inserted the lens of a young Urodele larva into the eye of an older larva from which the lens had been previously removed, the small lens was accelerated in its growth. For the second, as well as the first, we have an example in the work of Twitty (1930). Here cross-transplantation was made between larvae of *Amblystoma tigrinum* and *A. punctatum*. The latter species grows much more slowly than the former. Twitty removed the eye of a *punctatum* larva and replaced it by one of the same size from a *tigrinum* larva; owing to the higher growth-rate of *tigrinum*, the age of the donor was much less than that of the host. Even when the host was fed only minimally, the grafted eye now increased in size much more rapidly than that of the host: in one case it increased 50 per cent. in diameter while the host remained stationary in length. (See p. 197.)

The converse experiment consisted in removing the eye from a *tigrinum* larva and engrafting in its place an eye of the same size from a considerably older *punctatum* larva. In this case, the grafted eye made very slow growth. Here the rate of growth could be compared with that made by *punctatum* eyes grafted into *A. tigrinum* during the embryonic period.

[1] In some cases at least the change in growth-ratio will occur according to the law enunciated for Sphodromantis by Przibram, 1917: When Z is the normal final length of the regenerating limb, n the length after amputation, r its length at the beginning of a given period of time t, R the length at the end of the time t, V_a the normal coefficient of increase of the limb between one moult and the next, then $Z - n - r = \frac{K}{t.V_a} \cdot \frac{R}{r}$; and the growth-partition coefficient represents the limiting value of the growth-ratio when equilibrium is established.

It can at once be seen (Fig. 31) that the growth of the older eyes slowly approaches the normal growth-curve (see also Figs. 85 to 87, and especially 88).

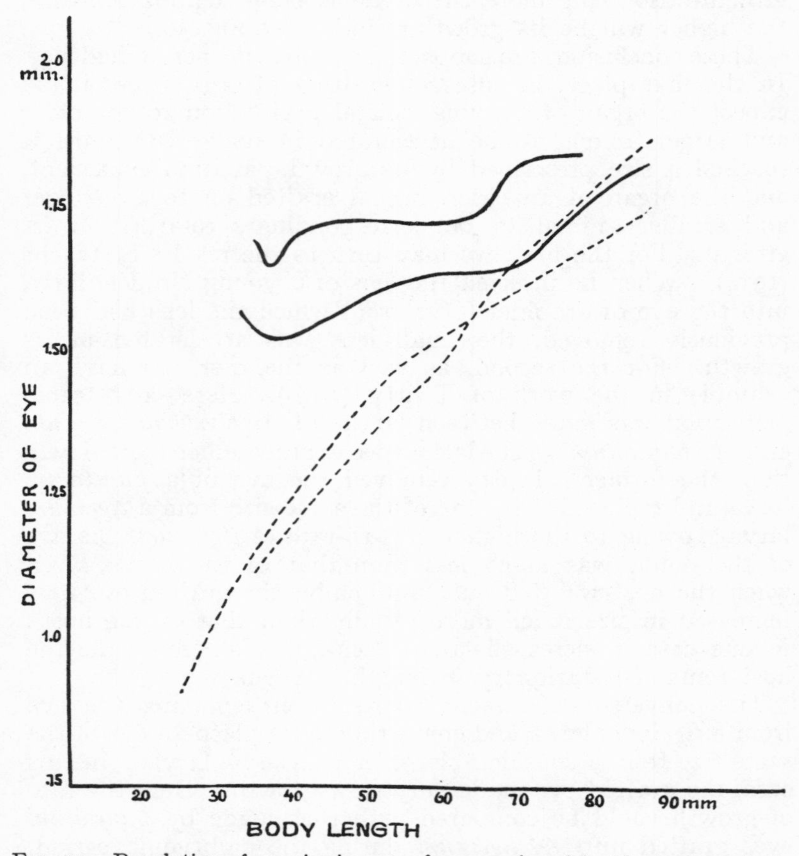

FIG. 31.—Regulation of eye-size in eyes of one species of Amblystoma grafted on to another.

Dotted lines, curves for two cases when embryonic eyes were grafted to a host of the same stage of development as the donor. Solid lines, curves for two cases when eyes were taken from an older larva and grafted on to a younger host larva; in this case the eyes hardly grow at all until they reach the correct relative size.

We thus are driven to the conclusion that though the eyes of the two species have different specific growth-intensities, and therefore different coefficients of growth-partition when both are present in the same body, there is for each body-size a characteristic eye-size towards the attainment of which the rate of eye-growth is regulated.[1]

In general, as is well known, the rate of regeneration is higher the more material is removed. This fits in with the ideas here presented and with the conceptions of Przibram (l. c.), but it throws into relief the very real difference between the idea of a constant differential growth-ratio and a constant coefficient of growth-limitation. Let us consider a Planarian worm, in which according to Abeloos (l.c.) the trunk grows heterogonically with reference to the head. During normal growth the relation is one of a constant differential growth-ratio. But during regeneration, the more is cut off, the more rapidly regeneration takes place: i.e. the smaller the fraction of the body left, the more rapid is the growth-ratio of the regenerate. What is constant is the final partition-coefficient between head-material and trunk-material; the normal constant differential growth-ratio is the special case of growth during which the partition-coefficient is always of this limiting value.

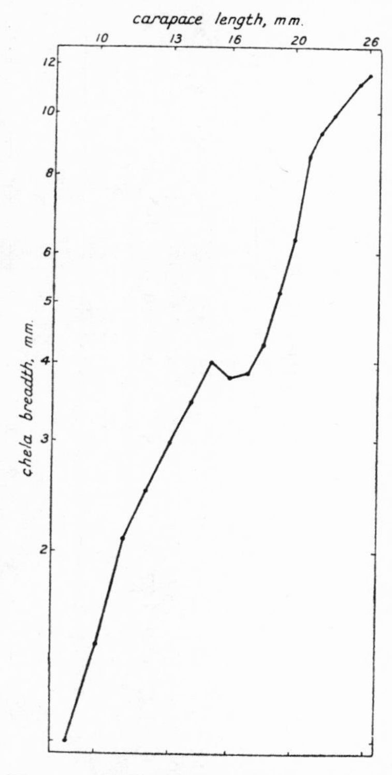

FIG. 32.—Chela breadth against carapace length in the male of the spider-crab, *Inachus mauritanicus*; logarithmic plotting. (See Table IIIA.)

The 'low' males (below about 14 mm. carapace length) and the 'high' males (above about 20 mm.) fall on a single curve with diminishing growth-coefficient (mean value of k over 2·3; for 'low' males, about 2·6; for 'high' males, about 1·4). Between these sizes the chela regresses to a narrow female type, then enlarging again with a very high growth-coefficient. (*Constructed from the data of G. W. Smith*, 1906A.)

Finally, the case of Inachus, investigated by G. Smith (1906A) whose data have been further analysed by me (unpublished) also supports this view-point. At Naples, *I. mauritanicus*, the species of Inachus studied by Smith, shows three forms—'low'

[1] Further details as to the specific growth-intensities of eyes and other organs when heteroplastically transplanted are recorded in Chapter VI.

TABLE IIIA

RELATIVE GROWTH OF CHELA-BREADTH OF 496 OF THE SPIDER-CRAB (*Inachus mauritanicus*)

(Constructed from Table 6, p. 97 of Geoffrey Smith, 1905)

(*a*) *Totals.*																		
car. l. class, mm.	9	10	11	12	13	14	15	16	17	18	19	20	21	22	23	24	25	26
No. of specimens	2	12	22	45	61	49	38	30	27	23	25	23	35	42	24	23	10	5
mean chela br., mm.	1·00	1·42	2·09	2·49	2·97	3·45	4·00	3·77	3·82	4·26	5·16	6·30	8·49	9·22	9·83	10·35	11·10	11·40
(*b*) *Female-type males in bimodal classes*																		
No. of specimens	—	—	—	—	—	—	—	36		17	14	9	—	—	—	—	—	—
mean chela br., mm.	—	—	—	—	—	—	—	3·00		3·41	3·64	3·89	—	—	—	—	—	—
(*c*) *Male-type males in bimodal classes*																		
No. of specimens	—	—	—	—	—	—	—	21		6	11	14	—	—	—	—	—	—
mean chela br., mm.	—	—	—	—	—	—	—	5·14		6·67	7·09	7·86	—	—	—	—	—	—
(*d*) *Chela-breadth frequency.*																		
Chela-br. class, mm.	1	2	3	4	5	6	7	8	9	10	11	12	—	—	—	—	—	—
No. of specimens	11	67	*131*	90	21	11	13	35	*46*	*46*	18	7	—	—	—	—	—	—

males, of small size with relatively small but male-type chelae, 'high' males of large size, with relatively large male-type chelae, and those of intermediate size, which have extremely small, female-type chelae.

When chela-size is plotted double-logarithmically against body-size, it is found that the means for the 'low' and 'high' males fall on two segments of a single simple curve, thus confirming Smith's view that these two types are merely breeding males in their first and second seasons respectively, and that those with female-type chelae are males in the non-breeding phase, during which the secondary sexual characters of their chelae have regressed to the female or neuter type.

Further, on the double logarithmic plot, the curve for these intermediate males first actually declines, and then mounts very steeply until it meets the prolongation of the straight-line curve for the low males, upon which it bends over and continues as the line for the 'high' males. In other words, after the regression period, the growth-ratio of the claw is much higher than normal, but becomes normal as soon as the theoretical equilibrium-size is reached.[1] (Table IIIA and Fig. 32). It is also seen that the frequency for chela-breadth is bimodal: this will be discussed in § 5.

§ 3. Heterogony in Holometabolous Insects

A somewhat different set of special cases is that provided by holometabolous insects. Many of these possess organs (usually of secondary sexual character, and these usually in the male sex), which increase in relative size with increase of absolute size of body. The most familiar of these are the mandibles of the stag-beetles (Lucanidae) and the 'horns', cephalic or thoracic or both, of various other beetles such as the Dynastidae; but Champy (l. c.) has collected numerous other examples, ranging from antennae (e.g. Acanthocinus: Champy, 1924, p. 167) and forelegs, to the 'tail' on the hind wing of Papilios and the swollen segments of the hindlegs in certain Hemiptera, such as Anoplocnemis (Champy, 1924, p. 173). See Figs. 33, 34, 91.

Analysis shows (Huxley, 1927 and 1931) that the relation between the dimensions of the organ and the body here too

[1] It is interesting to find that in I. dorsettensis, studied by Shaw (1928), the regression towards female type in the non-breeding season, though present, appears to be much less marked.

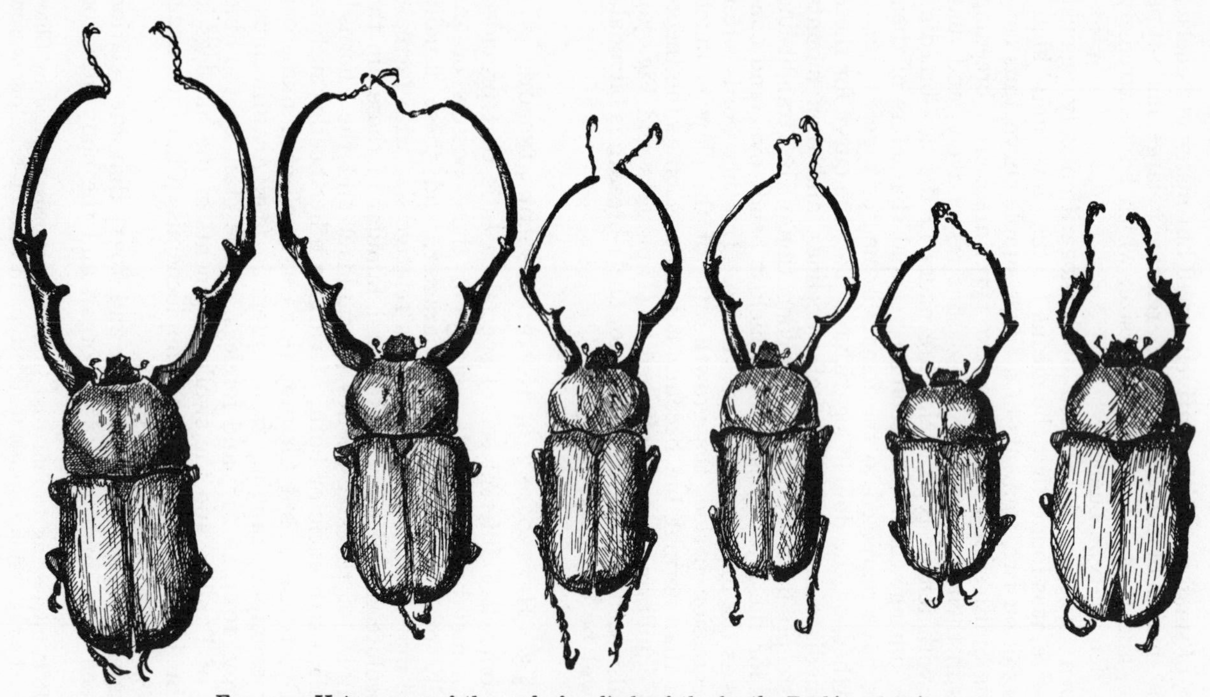

FIG. 33.—Heterogony of the male fore-limb of the beetle *Euchirus longimanus*.
The specimen on the right is a female; the rest are males of increasing absolute size.

approximates closely to the formula for a constant differential growth-ratio (Table IV, Fig. 35). Again, however, there can be no *growth-ratio* in the literal sense in which we have found

FIG. 34.—Heterogony of the 'tail' in the male of the swallow-tail butterfly, *Papilio dardanus* (the heterogony is stated by Champy not to occur in species in which the 'tail' occurs in both sexes).

it apply, e.g., to the partition of growth-potential between the large chela of Uca and the rest of the body. There cannot be, for the simple reason that in holometabolous insects the organ, as regards its imaginal characters, is not formed until the pupal instar, to emerge at the final moult in its definite shape and size. And as there are no further moults, it is incapable of further growth or form-change.

We are thus driven to suppose either that all the processes connected with the organ's heterogonic growth are confined to a very brief period, presumably just before and just after the moult from last larval in star to pupa; or else that, although the visible growth of the organ is confined to this short period, it depends for its amount on some substance whose chemical accumulation during the larval phase has had a constant differential growth-coefficient relative to body-weight (see e.g. Teissier, 1931, for a confirmation of this latter possibility).

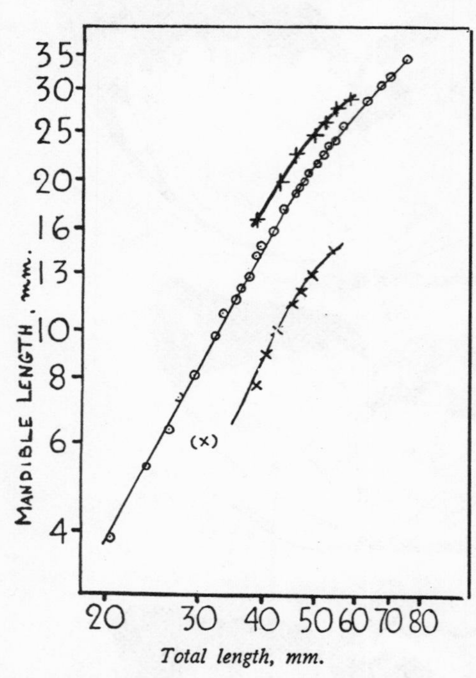

FIG. 35.—Relative growth of male mandibles in three species of stag-beetles (*Lucanidae*).

+, *Lucanus lunifer*; ×, *L. cervus*; ⊙, *Cyclommatus tarandus*. 'Total length' is true total length for Cyclommatus; for the others it is represented by (elytron length + mandible length). All the curves inflect at large absolute sizes (see text); for the remainder of the curves k is about 1·6 for *L. lunifer*, 2·3 for *L. cervus*, and nearly 2·0 for *C. tarandus*.

As we shall see later in considering dimorphism (p. 68) the former hypothesis is the more probable; but in any case we have, as in deer's antlers, the fact that the amount of growth attained is proportional to body-size raised to a power, the value of the power being equivalent to that of the constant differential growth-ratio in cases where visible growth is continuous over long periods. Thus the true growth-ratio of the organ during its short growth-period is far more rapid than indicated by the value found for the 'growth-coefficient' by the method of comparing organ and body at different absolute sizes. We are, in fact, again in the same predicament as with the deer's antlers, the chelae of male Inachus, or any heterogonic organ which is regenerating, and are driven to think of a limiting factor to growth, which we have defined as the growth-partition coefficient.

TABLE IV

MANDIBLES IN LUCANIDAE

(*a*) *Cyclommatus tarandus*

(data from Dudich, 1923: analysed Huxley, 1927)

x	y
20·38	3·88
24·01	5·31
26·38	6·33
27·76	7·32
29·65	8·17
32·20	9·73
33·11	10·71
35·01	11·49
36·13	12·08
37·32	12·73
38·44	14·11
39·26	14·70
41·34	15·84
43·22	17·39
45·51	18·83
46·32	19·19
47·28	19·92
48·40	20·79
50·04	21·53
51·50	22·54
52·50	23·25
54·23	23·96
56·01	25·38
62·06	28·49
66·06	30·69
69·00	32·00
74·00	34·50

$\Sigma = 178$

$k = 1{\cdot}97$, b = just over 0·01

(*b*) *Lucanus cervus*

(data from Bateson and Brindley, 1892: analysed Huxley, 1927)

x	y
(31·0)	(6·0)
38·65	7·75
40·50	9·00
42·55	10·00
45·00	11·20
46·93	11·86
49·18	12·82
53·60	14·40

$\Sigma = 48$

k = about 2·3

Lucanus lunifer

(data and analysis, Huxley, 1927)

x	y
(38·6)	(16·8)
42·4	19·9
45·6	22·4
49·3	24·5
51·9	26·0
57·3	28·8

$\Sigma = 18$

k = about 1·55

Figures in brackets ( ) indicate insufficient number of individuals in class.

In *L. cervus* and *L. lunifer*, x = 'total' length = (elytron length + mandible length), mm.

In *C. tarandus*, x = true total length = (body length + mandible length), mm.

y = mandible length, mm.

k = growth-partition coefficient: the values given do not hold at high values of x (see text).

There is a further point to consider in regard to heterogony in holometabolous forms. In other organisms—a fiddler-crab, for example—the growing system is an open one, in that it is continuously taking in food as it grows. The beetle or other holometabolous insect, however, during most of the period when the form of its adult organs is being laid down, is a closed system, taking in no further food, but depending on accumulated reserves and on the material derived from the breaking down of larval organs. This has two consequences for our problem. In the first place, in other forms the fairest comparison of relative size would seem to be between heterogonic organ and rest-of-body, since the size of the ingestive and digestive systems are functions of the size of the rest of the body, not of total size, as may easily be realized by reference to the fiddler-crab. Here two large specimens, male and female respectively, of equal rest-of-body weight and therefore presumably equal-sized alimentary systems, will be of very different total weight, since in the female either chela will weigh only 2 per cent. of the rest-of-body, while in the male the large chela may weigh up to 70 per cent. or more. But in a stag-beetle, for example, the conditions are quite different. Larvae of both sexes have jaws and guts of the same relative size. A male and a female larva of the same total size will have the same amount of reserve material, but during the pre-pupal and pupal period, say 1 per cent. of this is converted into imaginal female mandible, while perhaps 10 or 15 per cent. has to be converted into imaginal male mandible. Since the amount of reserve material is here the important factor for growth, it is total bulk, and not rest-of-body bulk, which should be here used as the standard against which to plot the bulk of the heterogonic organ (see Huxley, 1931c).

This is a minor methodological point ; but it has further consequences. To continue the example of stag-beetles, we should accordingly expect, in an exceptionally large male specimen where the theoretical relative size of the mandibles would be huge (as long as the rest of the body in some specimens of *Cyclommatus tarandus :* Dudich, 1923 ; see Fig. 91), that during the longer time necessary to lay down this large organ, the limited reserve-supply would come to an end, used up by other competing organs, and therefore that the organ would fall below the theoretical size expected on the formula for a constant coefficient of growth-partition. Thus, owing to the limitation of raw material due to the system being a closed one, we should expect, from a certain absolute size upwards, the actual values for the size of the heterogonic organ to fall progressively more and more below the theoretically expected value. And this is what we actually find. When organ-size is plotted against total size on a double logarithmic grid, the first part of the curve is a good approximation to a straight line, but the end portion curves over so as to be concave to the x-axis. This is so for all cases so far investigated, including the mandibles of three species of two genera of stag-beetles, the horn of Xylotrupes, the heads of polymorphic neuter ants, etc.[1]; Figs. 35, 37, 92.

§ 4. Heterogony and Polymorphism in Neuter Social Insects

We shall later note some further complications introduced into the situation by the fact of moulting. Here we may refer to the particularly interesting case, just mentioned, of polymorphic neuter ants. It is well known that in what are apparently the more primitive examples of such polymorphism, there is an unbroken array from smallest to largest neuters, the continuous series being quite arbitrarily divided up into 'worker minimae', 'worker mediae', and so on up to 'soldier mediae' and 'soldier maximae'; and some myrmecologists have introduced even more elaborate terms (see Wheeler, 1920). Now these series are invariably characterized by a relative increase of head- and especially mandible-size with an absolute increase of total size. Measurements of the weights of head and rest-of-body in species of two genera (the huge *Camponotus gigas*, from Borneo; and a driver ant of the genus Anomma from Africa) show that, over the major portion of the size-range, the formula for constant growth-partition coefficient is nicely adhered to[2] (Huxley, 1927; Huxley and Bush unpublished) (Table IVA and Fig. 37).

[1] Teissier, 1931 (p. 97), using weight and not linear measure, shows in his Fig. 20 no curving over of this type for the mandibles of *Lucanus cervus*. This might mean that my interpretation is wrong, and that mechanical reasons are interfering with great growth in length rather than nutritive reasons with growth in mass. On the other hand, the curvature in my material does not begin until elytron-length 33 mm., and Teissier has hardly any specimens as large as this.

[2] The Anomma curve bends over at high sizes, as described for stag-beetle mandibles, etc. This may presumably be accounted for as suggested earlier in this chapter. But the formula is also not obeyed

TABLE IVA

ANOMMA NIGRICANS

Data from Huxley and Bush (unpublished)

Analysed in Huxley (1927)

x	y
(138·3)	(165·3)
176·0	223·7
220·3	300·2
272·7	427·6
324·2	567·3
367·5	661·9

$\Sigma = 267$
k = about 1·55 (after first 2 points)
x = abdomen-length
y = head-breadth
(in arbitrary units)

Figures in brackets ( ) indicate insufficient numbers in class.

CAMPONOTUS GIGAS

Data from E. Banks (unpublished)

Analysed in Huxley (1927)

x	y
75·0	9·6
111·5	24·3
241·0	82·0
346·8	142·0

$\Sigma = 357$
k = about 1·6 (after first point)
x = total weight, mg.
y = head-weight, mg.

The obvious suggestion is that these series of workers and soldiers do actually represent nothing more than a series of size-forms of a single genetic type, possessing a mechanism for heterogony of the mandible and head. The difference from the other holometabolous cases hitherto considered is that the absolute size-range is much greater, and that the differences in size can only be supposed to be brought about by definite treatment of the larvae by their nurses, the largest types being fed to the limit, the smallest types being deprived of food, and so forced to pupate, while still quite small larvae. That enormous differences in size may be produced by cutting down or cutting off the food supply of insect larvae is established through experimental work on blowflies (Smirnov and

at very small absolute sizes, where the double-logarithmic curve is distorted in the opposite sense, with concavity upwards: the meaning of this, if not merely statistical, is unknown.

Zhelochovstev, 1926), houseflies (Herms, 1928), Drosophila (Gause, 1931), etc. And the postulated differential treatment of different worker larvae by their nurses is well within the known range of complexity of ant behaviour (see also Emery, 1921).

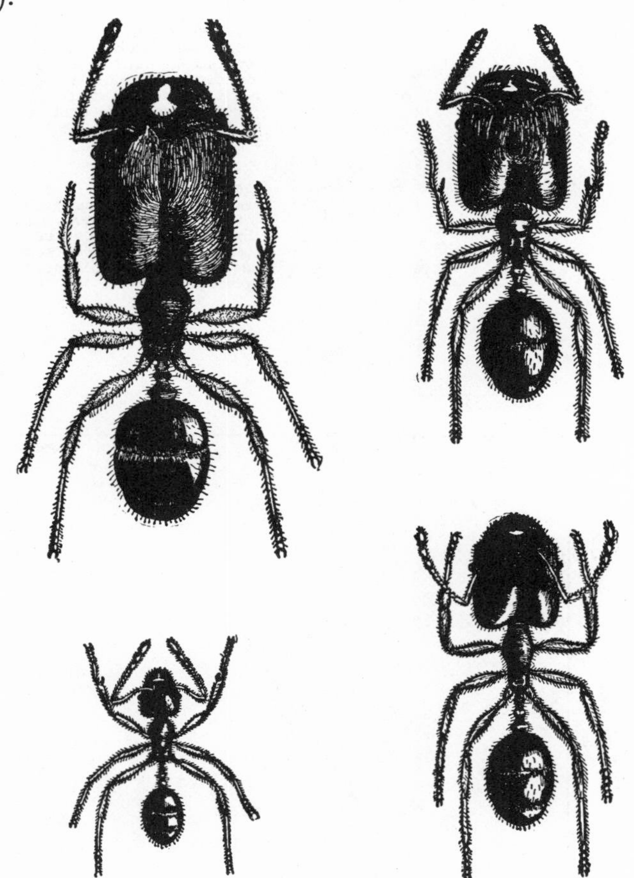

FIG. 36.—Increase of relative size of head with absolute size of body in the neuters of the ant, *Pheidole instabilis*.

That the effect of size-changes may be differential is also known; e.g. Eigenbrodt (1930) finds that the increase of total size in Drosophila caused by low temperature is accompanied by a decrease in wing-size, which is somewhat greater for breadth

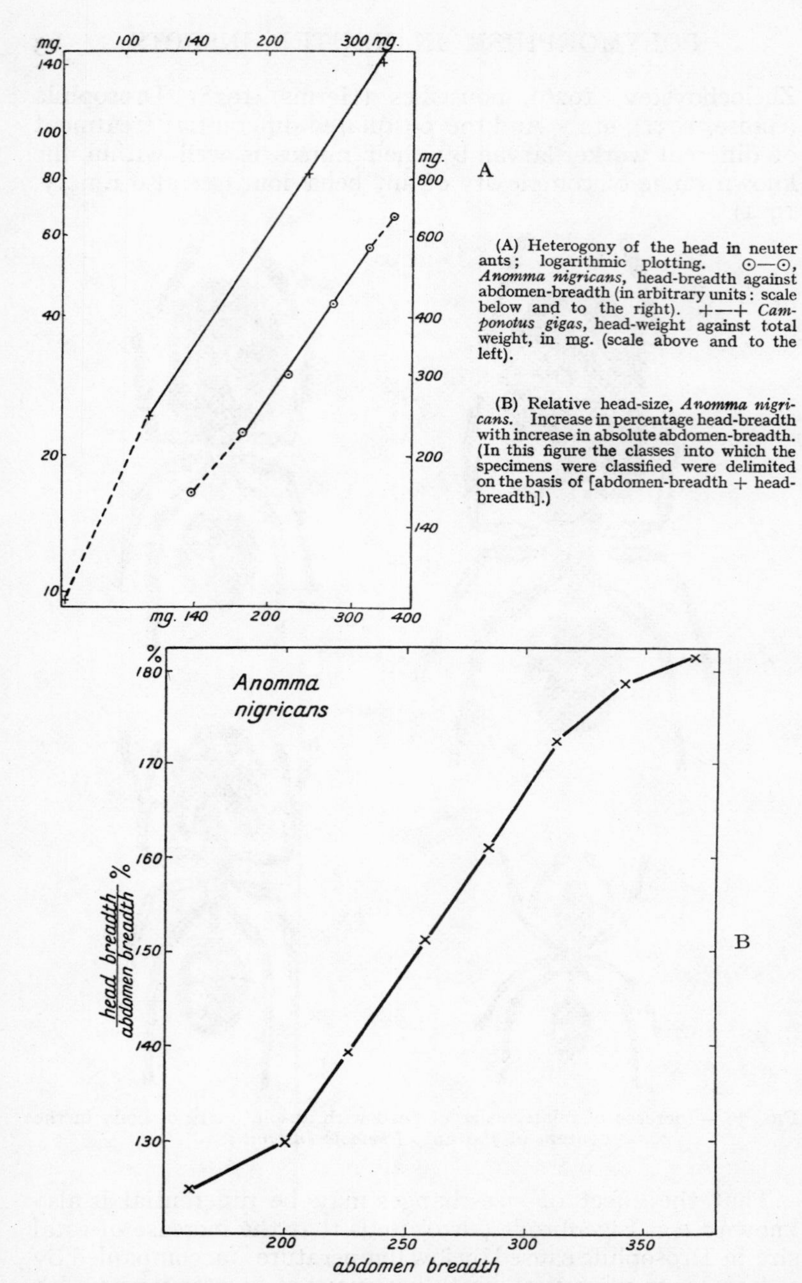

(A) Heterogony of the head in neuter ants; logarithmic plotting. ⊙—⊙, *Anomma nigricans*, head-breadth against abdomen-breadth (in arbitrary units: scale below and to the right). +—+ *Camponotus gigas*, head-weight against total weight, in mg. (scale above and to the left).

(B) Relative head-size, *Anomma nigricans*. Increase in percentage head-breadth with increase in absolute abdomen-breadth. (In this figure the classes into which the specimens were classified were delimited on the basis of [abdomen-breadth + head-breadth].)

FIG. 37.—Graph showing the increase of size of head with absolute size of body in neuter ants.

than for length; Alpatov (1930) finds the same phenomenon. Smirnov and Zhelochovstev (l. c.) for the blowfly Calliphora find a differential effect in different regions of the wing when total size is reduced by cutting off larval food-supplies after a given time. On the other hand, Alpatov (l. c.) in flies of the same species *reduced* in size by depriving larvae of food before they reached full growth, also finds a decrease in relative breadth of the wings, showing that absolute size is not the only factor determining proportions.

When, as in some more specialized forms, the neuter series is not continuous, but there exist only two or a few fairly sharply defined types—e.g. only large and relatively large-headed soldier and small and relatively small-headed worker—we need only suppose a slight specialization of the nurses' behaviour. An adumbration of this is seen in the fact, elicited by unpublished work of Miss Edmonds, of Sydney, that even in forms with a continuous series of neuters, the frequency curve for body-size is definitely multimodal (see also Palenitschenko, 1927).

It is to be hoped that those familiar with the technique of rearing ants in captivity will attack this problem experimentally. If my suggestion be verified, it will materially simplify the evolutionary problems connected with the polymorphism of ant neuters, for instead of having to postulate a large number of genetically distinct types, we need only assume one single genetic type of neuter, but provided with a heterogonic mechanism for head-growth, which will produce all the different head-types as secondary by-products of the animals' absolute size.

In termites, the case is somewhat different.[1] Within the soldier caste of primitive termites, we do sometimes meet with phenomena which appear to be quite parallel with what I have discussed for ants—considerable variation in absolute body-size accompanied by a progressive alteration in relative head- and jaw-size. This is well shown in the figures given by Heath (1927); see Fig. 38. Heath also (see his p. 402) demonstrates fairly conclusively the interesting fact that the polymorphism is due to the presence of forms which have gone through different numbers of moults; a fact which strengthens our hypothesis that there is but one genetic type with a heter-

[1] For much of the information concerning termites I am indebted to Professor A. E. Emerson, of Chicago University, to whom I should here like to express my thanks.

ogonic mechanism as regards head-size. The different soldier forms are discontinuous as regards size and head-proportions, which provides us with a further case of di- or poly-morphism due to a combination of heterogony with moulting (§ 5). Reference may also be made to some of the results given in the important paper by Kalshoven (1930), and in that by Hare (1931). Such a method of arriving at polymorphism of neuters would, of course, be impossible in the holometabolous ants. (See also the work of Light, p. 258.)

There are however other termite cases in which my hypothesis would not seem to apply. For instance, *Acanthotermes acanthothorax* (Sjostedt, 1925 : his Fig. 21, p. 61) has certainly two *qualitatively* different types of soldiers, which it would appear necessary to regard as different genetic types, or at least as produced by qualitative differences in feeding. Sjostedt himself figures no less than five kinds of soldiers, the largest number of types decribed for any species of Termite. Two of these, his Nos. 4 and 5, would appear to be growth-forms of one main type. To inspection, his Nos. 1–3 look as if they were growth-forms of another qualitatively different type ; but Dr. Emerson informs me that he has found that the smaller forms (Sjostedt's Nos. 2 and 3) are both infested with an insect larva which inhabits the head, and disproportionately reduces its size ; these forms are not found in nests from which the parasites are absent. The differential retardation of the head might be due to the existence of a true heterogony-mechanism for the head, which is not normally manifested, but is revealed when the parasite produces general size-reduction of the imago ; or it might be due merely to the fact that the head is the seat of the infestation.[1]

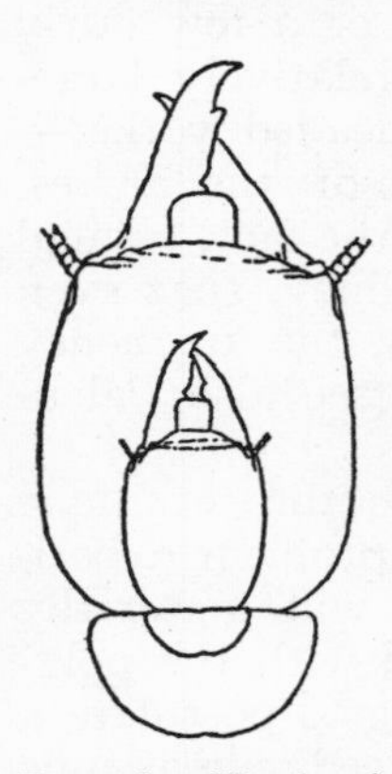

FIG. 38.—Heads of largest and smallest workers in a colony of the termite *Termopsis angusticollis*, showing heterogony and change of proportions with increase of absolute size.

The difference between worker and soldier would appear

[1] In passing, we may note that the effects of such parasites may be very striking ; e.g. in *Termes gilvus*, Silvester (1926) figures extraordinary qualitative changes in shape of head and jaws, as well as a general size-reduction and a highly disproportionate reduction in jaw-size, as result of the presence of a similar parasite in the head. And see, for a discussion of the corresponding problem in ants, Wheeler (1928) and Vandel (1930).

to be of another nature. Recent work such as that of Emerson (1926), John (1925), Heath (1927, 1928), etc., makes it highly probable that in all primitive termites, the forms which do the work of the colony are not a distinct caste, but are the juvenile forms of the soldiers. There is thus a marked heterogony of the head and jaws between the last worker instar and the first soldier instar (we have already seen that there may be more than one soldier instar: Heath, 1927: see also Hare, 1931.)

In more specialized forms, however, while some of the workers are juvenile soldiers, a true worker caste, distinguishable by large size and different proportions, also exists (Emerson, 1926, etc.) It would be extremely interesting to measure the head and body of these two types throughout their growth.

These results, combined with the fact that in some primitive forms, fertile soldiers are met with (Heath, 1928; Imms, 1920), and that soldiers with wing-buds are not unknown, whereas no true workers are known ever to be fertile or to possess wing-buds, indicates that workers have been derived from soldiers by a suppression of their final development into the normal big-jawed type—that, in fact, they are neotenic. This neoteny, however, possibly at first facultative, must at least in higher forms have been fixed as a constant caste-characteristic, either by differential feeding or perhaps more probably by some genetic mechanism (see Thompson, 1917).

This has a bearing on our problem, since the main feature in the evolution of the worker from the soldier would simply be the delay in the onset of the head's heterogony, a delay which, when it exceeded the time to the final instar, would eventually lead to the total absence of soldier characteristics. A less degree of delay would give rise to forms of intermediate soldier-worker type; these are known to exist in various genera, e.g. Armitermes and related genera.

Thus, although heterogony appears to play its part in the origin of polymorphism both in ants and termites, its rôle is different in the two groups. In ants, the control of absolute size through the feeding of the larvae appears to be the main method. In termites true heterogonic polymorphism is rare, and when present seems due to irregularity of moult-number; but neoteny due to postponement of the onset of heterogony also plays a rôle in the differentiation of castes.

§ 5. Heterogony, Moulting, and Dimorphism

Finally, certain quite different special problems arise from the interaction of constant differential growth-ratios with the characteristic Arthropod process of moulting. It will be clear, since growth in a typical Arthropod only occurs between the shedding of one exoskeleton and the hardening of the next, that in the life-history of any single specimen the theoretical growth-curve relating organ-size with body-size will never be realized. Instead, growth of both organ and rest-of-body will take place in a series of jumps, but the points thus arrived at will all lie on the theoretical curve.

Although moulting in crustacea and some insects tends to take place at each doubling of weight (Przibram, 1930), the large variations encountered, together with the variation in the initial post-larval weight, are often sufficient to obscure any recurrent modality in the sizes at which moulting occurs. In a large population of such species, moulting thus occurs at random at any size, and measurements of a heterogonic organ whose growth-ratio is constant over long periods accordingly fall on to a continuous curve when made on such a population.

The state of affairs, however, is entirely different if (*a*) the organ's high growth-coefficient is confined to one or a few instars and (*b*) is initiated in a more or less constant phase of an instar—e.g. always immediately after moulting. As the simplest case, let us take one in which the high growth-ratio lasts but one instar, as occurs with the female abdomen of Inachus (Shaw, 1928). When the data are grouped into classes by body-size, and then simply the means of the various size-classes taken, a curve is obtained which merely indicates two periods of approximate isogony, separated by a short phase of intensive heterogony (Figs. 22, 23). But when frequency-curves for relative abdomen-size are prepared for each body-size class, the true state of affairs is revealed (Fig. 39). All female Inachus are then seen to fall into one or other of two sharply non-overlapping groups as regards relative abdomen-size. Those below a certain absolute body-size all have narrow abdomens, those above another higher body-size all have broad abdomens. The curves for the short intervening range of body-size, however, are all bimodal, some individuals having narrow, others broad abdomens, but none being of intermediate type. When the curves for all the classes are summed, we

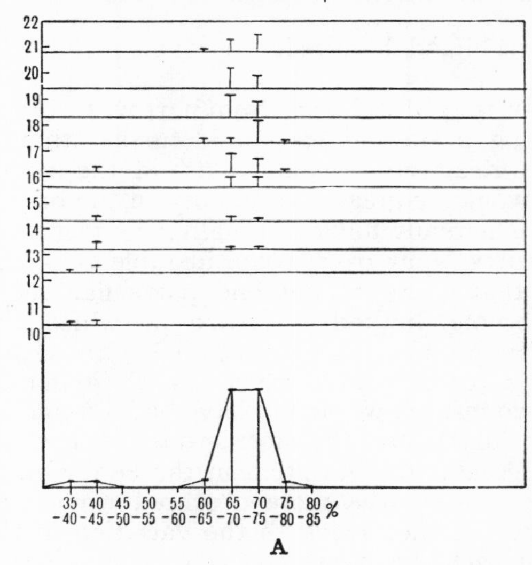

A

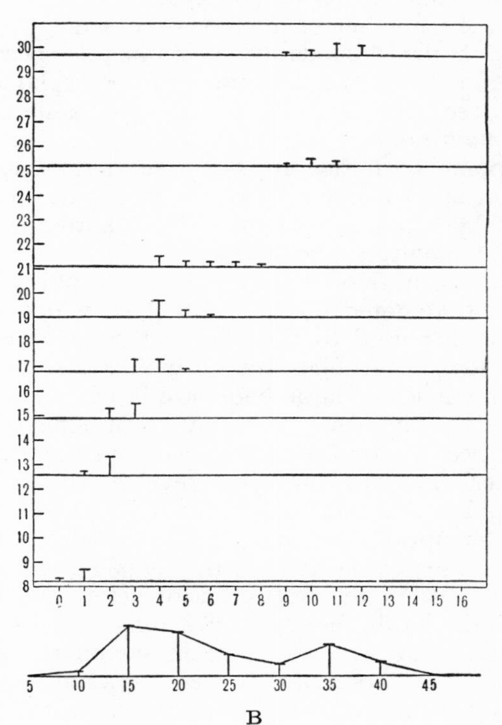

B

FIG. 39. — Dimorphism due to moulting in (A) female abdomen (breadth of 6th segment) and (B) male chela (breadth of propus) in the spider-crab, *Inachus dorsettensis*.

The values are relative, expressed as percentages of carapace length. The frequencies for each value of abdomen and chela are given for various values of carapace length, and also for all specimens taken together.

again of course obtain a bimodal curve for the whole population (Fig. 39).

This can only mean that the change from narrow to broad type is effected during a single instar; and further that it must be initiated at a relatively constant phase of the instar, for otherwise there would be greater variability in the broad-type abdomen than is actually found. Finally, the fact that the bimodal curves are found over a considerable range of body-size indicates that there is not one particular serial number of moult, or one limited body-size, at which the transformation (which presumably is associated with sexual maturity) is initiated. If the high growth-ratio of the abdomen had extended over two instars, we should have had a trimodal instead of a bimodal curve; but the more instars over which it extended and the greater the variation in the body-size at the initiation of heterogony, the more obscured would the curve's multimodality become, owing to the variation in the amount of growth at each moult, which becomes cumulative with the increase in the number of moults concerned.

A similar but less-marked bimodality occurs in this species as regards chela-size in males, and is doubtless to be explained in the same way; see also Table IIIA, p. 54, for a similar phenomenon in *I. mauritanicus*.

The important point to notice is that the unusual and apparently abnormal fact of dimorphism in one sex has here been brought about by a combination of the two normal and common processes of moulting and heterogony.

It will at once be seen that this type of dimorphism is quite different from that noted by G. Smith (l. c.) which we have just considered in regard to the chelae of males of the same genus. The 'low' and 'high' males in this case were simply the groups of small and large body-size in what would have been a case of continuous heterogony resembling that of the chela of male Uca, if it had not been for the intercalation of a non-breeding period in which the claw reverted to female type. This dimorphism has nothing to do with moulting, but is due to an interruption of a long-continued heterogony by a non-breeding phase; while that of the female abdomen of the same genus is due to the restriction of heterogony to a single moult-period. Both, however, are dimorphisms of developmental origin, and have nothing in common with genetic dimorphisms like those of 'diphasic' mammals or birds such as arctic fox, certain squirrels, herons, owls, etc., or the genetic polymorphism of the females of various mimetic butterflies. They also differ from the environmental dimorphisms, the best-known case of which is that of the wet-season and dry-season forms of certain tropical butterflies, where the difference between the two types is elicited by different environmental conditions bringing out different expressions of the same gene-complex.

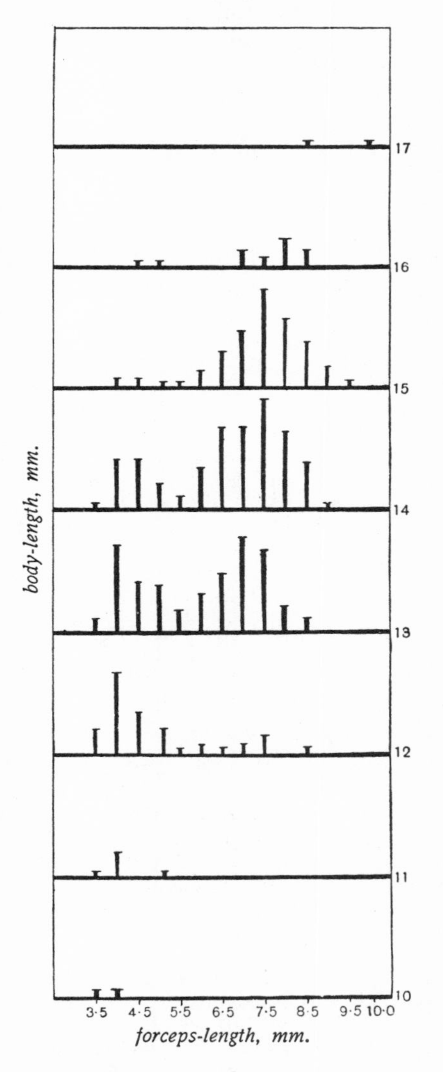

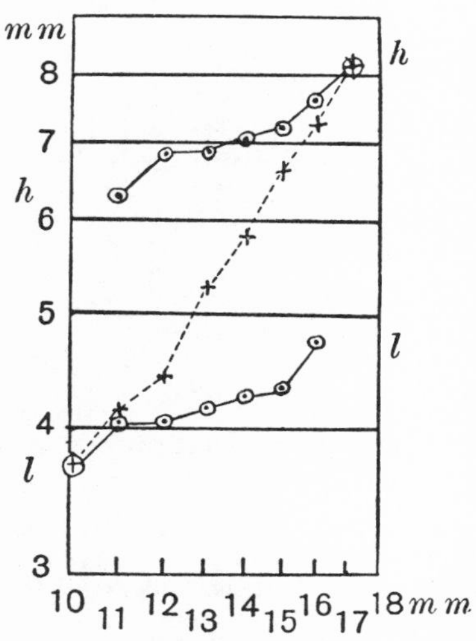

FIG. 40.—Bimodality and heterogony of the male forceps in the earwig, *Forficula*.

Left, absolute frequency of different forceps-length at various body-lengths in a random sample of 445 specimens. Right, logarithmic plot of means of 'high' forceps (h—h), 'low' forceps (l—l), and both together (dotted), against body-length in 1,519 specimens. k for all specimens, about 1·6.

Can we apply the results found in the abdomen of female spider-crabs to the classical cases of dimorphism in holometabolous insects—those of the earwig Forficula and the beetle

Xylotrupes (Bateson and Brindley, 1892)? I think that we can. Let us begin with the clearer-cut case of Forficula, in which there is a definite dimorphism of the male forceps, but no dimorphism of female forceps, or of the body-size of either sex.[1]

TABLE V

MEASUREMENTS OF 1,519 EARWIGS COLLECTED BY DJAKONOV: ANALYSED IN HUXLEY, 1927B, Table I (see Fig. 40)

Class by body-length, mm.	No. in class	Mean body-length, mm.	Mean forceps-length, mm.	Low type			High type			Per cent. high-type in class
				Total no.	Mean forceps-length, mm.	Relative forceps-length, per cent.	Total No.	Mean forceps-length, mm.	Relative forceps-length, per cent.	
10	5	—	3·70	5	3·70	37·0	—	—	—	0·0
11	40	—	4·15	38	4·04	36·7	2	6·25	56·8	5·0
12	160	—	4·41	139	4·05	33·8	21	6·81	56·8	13·1
13	363	—	5·24	219	4·18	32·2	144	6·86	52·8	39·7
14	481	—	5·80	216	4·26	30·4	265	7·06	50·4	55·1
15	326	—	6·61	69	4·32	28·8	257	7·23	48·2	78·8
16	129	—	7·28	15	4·77	29·9	114	7·61	47·6	88·4
17	15	—	8·17	—	—	—	15	8·17	48·1	100·0
Total	1,519	13·87	5·80	701	4·19	—	818	7·17	—	53·9

Analysis of Djakonov's results (Huxley, 1927B) has shown that when the data are tabulated by body-size, the mean

[1] Recently Kuhl (1928) has attempted to show that the dimorphism of the forceps of male earwigs is apparent only, due to unconscious selection of largest and smallest specimens in collecting. This, however, does not account for the monomorphism of male body-size or of female forceps; and in any case is quite unable to account for the degree of dimorphism found by Bateson and Brindley, Djakonov (1925), etc. If apparent dimorphism is so easily produced by such means, existing collections should demonstrate it for large numbers of species; instead it is quite exceptional.

It would appear that the absence of, or slight tendency to, bimodality shown in Kuhl's material (his pp. 362–3) is to be accounted for by the almost total absence of 'high' forms in his material, which again is to be correlated with low body-size. The *maximum* forceps-lengths in his four samples are 6·5, 6·5, 7·5 and 8·0 mm. respectively. As my figures show, the 'high' *mode* only occurs at 7·5 to 8·0 mm. In Djakonov's material the maximum forceps-length is 10 mm., and Brindley (referred to on my p. 313) records a maximum of 12·25 mm.!

values for forceps-length against body-length give a good approximation to the formula for constant differential growth-ratio between forceps-length and body-length. When, however, frequency curves for forceps-length are plotted for each class separately, a situation is revealed analogous to that for Inachus female abdomen. The smallest specimens show unimodal curves, their forceps being all of 'low' type; the largest, also unimodal curves, with forceps all of 'high' type; and the medium-sized show bimodal curves, the numbers of low forceps diminishing and high forceps increasing with increase of body-size. It should be noted that low and high types, as with Inachus chela and abdomen, differ only in size; further, that in this case there is slight overlapping of the two modal types (Figs. 40, 42.)

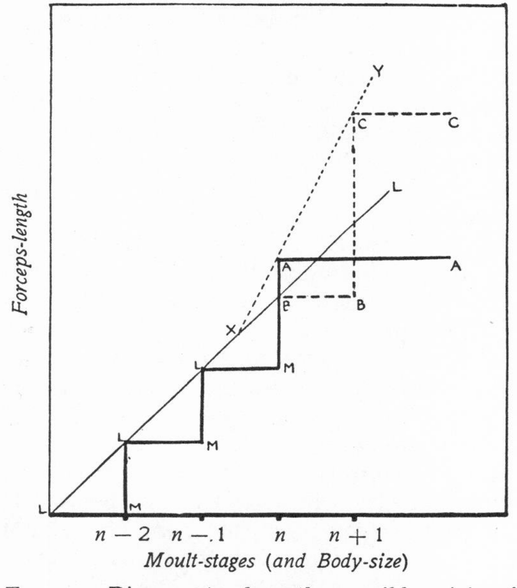

FIG. 41.—Diagram to show the possible origin of dimorphism of the male forceps in earwigs, *Forficula*.

LM, LM . . . sizes of larval forceps at successive moults; growth isogonic. X, point at which heterogonic growth of forceps is supposed to be initiated. If the change to the imago occurs at the *nth* moult, the modal level of adult forceps-size is at A—A ('low' males); if at moult ($n + 1$), the forceps, after continuing of larval type for one more instar (B—B), would reach the modal size-level, C—C ('high' males).

Przibram (1927) has suggested that the bimodality is due to some male earwigs having one or more extra moults; but his contentions, as they stand, will not explain the facts, for they should give bimodality of body-length as well as of forceps, and they do not take account of the overlap of 'low' and 'high' types, between which there is neither qualitative nor quantitative difference, but merely a modal distinction depending on their frequency distribution.

Furthermore, it is difficult to see how *more* than one extra moult can ever occur without leading to tri- or multi-modality.

The hypothesis of an extra moult may however be used to account for the facts, along the following lines. The processes responsible for the heterogony of the male forceps begin operating, we will suppose, late in larval life, and usually at a fairly definite phase of an instar. They cannot, however, be expressed in the form and size of male-type forceps until the imaginal instar is reached. This may be reached either at the next moult after the initiation of the heterogonic process, or only at the second moult. The result is that the heterogonic processes have either been operating for less than one instar, or for a period more than double as long, with resultant bimodality. Variation in the intensity of the heterogony, and in the time after moulting at which it is initiated, will bring about the overlap of high and low; there is, however, no qualitative difference to be expected between the two forceps-forms (Fig. 41).

The interesting fact discovered by Djakonov (l. c.) should be noted, namely that unfavourable nutritive conditions cause a decrease in mean and modal body-size, both for the population as a whole, and for the 'low' and 'high' groups considered separately. As regards forceps-length, however, the main effect is to cause a shift in the distribution of the forceps, there being fewer high-type and more low-type, but *without alteration of either modal value* (Fig. 42). Furthermore, though in unfavourable conditions there are fewer large individuals in both high and low series, yet for classes of the same mean body-size, whether we consider the population as a whole or the high and low groups separately, the mean forceps-size is actually *greater* for the animals which have grown up in the unfavourable conditions.

This apparently paradoxical fact may be explained in terms of our previous discussion. The unfavourable conditions reduce the total body-size. But the *initial* growth-ratio of the heterogonic forceps, starting at a late moult, will always be the same; it is only its final value which will tend to a limit imposed by the growth-partition coefficient. For most body-sizes therefore we should expect that the absolute size of the forceps would depend only on the time for which it had grown, not on the body-size at which it began to grow, and therefore with stunting of total bulk, a given body-size will show an absolutely as well as a relatively greater forceps-size. It is, further, quite possible that the equilibrium-size and final theoretical growth-ratio is in practice never attained, the imaginal stage with its cessation of growth supervening before this point is reached. (This is perhaps supported by the fact that the double-logarithmic plot for all forceps taken

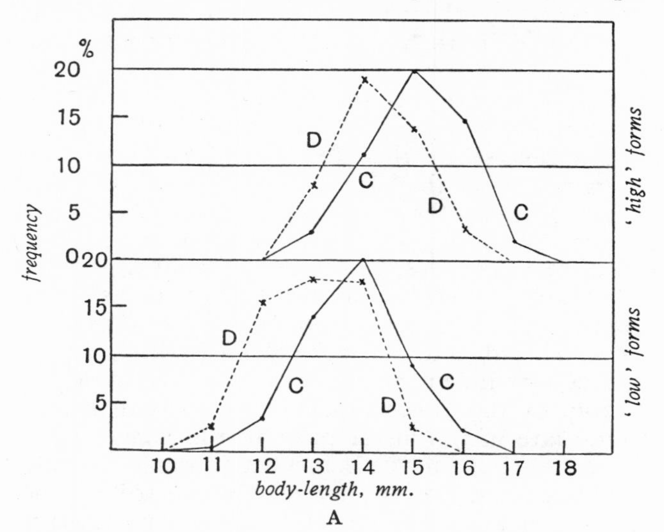

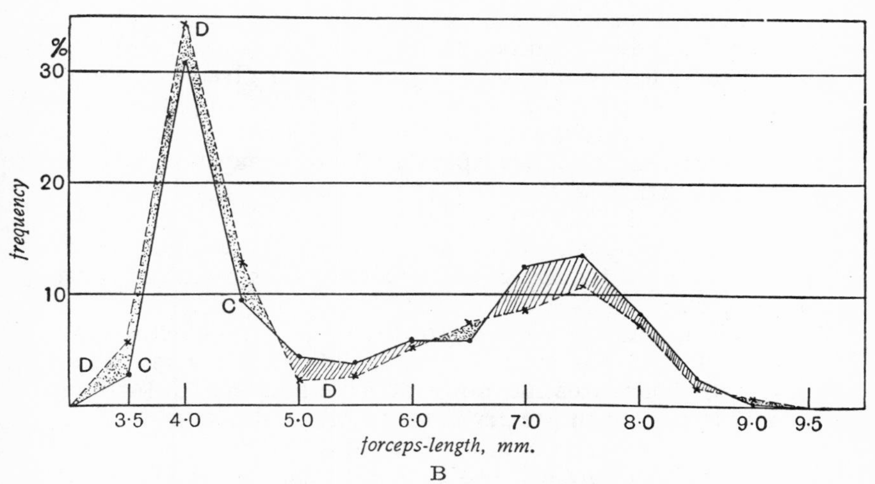

FIG. 42.—Different types of changes produced by unfavourable conditions (D—D) as against favourable conditions (C—C) on (A) body-length and (B) forceps-length in male earwigs. The modal body-length is reduced ; the frequency of forceps of low or high modal lengths is altered, but the modal length remain the same.

together does not curve over downwards at high body-sizes, as with various other holometabolous forms.) If so, then with earwigs within the known range of body-size, those with stunted body will always have larger forceps, though this should not hold if we could produce individuals of a much larger body-size.

A further peculiarity of the curves is that those for high and low forms taken separately (Fig. 40) both show a phenomenon the reverse of that found in most holometabolous insects—namely that they are concave upwards at their upper ends; in addition, they are concave downwards at their lower ends, indicating a rapid growth-coefficient at either end. This may, perhaps, be accounted for if we suppose that growth-ratio is accelerated during the period just before and just after a moult. The high-sloping early portion of the 'low' curve would be due to the normal initial high growth-ratio of the forceps-rudiment at the first onset of heterogony, and would comprise those individuals which started their forceps-heterogony late in the instar prior to the imago. The high-sloping end portion of the same curve would include those which started their heterogony relatively early in the same instar. But those in which heterogony was initiated towards the middle of the instar would only have one period of acceleration at the end of a period of slow growth-ratio, and therefore would show less change of forceps-size for a given increase of body-size. If an extra instar is added, the rapid growth-periods are repeated, with corresponding results on the curve. This is, of course, purely speculative, but may serve as the basis for further work.[1]

The case of Xylotrupes (Huxley, 1927c) has been less thoroughly analysed. It presents various complications, most notable being a tendency to trimodality over a certain range of body-size.

In Lucanidae (Huxley, 1931c) analysis of Dudich's paper (1923) and other data show several interesting facts. First, that in Cyclommatus the range of male body-length (without mandibles) is much greater than in other stag-beetles, being about as much as the mean body-size, whereas in *Lucanus cervus* and *L. lunifer* it is less than half the mean. (The ratio of largest to smallest body-length in Dudich's Cyclommatus specimens is 2·47; in Bateson and Brindley's *Lucanus cervus*,

[1] Most of the suggestions here advanced concerning Forficula modify or extend those put forward in my paper on the subject (l. c.).

the corresponding ratio for elytron-length is 1·6.) Secondly, the range of the male's body-size is much greater than that

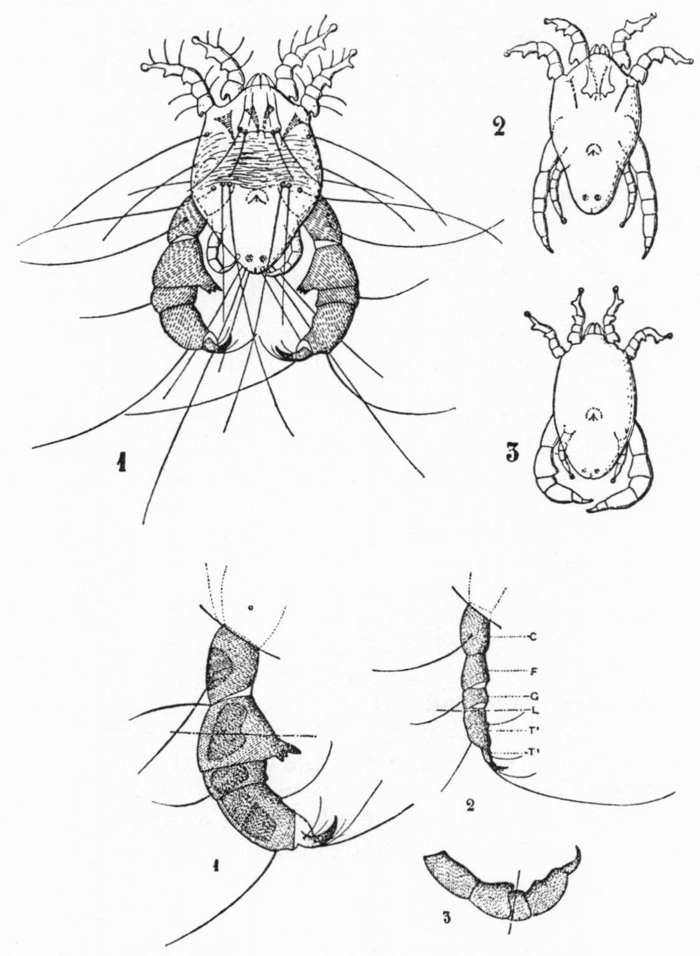

Fig. 43.—Heterogony and moulting of the 3rd pair of legs of the mite, *Analges accentorinus*.

Above: (1) large male with relatively huge 3rd legs; (2) small male with 3rd legs of the same relative size as in females; (3) small male in which the onset of heterogony has apparently taken place earlier, giving slightly enlarged 3rd legs. Below: the 3rd legs of the 3 forms, showing details.

of the female's, though both start at the same minimum size (16 mm.): the largest males have bodies 40 mm. long, the largest females only 25 mm. As the mandibles of the male are highly heterogonic (ranging up to 88 per cent. of body-length), the disparity in total size is even greater. This difference would appear only to be obtained on the basis of possible extra moults in the male. In this form the frequency curve for all mandibles is again multimodal, with two main and one or more subsidiary modes. But the multimodality is not so well defined when curves for separate body-size classes are plotted. There is also a slight tendency to multimodality or at least irregularity in the male body-length frequency curve. No multimodality is apparent in the Lucanus curves. Thus here again the insect with the presumption of a facultative extra moult is characterized by multimodality of heterogonic organ.

A case of male dimorphism in the Acarine mite *Analges accentorinus* which would appear to be due to similar causes has been described by Jucci (1924). Here the large adult males are characterized by a great hypertrophy of the third pair of limbs. There further exist small but also adult males whose third limbs are almost identical with those of the female. From the scale drawings given by Jucci, we can say that the size-difference between the two forms is very close to what we would expect (on the supposition that bulk is approximately doubled at each instar) if the larger had had one more instar in its development than the smaller (Fig. 43).

In addition, occasional small males are found with slightly enlarged third limbs: these would be specimens in which the onset of heterogony had taken place slightly earlier than usual. The case is thus very similar to the earwig, save that the normal 'low' type is more like the female than in Forficula.

Having concluded this survey of special cases, I shall in the next chapter pass to a more detailed analysis of the empirical laws of relative growth in heterogonic organs or regions.

CHAPTER III

GROWTH-CENTRES AND GROWTH-GRADIENTS

§ 1. Growth-gradients within Single Organs

So far we have only dealt with the question of relative growth in whole organs or regions of the body; and in our first chapter we have found an approximation to a simple mathematical formulation of relative growth, which we have called the law of constant differential growth-ratio. This, as we have further seen, is what we should have expected if we had worked on the problem *a priori*. It teaches us the striking fact that relative growth-rates of different parts of the body may stay constant over long periods of growth, which clearly is important as a contribution to the problem of form co-ordination and the orderliness of form-change, but it sheds little light upon any aspect of the growth-process itself.

In this chapter, however, we shall deal with certain further empirical laws or rules which will, I think, have to be taken into careful consideration in any future investigation of the biology and physiology of growth. It is of some interest that these rules, which to my mind constitute the most important part of any contribution made by me to the study of relative growth, emerged quite incidentally out of the investigations on differential growth-rates. In studying these latter, I had a perfectly clear-cut aim—to see whether change of proportions could be envisaged as the result of any simple laws of relative growth. But of the growth-gradients to be discussed in this chapter, I had no suspicion: their existence thrust itself upon me as a new empirical fact, any explanation of which is for the moment entirely problematical. I say a new empirical fact, for although D'Arcy Thompson (l. c.) had already adumbrated a similar view, for one thing he had not fully generalized it or pursued its consequences to their limit, and for another, it was new to me, as I had not at first grasped the full implications of his ideas, which only became clear on re-reading his book after obtaining certain empirical results for myself.

The starting-point of these investigations was afforded by the fact, obvious to simple inspection, that whereas the proportions of the separate joints of the female-type chelae in the sexes of Uca do not change appreciably during growth, those of the joints of the large or male-type chela do change, and very markedly. The most obvious alteration is a relative increase of the size of the propus with absolute increase of the size of the whole chela. However, when the weights of different chela-regions were accurately determined, it was found that there existed within the limb what we may call a *growth-gradient*, the distal region (chela + propus) having the highest relative growth-rate, the central region (carpus) the next highest, and the basal region, nearest the breaking-joint, (merus + part of ischium) the lowest, although its relative growth-rate was still above that of the body.

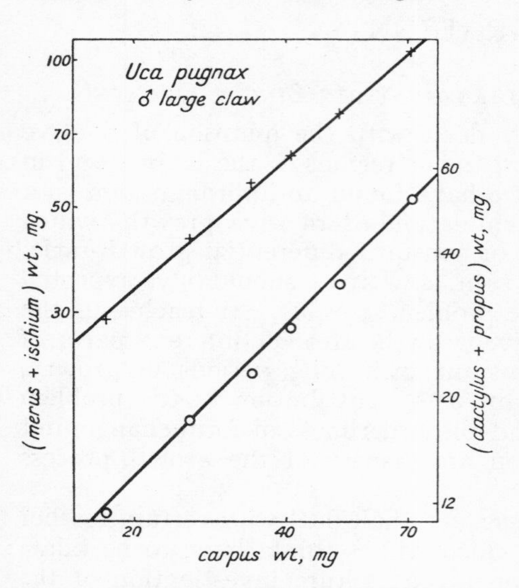

FIG. 44.—Graph to show different relative growth-rates of different parts of the large claw of the fiddler-crab, *Uca pugnax*.

Weights in mg. of dactylus + propus (o, scale on right) and merus + ischius (+, scale on left) against weights of carpus; logarithmic plotting. The distal region shows positive heterogony (k about 1·05) relative to the intermediate region (carpus); the proximal region shows negative heterogony (k about 0·9). (See Table VI.)

If we put this crudely into graphic form, using growth-coefficients as ordinates and spacing the different regions arbitrarily along the x-axis, we obtain a curve representing the distribution, along the main axis of the limb, of what we may for brevity's sake, without introducing any theoretical ideas, speak of as *growth-potential*. This curve is inclined to the horizontal, and is therefore the graphic representation of a *growth-gradient* within the appendage, the inclination of the curve representing the steepness of the gradient, or in other words the difference in absolute growth-potential between the two ends of the gradient (as measured by growth-coefficients, which for our present purpose afford the only comparable standard for measuring intensity of growth-potential in a number of different regions or forms) in relation to length of the gradient—i.e. the relative length of the appendage. Such a graph, however, is as I say only a crude representation of the true growth-gradient. The most obvious reason for this is the impossibility of assigning fixed points along the abscissa-axis to the several joints, since the very fact of their differential growth is causing their centres (or ends) to shift differentially with increase in absolute size.

And secondly, we have the difficulty that the values we have obtained for the growth-coefficients are merely mean values for large regions of the organ, whereas if the idea of a growth-gradient be really justified, we should expect a progressive change of the growth-coefficient from point to point along the axis, even within the limits of a single joint—a theoretical consideration supported by certain actual evidence in other forms (pp. 98, 261–2).

The full solution of the problem, so as to obtain a quantitatively accurate picture of the graded change in growth-potential along an organ, will be a matter of considerable difficulty, partly owing to the formal difficulties arising from the constant change of the relative size of the parts measured, partly owing to the practical difficulty of finding sufficient distinctive points within the limits of a region such as the segment of a limb, on which to take measurements to determine the detailed form of the gradient empirically and not by mere extrapolation from a few mean values.

Finally, there is still another difficulty. As pointed out to me by Mr. J. B. S. Haldane, if y be the value of the weight (or linear measurement) of the organ as a whole, and if $y_1, y_2, \ldots y_n$ be the corresponding values of its constituent joints or segments, then if $y = bx^k$ be a correct expression for the limb as a whole, then $y_1 = b_1x^{k_1}$, $y_2 = b_2x^{k_2}$, and so forth cannot be accurate expressions for the separate parts (or vice versa), since the sum of the several expressions for the parts will not exactly fit the expression for the whole. On the other hand, within certain limits of the value of k, the discrepancy will only be slight, and we are justified, from the actual figures obtained, in taking an expression of the above general form as giving a close approximation to the truth, and therefore in using the values of the growth-coefficients (k) as obtained from this type of expression as standards of growth-intensity.

TABLE VI

UCA PUGNAX ♂. MEAN WEIGHTS (MG.) OF THREE REGIONS OF LARGE CHELA (57 SPECIMENS)

Region	Class 1	2	3	4	5	6
Distal (dactylus + propus)	114	179	222	280	344	515
Intermediate (carpus) .	17·5	25·7	33·9	40·6	50·7	70·8
Proximal (merus + ischium to breaking-joint) . .	28·8	42·2	55·7	63·1	76·9	103

TABLE VII

GROWTH-COEFFICIENTS OF DIFFERENT REGIONS OF THE LARGE MALE CHELAE IN DIFFERENT CRUSTACEA (BASED ON HUXLEY, 1927 AND UNPUBLISHED, AND DEAN, UNPUBLISHED, ANALYSIS OF KEMP)

Units of Measurement.	Species	Merus + ischium	Carpus	Propus + Dactylus
Weight, relative to total chela weight (distal to breaking-joint)	Uca pugnax	0·89	0·97	— — 1·03
,, ,,	Maia squinado	0·81	0·93	1·20 1·14 1·19
Weight, relative to body-weight	,,	1·51	1·72	2·22 2·10 2·19
Length, relative to total cheliped length (distal to breaking-joint)	Palaemon rudis	ischium 0·83 merus 1·02	1·04	1·11 1·05 —

Investigations are now in progress which have for their aim the clearing up of some of the practical and theoretical difficulties in the way of obtaining a quantitatively accurate picture of the growth-gradient within an organ. Until these have been completed, I shall here content myself with establishing the fact that a gradient of some sort exists; and in the graphic representation of growth-gradients I shall arbitrarily represent the centres of homologous regions as equidistant along the abscissa axis, and shall use the k-values, as obtained from the formula for constant differential growth-ratios, as reasonable approximations for the values of growth-intensity.

§ 2. Steepness of Growth-gradient within an Organ and Growth-intensity of the Organ as a whole

The fact of a growth-gradient once established for the heterogonic chela of Uca, the next step was to see if similar growth-gradients occurred in other heterogonic organs. This proved to be the case. We will first take other examples from Crustacean appendages. The weights of the separate joints of the chela (five of them distal to the breaking-joint) were taken for both male and female *Maia squinado* (Huxley,

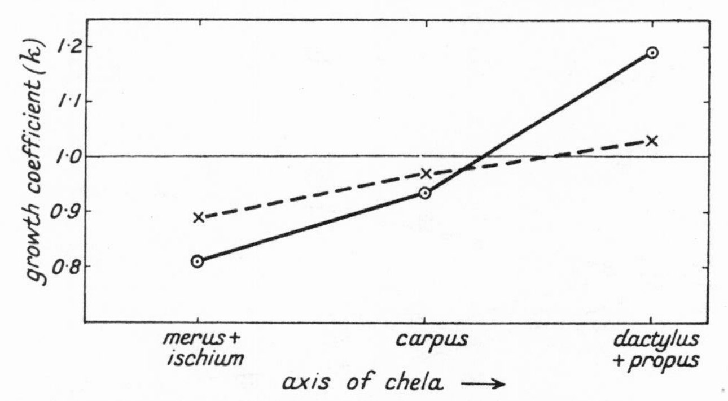

Fig. 45.—Growth-gradient in the large chela (×) of the fiddler-crab, *Uca*, (⊙) of the spider-crab, *Maia*.

The growth-coefficients of the different regions are here taken relative to the total weight of the chela distal to the breaking-point, not, as in Fig. 44, to the carpus-weight.

1927, and unpublished). It was found that while the proportionate weight of the joints of the female chela remained approximately constant within the limits of variation at all absolute sizes,—i.e. their growth-coefficients relative to the chela as a whole, like that of the chela relative to rest-of-body, were all = 1·0,—those of the male chela during its period of heterogony were all greater than unity and were arranged in a regular growth-gradient. This was double in form, with high point in the propus, a slight fall towards the tip (dactylus), and a more rapid and more prolonged fall towards the body. The high point of the gradient we will call the *growth-centre*.

Presumably the growth-gradient in the large chela of Uca was of the same form, but this was not apparent, owing to the propus and dactylus having been lumped together for purposes of measurement. (See Table VII; Fig. 45.)

The gradient is steeper in the chela of Maia than in that of Uca. This appears to be due to the fact that though the ♂ chela of Maia begins its period of heterogony much later

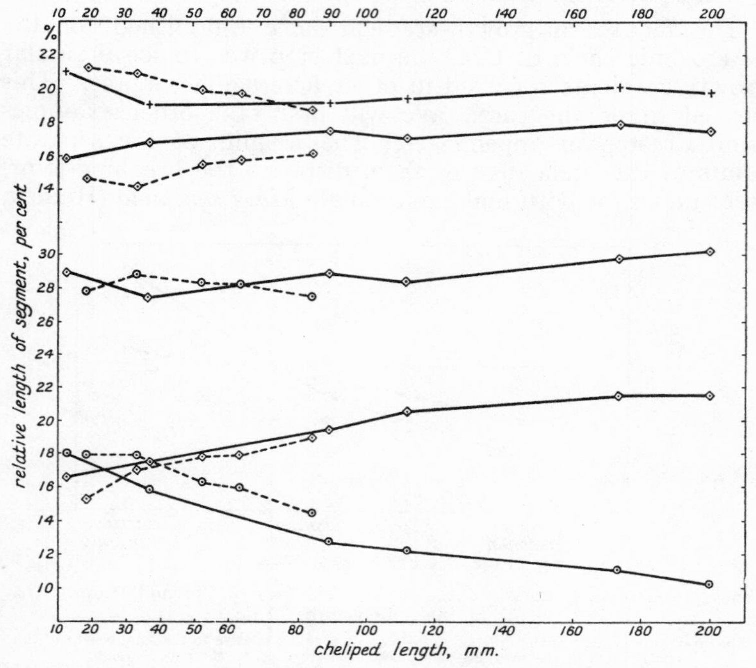

Fig. 46.—Changes in relative length of different segments of the cheliped of the prawn, *Palaemon rudis*, with increase of cheliped length.

Solid lines, males; dotted lines, females. From above downwards: merus; dactylus; carpus; ischium (⊙), and propus (◇). The ischium decreases markedly, the merus and the carpus are almost constant; the propus increases markedly (growth-centre), the dactylus slightly.

in life than that of Uca, and therefore never attains the same enormous relative size, yet during this period, its growth-coefficient is higher than that of Uca (about 1·85 as against 1·6 in Uca's first phase and 1·3 in its second). It would appear natural that the greater is the growth-coefficient of an organ as a whole, the steeper will be the growth-gradient of its parts; this is confirmed by all the evidence so far collected.

The only references I can find to these striking changes in the proportions of heterogonic Crustacean limbs are those of Kemp and his fellow-workers (Kemp, 1913, 1914, 1915; Henderson and Mathai, 1910). But they only draw attention to the change in percentage length of the joints, and have not proved the existence of constant differential growth-ratios, or propounded the idea of a growth-gradient. The figures for *Palaemon rudis* have, however, been analysed by Miss I. Dean (unpublished) and in both series show a definite growth-gradient with growth-centre in the propus (Table VII; Fig. 46).

In general, measurements of crustacean limbs show that wherever there is marked heterogony, there is a comparatively steep growth-gradient within the limb, with well-marked growth-centre near the tip (apparently always in the propus); when the heterogony is only slight, the growth-gradient is far less steep, and its centre usually near the middle of the limb: in most cases examined either in the merus or carpus. Thus, although the joints of the small (female-type) chela of male Uca do not, to simple inspection, appear to alter in proportionate size, measurement shows that in respect at least of linear dimension, they do so, albeit slightly; the growth-centre here is in the carpus (Huxley and Callow). (It will later be shown that precisely similar relations hold for those brachyuran abdomens which have been measured.)

Benazzi (1929) has given results on the regeneration of the limbs of the larva of the dragonfly *Aeschna grandis* from which it can be calculated that for regeneration during two instars, the growth-coefficients (k) of femur, tibia and tarsus, relative to the sum of the three parts, are as follows: femur, 1·12; tibia, 1·02; tarsus, 0·72. There is thus a regeneration-gradient in the limb with high point proximally.

Whereas the gradients of most organs appear to be of the simple form above described, there are some of unusual type in which part of the organ has growth-coefficients above unity, the rest below unity. This is the case with the first antennae of certain copepods (Seymour Sewell, 1929) (see Fig. 47). There is a positive growth-centre at the eighth or ninth segment, and a negative growth-centre at the extreme tip. The change from positive to negative heterogony of the segments, relative to total antenna-length, occurs close to the joint between the eighteenth and nineteenth segments. It is worth noting that when a hinge is developed in the male's grasping antenna, it is formed at this joint. In almost all cases relative growth again falls off steadily on the proximal side of the positive growth-centre.[1] The first and sometimes a few more of the basal segments usually exhibit negative heterogony, but occasionally show a very low positive heterogony. The first antenna as a whole shows a slight negative heterogony relative to total length (l. c., p. 9).[2]

We may suggest that it is biologically desirable for the terminal portion of the antenna to decrease, the proximal

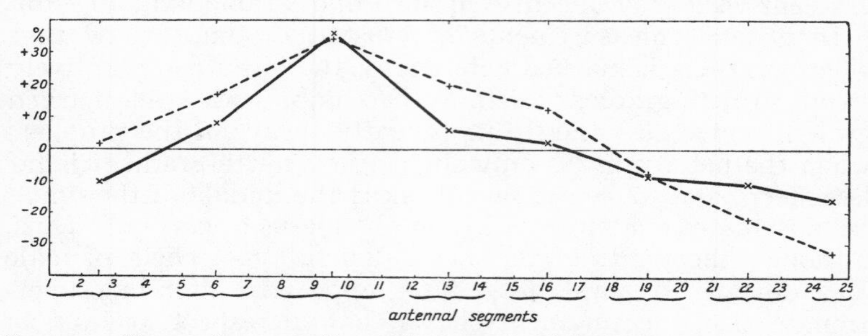

FIG. 47.—Change in proportions of segments of first antenna of Copepods during growth. Constructed from the data of Seymour Sewell, 1929. He gives the proportionate sizes of the antenna segments relative to total antenna length at various sizes.

The graph gives the percentage change in proportionate size between the smallest and largest stages measured (the segments have been grouped as indicated, and the means taken for the groups). In every case the growth-rate of the basal region is low, usually negatively heterogonic; there is a centre of maximum growth at the 8th or 9th segment, and a centre of minimum growth at the distal end. The transition from positive to negative heterogony occurs at about the 18th or 19th segment.

× *Nannocalanus minor*. + *Eucalanus subcrassus*.

region to increase in relative size. Since changes in relative size appear to operate by means of growth-gradients, the negative growth-centre in the tip of the antenna will be connected

[1] Seymour Sewell's data for *Undulina vulgaris* indicate that in this species, after a low point of no change in proportions in the fourth segment, the growth-gradient again turns upward as we pass towards the body, which would give a still more complex growth-gradient.

[2] The increase in total body-length, however, is due partly to the formation of new segments in the growing zone in the sub-terminal region of the abdomen. As I shall attempt to show in a later chapter, growth during early stages of the process, during which differentiation from embryonic tissue is actively proceeding, obeys different laws from those concerned with heterogony of parts which are already differentiated. It would be better to compare the growth-rate of the antenna with some definitely-formed part of the body, e.g. cephalothorax-length, in which case it would probably show slight positive heterogony.

by a continuous growth-gradient with the positive centre in the region of the ninth segment. Further, we may safely assume that the copepods, in common with almost all other animals, show a negative heterogony of the head region: accordingly this centre of low growth-intensity will again be connected via a continuous growth-gradient with the positive growth-centre of the antenna. Since the gradient is of the same type in both sexes, the functional differentiation of the terminal region of the male antenna as a clasping organ cannot have any causal significance in determining the low growth-rate of this region. On the other hand, the fact that the heterogony passes from positive to negative at about the eighteenth or nineteenth segment may have had something to do with the fixing of the hinge-joint between the clasping region and the rest of the male antenna at this spot; such a suggestion must, however, be regarded for the moment as purely speculative.

§ 3. Reversal of the Sign of the Growth-gradient in Negative Heterogony

Those pereiopods which are used as walking legs appear usually to show slight but distinct positive heterogony, and to have a definite but slight growth-gradient with centre in the merus (Bush, 1930). It is of interest that in the actively-running shore-crab Ocypoda, the young (like the active young of Ungulates) must be provided from the start with relatively large legs if their speed is to be sufficient, so that their pereiopods show a definite negative heterogony, or *decrease* in relative size with increase of absolute size: and that here the low point of growth, or 'negative growth-centre', is also in the merus [1] (Cott, l. c.; Huxley, 1931B).

A similar reversal of gradient-sign appears to occur in the individual development of Ungulates. D'Arcy Thompson (l. c.) gives a figure (Fig. 48), of the proportions of the foot in ox,

[1] This is from length-measurements kindly supplied in answer to a query of mine by Mr. Cott; unfortunately, he only had a few specimens available for measurement, and the results, while clearly showing the merus as the joint of lowest growth-ratio, are not sufficient to construct a growth-gradient. The indication is that the gradient is complex, first rising above the level for the body-standard (carapace length), then sinking well below it in the merus, then rising again. It would be of great interest to establish this by obtaining statistically adequate data, as this is the only indication so far obtained of a complex growth-gradient with two points of inflexion within a single limb.

sheep and giraffe which together with inspection of skeletons makes it fairly clear that in the phylogenetic elongation of the giraffe's leg, the growth-centre has lain in the cannon-bone, with a steep gradient distally, a less steep one proximally. Meanwhile actual weight (and length) measurements made by Hammond (1927, 1929) and analysed by Huxley (1931B) on the individual growth of the hind-limbs in sheep, show that in regard to the pelvis and the three segments femur, tibia and cannon-bone (unfortunately the digits were not measured) there is, correlated with the negative heterogony of the whole limb relative to the body, a reversed growth-gradient with low point distally (Table VIII; Fig. 49). Though these constitute but two isolated bits of evidence, they indicate, so far as they go, that the growth-mechanisms underlying all heterogony are similar, and that when heterogony is negative, the sign of the gradient is simply reversed.

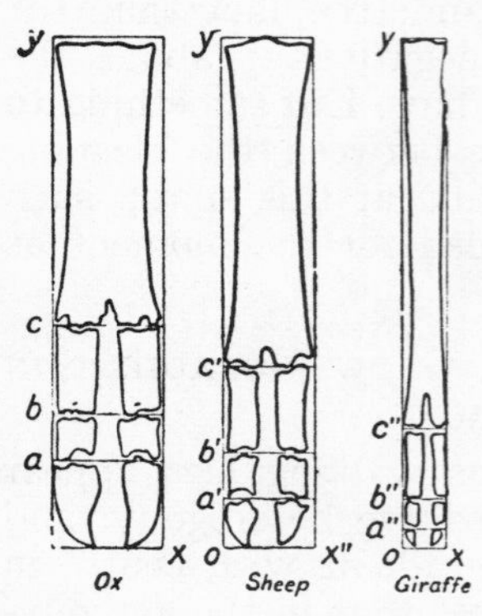

FIG. 48.—Comparison of the skeleton of the foot in Ox, Sheep and Giraffe, to show graded alteration in proportions of parts.

To effect the transformation form a typical (e.g. ox) form to that in the giraffe, *y—c* has been enlarged, *c—b* has remained nearly constant, *b—a* has been decreased, and *a—o* markedly decreased. In addition, the length : width ratio has been increased.

Hammond (1928, see also 1921) has also shown that the growth-gradients in the limbs and elsewhere affect the muscles as well as the bones, so that the study is of practical as well as theoretical importance. An important point made by Hammond may be given in his own words.

"As the animal grows, it changes its conformation; at birth the calf or lamb is all head and legs, its body is short and shallow, and the buttocks and loin are comparatively underdeveloped; but, as it grows, the latter—buttocks, loin, etc.—grow at a faster rate than the head and legs, and so the proportions of the animal change. . . . The extent to which these proportions change determines its conformation; those which develop most for their age have the best meat conformation, while those which develop least have the worst . . . Breed improvement for meat, therefore, means pushing a stage further the natural change of proportions as the animal matures. . . . The adult wild Mouflon ewe is in its proportions but little in advance of the improved Suffolk lamb at birth, although it is much larger.

What this means to the butcher and consumer is that of 100 lbs. live weight of an animal shpaed like the Suffolk lamb four days old the butcher can hang up as carcase in his shop 53 lbs., and the customer can eat as flesh only 30 lbs.; on the other hand, when the animal is shaped like that of the adult Suffolk ram, from 100 lbs. live weight 67 lbs. of carcase is obtained, and of this 61 lbs. is flesh which can be eaten—more than double that from of the badly shaped animal."

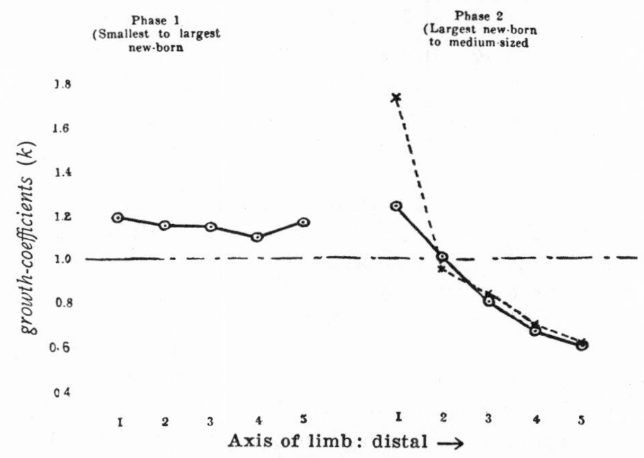

FIG. 49.—Reversed growth-gradient in organs showing negative heterogony (limbs of sheep).

The figures on the abscissa represent: 1, limb-girdle; 2, humerus, or femur; 3, radius + ulna, or tibio-fibula; 4, carpals, or tarsals; 5, metacarpals, or metatarsals. (solid line, forelimb; dotted line, hind-limb). The ordinates denote growth-coefficients (k), those below 1·0 signifying negative heterogony: they are taken relative to vertebral column weight. Size-phase 1 includes smallest to largest new-born specimens (100 to 256 g. vertebral column weight); size-phase 2, largest new-born to half-grown (256 to 690 g. vertebral column weight). The growth-gradient is at first flat, with slight positive heterogony; then steeply tilted downwards distally, upwards proximally.

TABLE VIII

Relative weights of parts of the skeleton of Suffolk sheep at three different ages (♂ Hammond, 1929; ♀ Hammond, 1927), reduced to proportions of weight of cannon-bone taken as 100. From Huxley, 1931B.

	Cannon-bone (hind-leg).	Tibio-fibula	Femur	Pelvis	Cannon-bone (fore-leg)	Radio-ulna	Humerus	Scapula	Skull and Jaws	Vertebrae				Ribs and sternum
										Cervical	Thoracic	Lumbo-sacral	Total	
Birth . .	100	286	212	220	100	187	150	75	1097	404	489	339	1230	722
♂ 5 months	100	352	320	420	100	213	241	169	1179	581	564	633	1778	1205
♂ 4 years .	100	380	361	619	100	276	289	257	1235	854	775	794	2423	1706
Ratio 4 yrs.: birth	1·00	1·36	1·70	2·81	—	1·47	1·93	3·43	1·13	2·11	1·58	2·34	1·97	2·36
Birth .	100	147	217	142										
♀ 5 months	100	245	285	430										
♀ 4 years .	100	272	324	569										
Ratio 4 yrs.: birth .	1·00	1·38	1·50	4·07										

Thus it would appear that one of the chief advances made by man in creating improved breeds of sheep and other meat animals has been simply to steepen growth-gradients which already operate during post-natal development in the wild ancestral forms. Hammond himself (1927) has expressed a similar idea. 'The improver of meat-producing animals has apparently not chosen mutations occurring in isolated points independently, but rather has based his selection on the generalized correlated changes of growth'. (See also Fig. 96, p. 223.)

In consequence of the gradient, there will be much less difference in the size of the metatarsal between a semi-wild and an improved breed than in the size of the femur. This is well brought out by Hammond (1927) in his Fig. 4.

In this connexion, it is well to remember that during embryonic life, the limbs of sheep must show a growth-gradient precisely opposite in sign to that of their post-natal period. Lambs are born with relatively long legs, as an adaptation to accompanying their dams almost from birth. To achieve these unusual proportions, the leg must have exhibited positive heterogony during foetal life; and to allow for the fact of the later centre of negative heterogony in the cannon-bone, this same region must have been the positive growth-centre in the earlier period. The same reasoning applies to Ocypoda, whose young are similarly precocial.

§ 4. The Form of Growth-gradients

Analysis of the data of Kemp and his co-workers on Palaemon spp. undertaken by Miss I. Dean (unpublished) gives a further interesting result. In these prawns, both male and female have obviously heterogonic chelae, but the male's heterogony is considerably higher. Thus a male and a female of the same absolute size will possess chelae of very different sizes, the female's being considerably the smaller. But if we take a male chela and a female chela of the same absolute size (which will of course be borne by a small male and a large female body) the proportions of the separate joints will be found to be fairly similar. This indicates that whenever marked heterogony, or at any rate heterogony designed to give rise to a large chela, is present, it must operate by essentially the same growth-mechanism within the limb (and a mechanism quite different from that in a slightly heterogonic pereiopod), whether the growth-coefficient of the whole limb relative to the body be moderate or high. The growth-gradient of the female is not quite so steep as that of the male, a fact also brought out by Tazelaar on *P. carcinus* (p. 92); but the male and female chela-gradients are much more like each other than they are to the gradients of any of the pereiopods.

Still further proof of the radical difference of the growth-gradients leading to pereiopod and to large chela is afforded by the male hermit-crab Eupagurus (Bush, 1930; Bush and Huxley, 1930). Here the right chela during early life is not much enlarged, and its growth-coefficient is no greater than that of the pereiopods; only later does it begin the marked

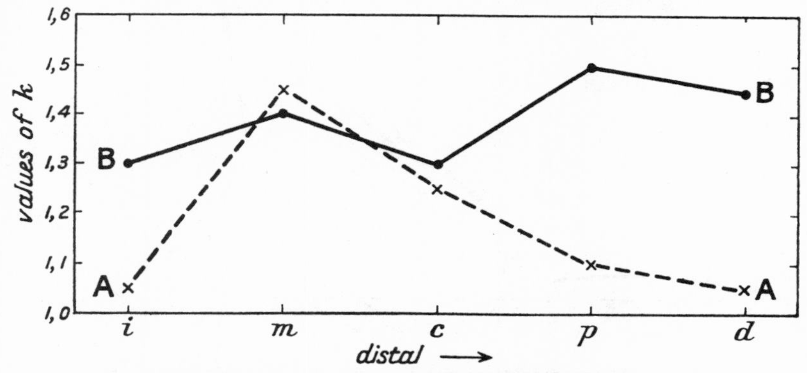

FIG. 50.—Change in form of growth-gradient with increase of growth-rate in large (right) male claw of the hermit-crab, *Eupagurus*.

i, ischium; *m*, merus; *c*, carpus; *p*, propus; *d*, dactylus. A—A, juvenile phase; the growth-gradient resembles that of a pereiopod. B—B, phase of heterogony of right chela; the main growth-centre shifts distally.

heterogony which provides its definitive enlargement. And during the earlier period its growth-gradient is similar to that of a pereiopod, with centre in the merus; while so soon as the final heterogony becomes marked, the main growth-centre shifts to the propus (Fig. 50).

Tazelaar (unpublished) has also collected facts bearing on this subject. In *Palaemon carcinus*, there is a change in the growth-coefficient of the chela in both sexes at about 4·5 cm. carapace length. In the female, before this, the chela has been growing less rapidly than the neighbouring pereiopods; after this it exhibits a considerable heterogony. During the first of these phases its growth-gradient is almost flat, like those of the pereiopods, but with a slight growth-centre in the propus. Later it exhibits a marked growth-gradient with centre in the dactylus.

In the male, the chela in the first phase shows definite heterogony, about the same as the female chela in the second phase. During this phase it shows a growth-gradient with

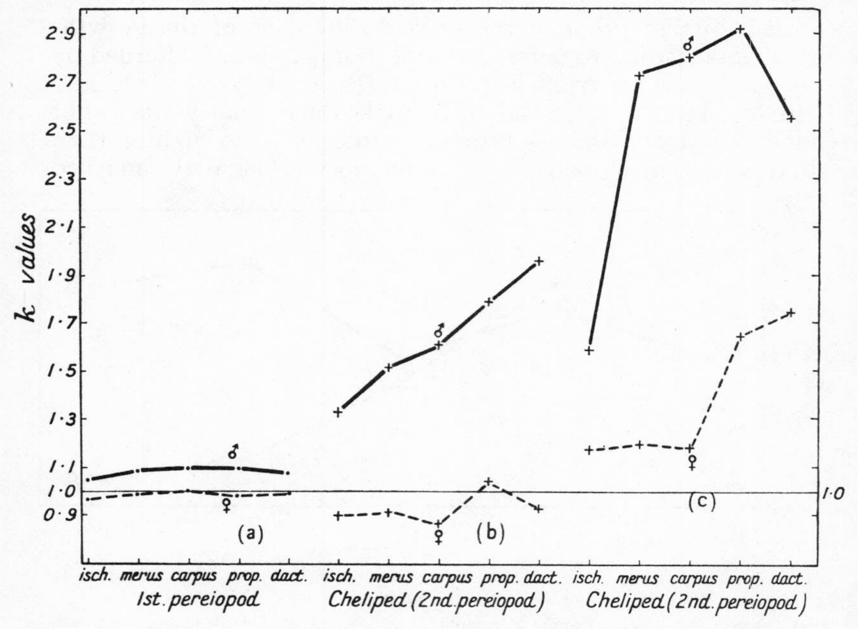

FIG. 51.—Growth-gradients in the 1st pereiopod, and the chela of the prawn, *Palaemon carcinus*.

——, male ; – – – –, female. (*a*) 1st pereiopod ; the gradient is flat and close to unity throughout , (*b*) and (*c*) 2nd pereiopod (chela). (*b*) 1st phase ; female with slight initiation of growth-centre distally ; male with regular growth-gradient (distal growth-centre). (*a*) 2nd phase ; female with definite growth-gradient but incomplete proximally ; male with very marked growth-gradient (subtermunal growth-centre).

centre in the dactylus. During the second phase, the male chela shows extremely marked heterogony ; and it now possesses a striking growth-gradient, with centre in the propus (Fig. 51). It would seem as if the steepening of the gradient began near the top, and then gradually extended centripetally (cf. p. 168).

§ 5. GROWTH-GRADIENTS IN REGIONS OF THE BODY

Precisely similar gradients to these found in appendages may be traced in the growth of whole regions of the body.

The most clear-cut examples concern the abdomen of crabs, which in all cases are narrow in the male, broadly expanded in the female. A large series of measurements has been made by Sasaki (1928) on both sexes of the Japanese species *Telmessus cheiragonus*. Analysis of these data shows that whereas the growth-gradient for breadth in the male abdomen is nearly

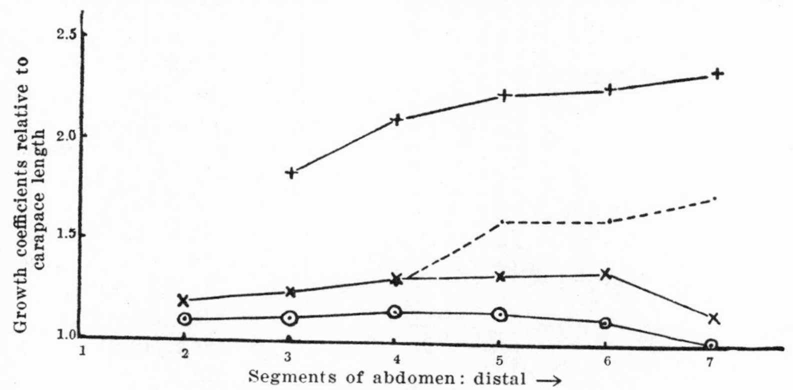

FIG. 52.—Growth-gradients in the abdomen of crabs.

Solid lines, for breadth of abdominal segments: ⊙, *Telmessus cheiragonus*, ♂; ×, *Telmessus cheiragonus*, ♀; +, *Pinnotheres pisum*, ♀. Dotted line, for length of abdominal segments in *Pinnotheres pisum*, ♀.

flat, with its growth-centre, if so it may be called, near the centre of the region, in the female it is steeper, with its centre (as in the typical male chela) in the penultimate segment (Fig. 52).

These figures may be compared with those cited for the edible crab, *Cancer pagurus*, by Pearson (1908, p. 21) in two large specimens of the same size but opposite sex (Table IX).

TABLE IX

CANCER PAGURUS; FROM DATA OF PEARSON, 1908.

Abdomen segment	Carapace-breadth 235 mm.				♀/♂ ratio per cent.	
	♂		♀			
	Length mm.	Breadth mm.	Length mm.	Breadth mm.	Length	Breadth
1	17	22	17	25	100	114
2	8	17	8	22	100	129
3	7	23	7	30	100	130
4	8	20	8	32	100	160
5	9	17	10	35	111	206
6	13	16	20	35	154	219
7	13	13	18	21	138	162

I have (last column) calculated the ratios of ♀ to ♂. These indicate clearly that both for length and breadth there exists a growth-gradient in the ♀ abdomen, with high point in the sixth segment, though the two gradients must be of very different slope and shape.

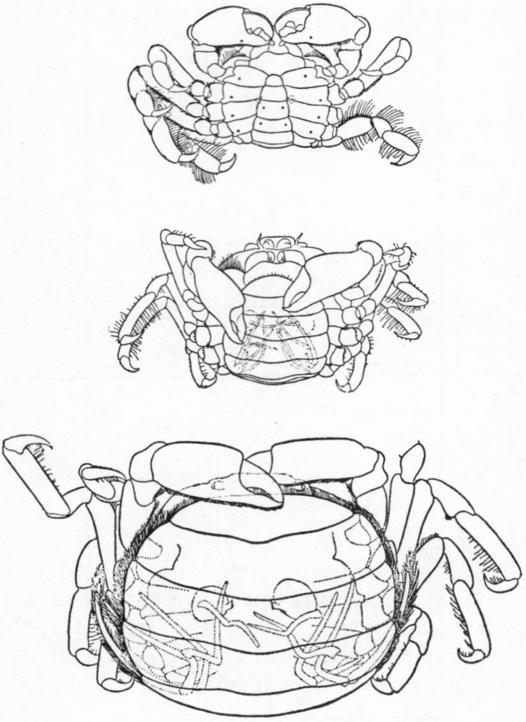

Precisely similar results have been obtained for the female abdomen of the common *Carcinus maenas* (unpublished). That the growth-centre need not be in the penultimate segment, however, is shown by the Pea-crab, *Pinnotheres pisum*. Scale drawings of this at different stages of its growth are given by Atkins (1926); I have measured these, and although the number of specimens is very small, it is quite enough to demonstrate the gradient. The gradient is first of all interesting because of its steepness. The growth-coefficients of the separate segments, relative to carapace length, range up to 2·3 and over, a very high figure. In passing, this remarkable heterogony is doubtless correlated with the small absolute size to which the crab is restricted within its host's shell; the animal has to attain its full female proportions at a much

FIG. 53.—Changes of shape in the female abdomen of the pea-crab, *Pinnotheres*, during growth.

Above, stage I, late (carapace width about 3 mm.); centre, stage III*b* (carapace width 6 mm.); below, final form, stage V (carapace width 11·5 mm.). The abdomen is at first male-type, but shows marked heterogony, with terminal growth-centre, until it overtakes the bases of the legs.

smaller absolute size than is necessary in most crabs. (Fig. 52.) (See Huxley, 1931B.)

During growth, the young female type of abdomen, which with its flat or concave margins resembles the male's, is converted into an almost circular structure, with the fourth and fifth segments the broadest. Casual inspection would indicate that one of these segments must contain the growth-centre; but casual inspection is wrong—the terminal segment (telson) is so small in the young female that for it to achieve its only moderate definitive size it must, and does, contain the growth-centre (Figs. 52, 53).

In the heterogonic growth of the face of mammals, which we have noted in Chapter I, it would appear from the work of Todd (1926) that in the palatal region the centre of maximum growth lies in the palatal processes of the maxillae; there is a moderate amount of growth in the premaxillae and very little in the palatals. It is impossible to arrive at quantitative expression, but the facts are consonant with the idea of a double gradient culminating within the maxillae.

§ 6. Graded Growth-intensity in the Different Planes of Space

So far, we have been considering growth-gradients referring either to weight, or to linear measurements in one axis.

In Pinnotheres, measurements were taken both for the length and breadth of the abdominal segments (Fig. 53); when we consider the two sets of measurements in relation to each other, we find that the ratio between growth-coefficient for breadth and that for length is as follows for the distal half of the female abdomen:

Segment	4	5	6	7
Ratio, $\frac{\text{growth-coefficient for breadth}}{\text{growth-coefficient for length}}$	1·62	1·46	1·40	1·36

Thus the difference between male-type and mature female-type abdomen is brought about (1) by greater growth in the female, both in length and in breadth; (2) by the breadth-growth being throughout higher than the length-growth; (3) by the excess of breadth-growth over length-growth being greatest, in the segments measured, in the fourth segment, and decreasing distally. Without further measurements it is impossible to state whether the high point in breadth-growth predominance is really in the fourth segment, or further basally; but at least the orderly and graded relation between the intensity of growth in the two planes of space is clearly brought out.

But a more complex, and perhaps more interesting problem is afforded by the two chelae of markedly heterochelous Crustacea of which the lobster (Homarus) is the most familiar, and the pistol-crab (Alpheus) the most extreme example. In all of these, the heavier (crusher) claw is broader and altogether bulkier in build; and the lighter (nipper) claw *always* has a relatively and often an absolutely longer dactylus (Huxley, unpublished: see figures in Przibram, 1930). It would seem to mere inspection that the fundamental difference between the two claws lies in the difference of their growth-gradients in the three planes of space, the growth-coefficients for breadth and depth being relatively as well as absolutely much higher in the crusher, and the growth-centre as regards length being more distal in the nipper.

A few measurements have been made on the chelae of lobster (Homarus) to check this (Huxley, unpublished). So far as they go, they confirm the impression made by inspection. The (only very approximate) value of the growth-coefficients are as follows (relative to carpus length as standard):

TABLE X

APPROXIMATE GROWTH-COEFFICIENTS (k), RELATIVE TO CARPUS-LENGTH, OF THE LINEAR DIMENSIONS OF THE SEGMENTS OF THE CHELAE OF THE EUROPEAN LOBSTER

	Crusher			Nipper		
	Length	Breadth	Depth	Length	Breadth	Depth
Dactylus .	0·6	1·2	1·1	0·85	1·05	1·0
Propus . .	0·95	1·2	1·1	0·8	1·05	0·9
Carpus . .	(1·0)	0·95	1·1	(1·0)	1·05	1·15
Merus . .	0·85	0·9	1·0	0·85	1·15	1·1
Ischius . .	0·85	0·85	0·9	0·8	1·05	1·05

It will be seen that the chief difference between the two claws is in the distal region. In the crusher dactylus, the length-growth falls off notably to a much greater extent than it does in the nipper dactylus. In the nipper, breadth-growth is approximately constant in all joints, while depth-growth decreases distally. In the crusher, on the other hand, depth-growth increases steadily, and breadth-growth markedly, from ischium to propus, to fall slightly in the dactylus ; (it should be recalled that the rates of growth are taken relative to the claw's own carpus-length, not to an independent standard in the body). Perhaps the most interesting point is the change

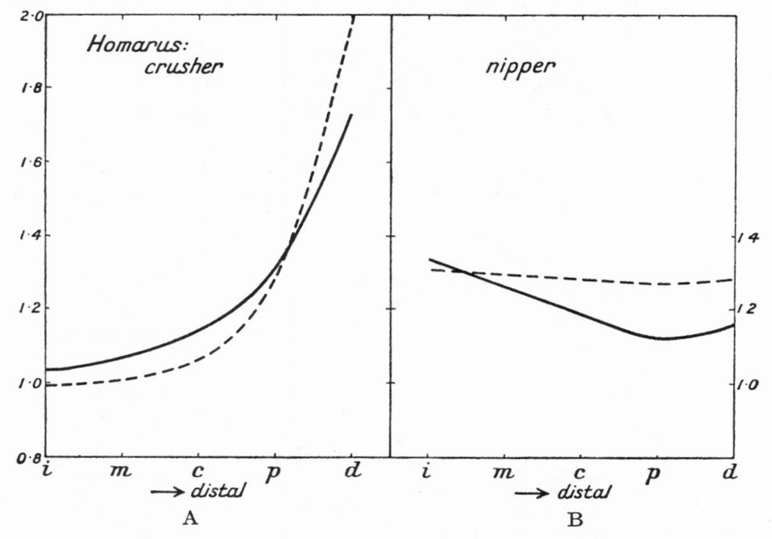

FIG. 54.—Probable change in the ratio between growth-intensity in depth and that in length (solid line) and the ratio between growth-intensity in breadth and that in length (dotted line) along the axis of (A) the crusher-claw and (B) the nipper claw (right) of the Lobster.

i, ischium ; *m*, merus ; *c*, carpus ; *p*, propus ; *d*, dactylus.

in the ratios between the growth-intensities in breadth and depth to that in length, as we pass along the axis of the limbs.

The present figures are based only on a few specimens, and curves constructed directly from them are rather irregular. I have therefore ventured to prepare smoothed curves which represent to my mind the most probable state of affairs (Fig. 54).

Thus the main difference between crusher and nipper is not merely one of total growth-potential, but also of the different distribution of growth-potential in the three planes of space. The minor differences, e.g. number and form of 'teeth' are presumably due to specific gene-differences, not correlated directly with growth.

A further point of interest was elicited in these chelae by making measurements not only of whole segments, but to intermediate points marked by spines, etc. (notably the four large spines on the median side of the propus). These measurements indicate that, as suspected, the growth-coefficient of a given whole segment merely represents a mean value, the growth-gradient being real and continuous, and that the values of the growth-coefficients are altering continuously and regularly along the segment. Many more measurements, however, would be necessary before the true gradients could be plotted in detail, and to obtain these would be a very difficult task, as fixed points to measure are not numerous enough. (See also Locket, p. 261.)

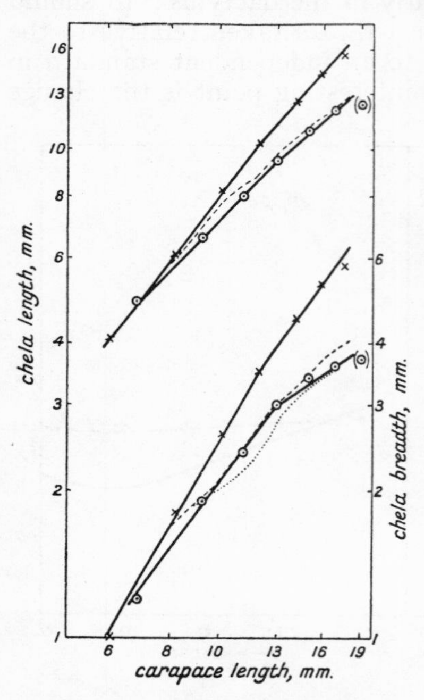

FIG. 55.—*Upogebia littoralis* (*Decapoda, Reptantia, Anomura*). Chela (dactylus + propus) length and chela (propus) breadth against carapace length; logarithmic plotting. See Table XI.

×, males; ⊙, females; ----, parasitized males; ......, parasitized females. The male chela, as in *Uca*, begins by being distinctly heterogonic and then becomes less heterogonic. For details, see text.

(*From data of Tucker*, 1930.)

Tucker (1930) has made measurements on the chela of the anomuran *Upogebia littoralis*, for normal specimens of both sexes as well as for those parasitically castrated by the Bopyrid *Gyge branchialis*. The length measurement chosen was (propus + dactylus) while the breadth measurement was propus-breadth. The results for normal specimens are shown in Fig. 55. For both males and females, two distinct phases of growth are found, the alteration occurring at 12–13 mm. carapace length. In the first phase of both sexes there is a good approximation to constant differential growth-ratio; this exists also in the males' second phase, with a slight falling off in the last point. The male chela shows positive heterogony in both phases. In the female, there is a progressive falling off in the growth-coefficient during the second phase, more marked for breadth than for length. The female chela is positively heterogonic in the first, negatively heterogonic in the second phase. If the growth-coefficients are calculated for the whole of the first phase, and between the first and third points of the second phase, we obtain the following result:

TABLE XI

GROWTH-COEFFICIENTS FOR LENGTH AND BREADTH OF CHELA IN UPOGEBIA LITTORALIS, CALCULATED FROM THE DATA OF TUCKER, 1930

	k for chela breadth	k for chela length	Ratio of $\frac{k \text{ breadth}}{k \text{ length}}$
Males, first phase . .	1·82	1·37	1·33
Males, second phase . .	1·47	1·13	1·30
Females, first phase . .	1·37	1·05	1·30
Females, second phase .	0·72	0·90	0·80

I.e. the ratio of growth-intensity for breadth to that for length remains the same within the limits of experimental error so long as the chela is positively heterogonic. When, however, as in the females' second phase, it becomes negatively heterogonic, the situation is reversed, and growth-intensity for breadth falls below that for length. This reversal would appear to be analogous to the reversal of sign of the growth-gradients in negatively heterogonic organs (p. 87). Logarithmic plots (see Fig. 55) as well as Tucker's graphs indicate some interesting points with regard to the action of the parasite upon chela growth. The parasite has a general feminizing action, very similar to that of Sacculina. In males, the effect in reducing the growth-coefficient of chela-length is slight up to 10·5 mm. carapace-length, for chela-breadth up to 8 mm. carapace-length. From then on the growth-coefficient for length remains just above that for females throughout; that for breadth, after approximating closely to that for females until the second phase, fails to be as much reduced as the females during the remainder of life (k = about 0·95 instead of 0·72).

In the parasitized females, k for length is throughout just below that of normal specimens, whereas k for breadth descends well below the normals between 10·5 mm. and 13·5 mm. carapace-length, then gradually ascending almost to the level of the normals. This latter fact would seem to indicate that in females the parasite (as in Sacculina) acts as would a precociously functioning ovary, but that the value for normal chela-breadth in the larger normal specimens represents a final partition-coefficient towards which parasitized as well as normals tend to approach. It is further clear that breadth-growth is more sensitive than length-growth to ovarian influence and to the parasite's pseudo-ovarian influence, and reacts to them both earlier and to a greater degree. The failure of the chela of parasitized males to respond in breadth-growth to the influence of the parasite as much as the normal female chelae to the influence of the ovary may be due to a specific difference in male and female chela-tissue, but it is more probably due to the fact that the degree of feminization effected by the parasite is variable. As breadth-growth is more sensitive than length-growth to feminizing influences, this variability will be reflected more markedly in the curves for breadth-growth.

Finally, the work of Hecht (1916) on the growth of Teleost fish is of some interest here. He finds in a number of species of markedly differing body-form that the relation of the maximum width of the body to total length remains fairly constant (to about $\pm$ 16 per cent.) whereas the relation for depth is highly variable (by over $\pm$ 55 per cent.). The form of fish appears thus to be determined much more by changes in the growth-coefficient for depth than by changes in that for width.

§ 7. Gradients in Growth-rate of Epidermal Structures

Another extremely interesting case of growth-gradients is provided by the work of Juhn, Faulkner and Gustavsen (1931). Working on the domestic fowl, they find in the male (normal and castrated) definite regional gradients in the rate of growth of feathers which are regenerating after plucking. The most marked of these regional gradients is in the breast, and has its high point posteriorly (Fig. 56).

In addition, there is in normal and castrated males a dorso-ventral gradient in breast, back, and saddle, with high point dorsally, e.g. the regenerating saddle feathers of capons grow more slowly than the breast feathers. A similar gradient exists in the breast, but not in the saddle.

Two further points demand special attention : one is a quite new one, of obviously first-class physiological importance—viz. that in all feathers low rate of growth is associated with a low threshold of sensitivity to the sex-hormone, as our authors have demonstrated by means of an ingenious series of experiments. Accordingly, saddle feathers respond by

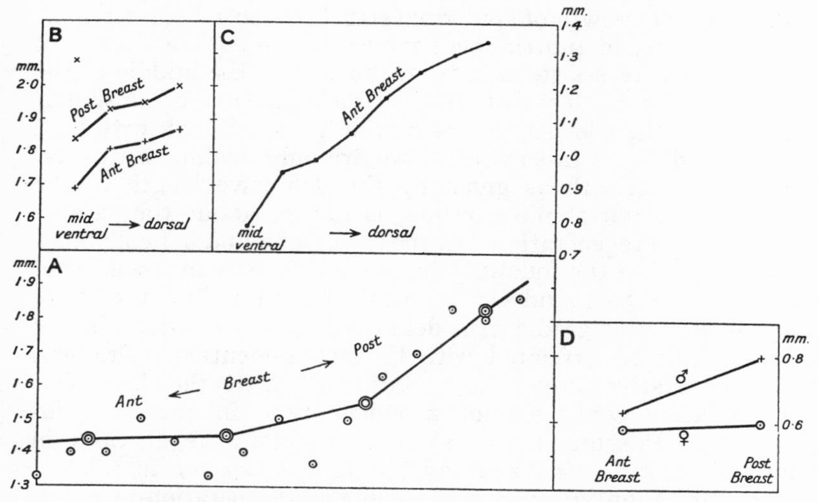

FIG. 56.—Gradients in feather-growth in fowls ; constructed from the data of Juhn, Faulkner and Gustavsen, 1931. The ordinates represent rates of feather-growth per day, in mm.

In A, the abscissa axis represents the antero-posterior axis of the breast region in a capon (⊙, points for single feathers ; ◎, means). In B and C, it represents the gradient downwards from the mid-ventral line (in B, in both anterior and posterior breast regions of a cock ; in C in the anterior breast region of a capon). In D, the rates of growth of anterior and posterior regions of the breast of a cock and a hen are compared.

feminization to small doses of female hormone which have no effect upon breast feathers : or again, a single injection of female hormone capable of inducing a large feminized pattern on the slowly regenerating feathers of the anterior breast produce smaller bars of feminized pattern on the more rapidly-growing feathers of the posterior breast. (And see p. 260.)

The other point is a confirmation from this quite new quarter of the principle we found to hold in crustacean limbs, etc., that increase in relative rate of growth in an appendage or region was accompanied by a steepening of the growth-gradient. This holds even for the type of growth-gradient affecting feather-growth only, which we are here considering. In females and feminized males, the absolute rate of growth of, e.g., the breast-feathers is reduced, and the feathers, growth-gradient in this region is at the same time almost flattened out (Fig. 56D).

Mr. Miller, of the Animal Breeding Research Department, Edinburgh, has pointed out to me that similar gradients affecting the growth of epidermal structures occur in mammals. For instance, in (untrimmed) manes of horses, the length of the hairs increases steadily from the ears to the middle of the mane, then sinks steadily to the hind end of the mane. A similar gradient in hair-length occurs in the human beard, with high point medially. In such cases we are only dealing with total amount of growth as given by the definitive length of the hair: I do not think anything is known about the rate of growth or of regeneration. In the fowl, Juhn and her collaborators (l. c.) find the following interesting facts. In breast, and also in the back-and-saddle region, there exists a distinct antero-posterior gradient in definitive feather-length. In the breast, this is correlated with the above-mentioned gradient in regenerative growth-rate, but also with the fact that regeneration continues for a longer time in the posterior feathers. Presumably the same two factors are at work in normal growth. In back and saddle (information in a letter from Dr. Juhn) there is no gradient in regeneration-rate, but only a gradient in the length of time for which regeneration proceeds. We thus have two distinctive methods of growth, both capable of gradation, affecting the size of epidermal structures.

§ 8. Conclusion

The chief points brought out in this chapter are the following. Differential growth of a limb or appendage or a well-marked region of the body appears never to be brought about by an equal distribution of excess growth-potential throughout the organ or region. On the contrary, the growth-potential of the organ or region is distributed in the form of a growth-gradient, normally with a single high point or growth-centre, from which growth-intensity grades downwards in both directions (or in one, if the growth-centre be terminal). In general, the less the difference between the growth-coefficient of the organ or region and that of the rest of the body, the less marked and flatter is the gradient, so that in isogonic organs there is scarcely any gradient, but all the parts grow at approximately the same rate as the body as a whole. There are, however, some unusual gradients, as in the first antennae of certain copepoda, in which the growth-coefficients of part of the organ are above, the rest below unity.

When one of two corresponding organs is positively, the other negatively, heterogonic, the growth-gradients of the two appear to be similar but reversed in sign, the same joint being in one a centre of maximum, in the other a centre of minimum growth.

In most markedly heterogonic organs so far investigated (chelae of crustacea, abdomens of female crabs, limbs of ruminants) the growth-centre is terminal or sub-terminal.

When an organ is markedly heterogonic in both sexes, then, even though the growth-coefficient of the whole organ may differ considerably in male and female, the growth-gradient within the organ appears to be essentially similar—i.e. there is a certain qualitative type of growth-gradient needed to produce an organ of a certain morphological type, though quantitative details may be different. This is confirmed by the change from a growth-gradient with sub-basal centre (as in the pereiopods) to one with a sub-terminal centre (as in other large chelae), when the chela of the male Eupagurus passes from a slight heterogony, no greater than in the walking legs, to the marked heterogony of maturity.

The gradients for growth in the three planes of space may be different; applications of this are seen in the abdomen of male and female crabs, and still more in the crusher and nipper claws of heterochelous crustacea.

The existence of growth-centres and growth-gradients is an empirical fact, whose physiological explanation is quite unknown but may prove to be of importance for the study of growth in general.

CHAPTER IV

GROWTH-GRADIENTS AND THE GENERAL DISTRIBUTION OF GROWTH-POTENTIAL IN THE ANIMAL BODY

§ 1. General Growth-gradients: D'Arcy Thompson's Graphic Method

THE next step is to inquire whether the growth-gradient mechanism may not underlie the general growth of the body as well as the growth of specialized appendages or regions. In other words, we want to see whether the sharply-marked growth-gradients of a chela or an abdomen are not merely special cases of a more general but still orderly mechanism underlying the distribution of what we may, for want of a better term, call growth-potential, throughout the whole organism.

A strong indication that this is so was afforded by D'Arcy Thompson (l. c., chap. 17) through his ingenious application of the principle of Cartesian co-ordinates to the problem of animal form. Let us take the most spectacular and at the same time one of the simplest of his instances. The strange fantastic sun-fish, Orthagoriscus, is a close relative of such types as Diodon; but its adaptation to an almost planktonic life at the surface of the sea has led to a change of form so radical that it is at first sight difficult to see how it could have been brought about within any short space of evolutionary time. D'Arcy Thompson, however, pointed out that if you inscribed the outline of a Diodon in a framework of rectangular co-ordinates, and then distorted this in a certain perfectly regular way, you would obtain a very close approximation of the outline of an Orthagoriscus (Fig. 57). From the figure it will be immediately obvious that the essence of the transformation, considered biologically and not merely as an exercise in higher geometry, must have been the origin of a very active growth-centre in the whole of the hind-region of the body, whence the intensity of growth diminished regularly towards the front end. In other words, superposed on whatever growth-mechanisms may be necessary to generate a form similar to that of Diodon, there has arisen a steep and unitary postero-anterior growth-gradient extending throughout the entire body, with high point almost or quite at the extreme hind end.

The diagram, however, also illustrates the serious limitations of the method. In the first place, two adult forms are contrasted. What should rather be contrasted are, phylogenetically, the young form of the presumed Diodon-like ancestor, and the adult Orthagoriscus; or, ontogenetically, a young Diodon-like stage of Orthagoriscus with the adult condition.

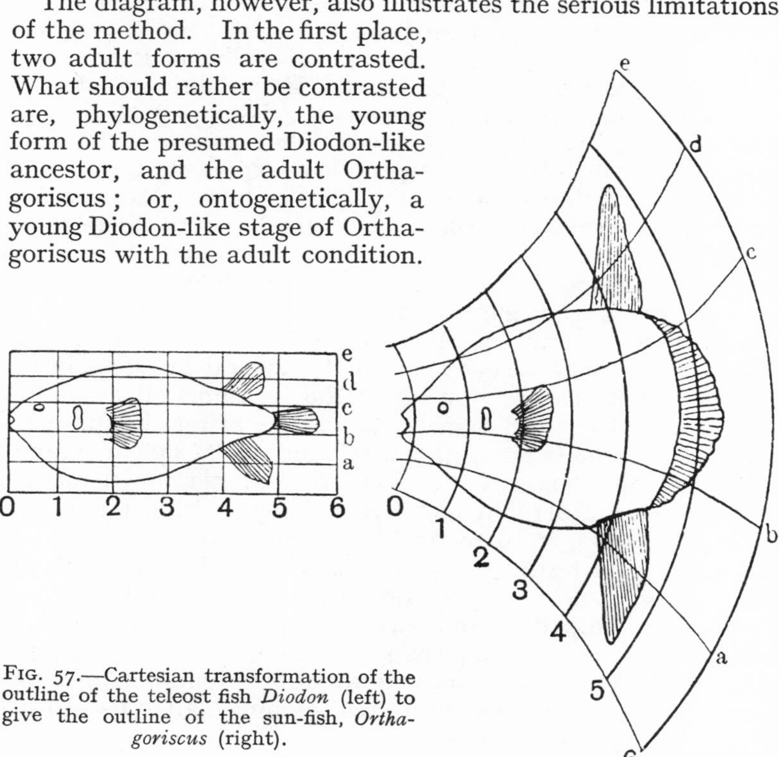

FIG. 57.—Cartesian transformation of the outline of the teleost fish *Diodon* (left) to give the outline of the sun-fish, *Orthagoriscus* (right).

But this is not so serious as the following point: that even should you compare the correct two stages, this graphic method, if interpreted in terms of growth, can only give a general and qualitative picture of the mechanism at work, in place of a specific and quantitative one. For if, as D'Arcy Thompson points out, the transformation, so difficult to understand at first sight, becomes readily comprehensible on the idea of an orderly change in the distribution of growth-activity along the axis of the body, then clearly the proportions of the animal must be continually changing so long as it is increasing in absolute size, or at least over a long space of time. But the fish's outline and the system of co-ordinates drawn to fit it, represent the state of affairs only at one particular moment of its life-history. If the fish had grown to twice the bulk, its proportions would have changed, and the co-ordinate grid would have to be altered; yet the underlying growth-gradient might have remained wholly unaltered.

An improvement would be effected if the absolute sizes of the two outlines which mark the onset and stoppage of orderly differential growth could be given: but even so, the real invariable, namely the growth-gradient expressing the values of the growth-coefficients along the body, could only be obtained from this by calculation, and is not deducible by inspection.

For this reason, the co-ordinate method, while of the utmost importance as affording a graphic and immediate proof of the need for postulating regularities in the distribution of growth throughout the body, is of little use for detailed analysis, because by its nature it neglects the fundamental attribute of differential growth, namely the change of relative proportions with absolute size: it is static instead of dynamic, and substitutes the short cut of a geometrical solution for the more complex realities actually underlying biological transformation.

None the less, it is invaluable as demonstrating the need for thinking in terms of growth-gradients and, in general, of an orderly system in the distribution of growth-activity throughout the body: and this whether we are considering individual or evolutionary change of form. We have only to glance at D'Arcy Thompson's figures of brachyuran carapaces, ungulate limbs, and so forth, to realize immediately its utility in this respect (Figs. 48, 58).

Another interesting use has been to deduce the course of evolution over gaps where actual fossil data are missing. It is easy to say that we can do this by common-sense, simply inserting hypothetical intermediate stages between known end-points. But this is insufficient. There are, for instance, hundreds of possible ways of bridging the gap between the pelves of Archaeopteryx and the cretaceous bird Apatornis, to take an example worked out by Thompson; but if the evolutionary modification of such a structure be due to growth-changes, and if orderly and regular growth-gradients be the mechanism by which growth-changes must operate, then you can deduce the precise course of evolution within comparatively narrow limits (see Fig. 59).

As showing the validity of the method, D'Arcy Thompson has applied it to the evolution of the skeleton of the horse,

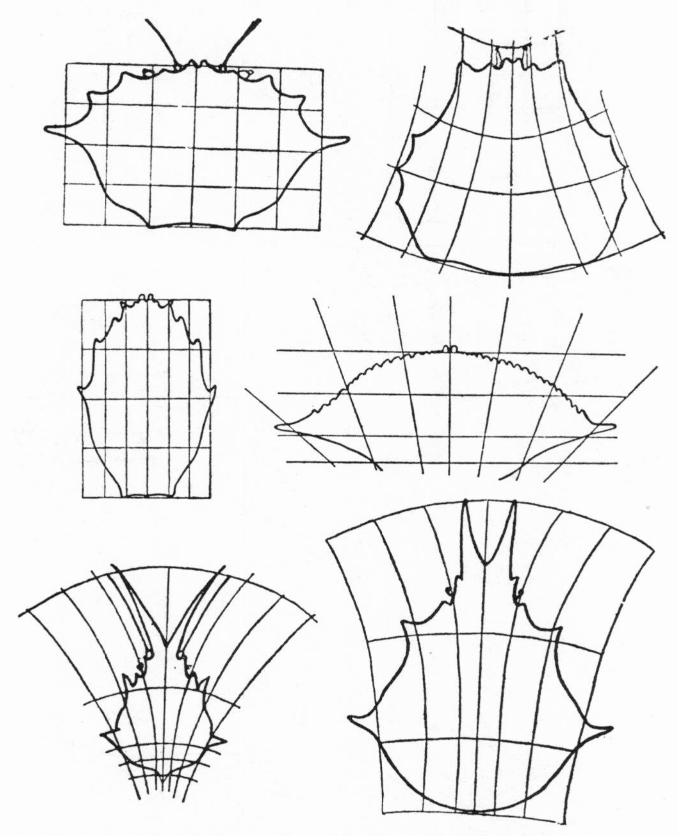

FIG. 58.—Carapaces of various crabs to show how they may be readily derived from one form by simple Cartesian transformation.

and then compared his deduced results with the known fossil forms (Fig. 60).

The agreement of actual with deduced skull-form is very close for Protohippus, Miohippus and Mesohippus. But the species of Parahippus figured by Thompson, as he points out,

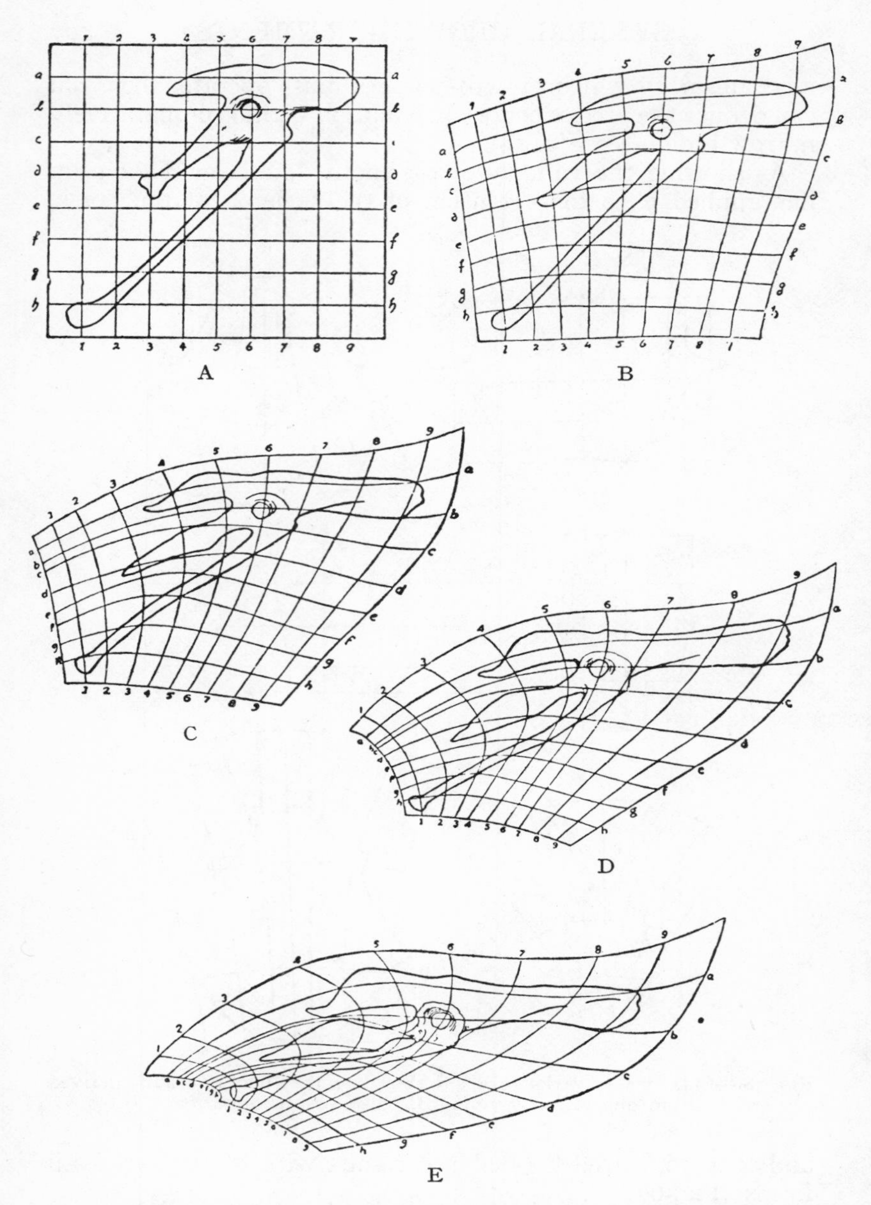

FIG. 59.—B, C, and D, Hypothetical intermediate stages, constructed by applying the method of Cartesian transformation, between A, pelvis of Archeopteryx and E, that of Apatornis.

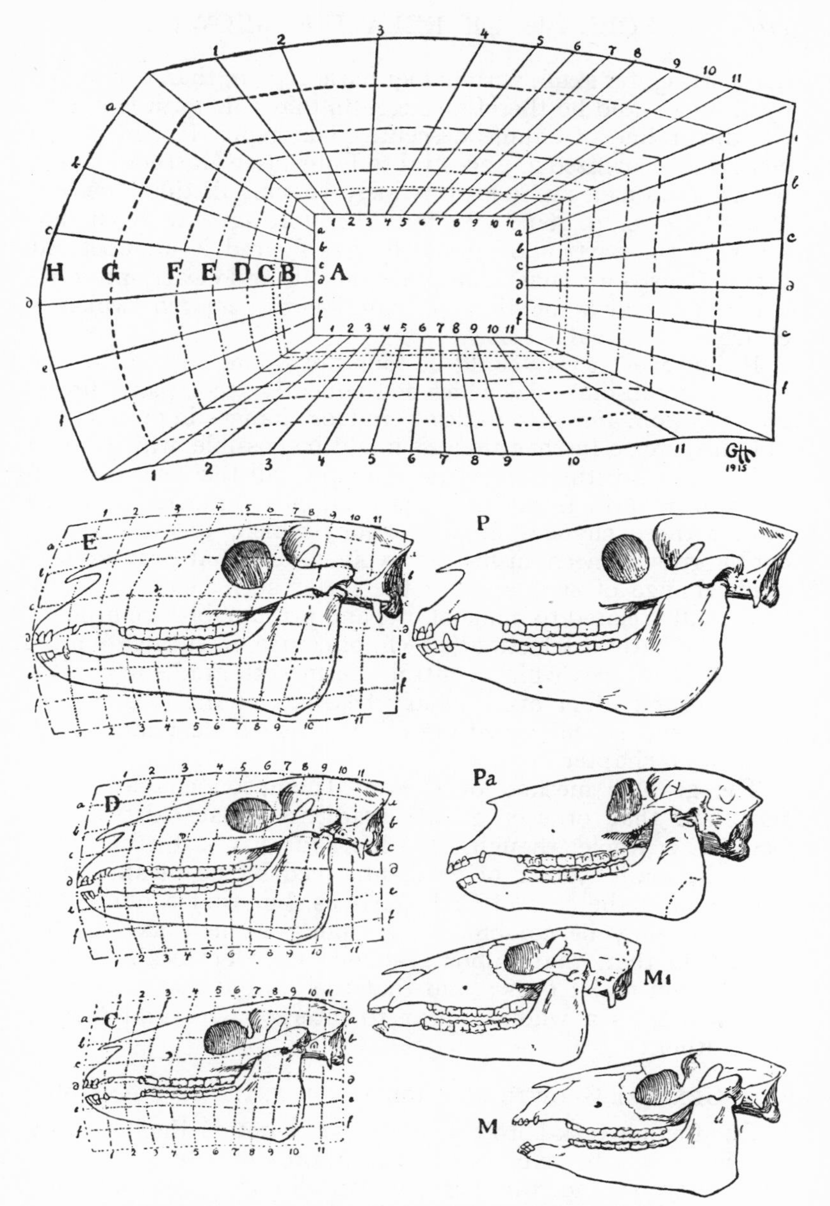

FIG. 60.—Comparison of hypothetical forms of the skull in the ancestry of the horse (C—E), constructed by applying the method of Cartesian transformation (A—H) with actual forms; (M) Mesohippus; (Mi) Miohippus; (Pa) Parahippus; (P) Protohippus.

has a straighter skull, with higher nasal region, than demanded by theory: and he therefore suggests that this form is not on the direct line of equine descent. The genus Parahippus is generally recognized as ancestral to Equus (see Matthew, 1926), but it contains a number of 'widely varying distinct species' (Matthew, l. c., p. 160). We may therefore agree with Thompson that the particular species he has figured is an aberrant type. This is a remarkable achievement, and clearly provides a new method in detailed paleontological research, which is clearly of real value.

If, however, a way could be found of taking account of the changes in absolute size which so frequently accompany directive evolution, and automatically induce changes in proportions of limbs, etc., a further analysis might be possible, which would enable us to distinguish what we might call the consequential changes in form from the strictly adaptive, meaning by the former those changes in proportions which, unless counteracting growth-mechanisms are evolved, automatically accompany change of size, and by the latter such changes as are specifically related to mode of life, and presumably are brought about by natural selection (which, of course, in such case must act via the genes which control the growth-gradients). But of these and other evolutionary bearings of the existence of growth-centres and growth-gradients I shall deal more fully in a later chapter.

The graphic method of D'Arcy Thompson enables us to postulate that orderly growth-gradients must exist in the body as a whole, though from his figures it is further clear that the main system of gradients need by no means be so simple as in the case of Orthagoriscus, but that a number of gradients may be combined even along a single axis of the body; further, that minor gradients may be locally superposed upon major ones; and that the various components of the system as a whole appear to interact with and modify each other.

§ 2. General Growth-gradients: Quantitative Analysis

Our further task is to try to find out whether these growth-gradients can be expressed, even approximately, in quantitative terms, and whether we can discover any empirical rules concerning the nature of the influence exerted by one gradient upon another. To do this is no easy task, for it demands quantitative measurements, either of mass or linear dimensions, of numerous parts of the body; and these must, for one thing, be made at a considerable number of absolute sizes, to make sure that the gradients, etc., do not change with age, and for another be made on a considerable number of specimens within each size-group, to exclude the effects of random sampling. Such a body of measurements only exists for very few animals, and even there not for sufficient points or organs to give a complete picture of the growth-gradients; but in spite of these limitations, the results are of considerable interest. We will begin with the case of the hermit-crab, Eupagurus, analysed under my direction by S. F. Bush (Bush, 1930; Bush and Huxley, 1930). Measurements were made of the eyestalks, first and second antennae, third maxillipeds, chelae (=first pereiopods), pereiopods 2–5, and uropods, as well as of median pre-thorax length as standard; (in some appendages the length of an arbitrary portion was taken in place of total length). When groups of small and large body-size were compared, the following significant facts emerged. We will first consider males, and in them the left side of the body, where matters are not complicated by the pronounced asymmetric heterogony of the right chela. Here we find a quite definite double growth-gradient, with growth-centre in the third pereiopod, whose growth-coefficient is considerably higher than that of the body (pre-thorax); from this point growth-activity falls off both anteriorly and posteriorly, until it is about equal to that of the body in the third maxillipeds and fourth pereiopods respectively, while less in the cephalic appendages and uropod (Fig. 61). There is a slight irregularity in the region of the last pereiopods; otherwise the gradient is quite regular.

This is remarkable, since the organs measured are not only of very different shape, but differ vastly in absolute size; yet the short eyestalk, first antenna and third maxilliped, the long second antenna, the large first to third pereiopods and the small fourth and fifth pereiopods, all, as regards their growth-ratios, fall into this regular graded series.[1]

The further significant fact emerges that although we are only considering the growth of localized appendicular organs,

[1] This appears doubly surprising, since the size of an appendage must in the long run depend upon its relative growth (growth-ratio); we shall meet with the probable explanation of this paradox later (§ 3, p. 118).

the gradient appears to be a continuous one. In other words, although appendicular growth can only take place in relatively small, pre-localized regions, yet the gradient determining the amount of that growth is continuous. The growth-gradient is thus probably something of a very fundamental nature, akin to those gradients (of equally recondite nature) which Boveri, Child, von Ubisch, Weiss and others have found it necessary to postulate in the early development of the egg and in regeneration to account for polarity and certain orderly phenomena of morphogenesis; the gradients happen to exist, and where growth is possible, they influence the amount of that growth. We shall be forced to draw similar conclusions in Chap. V (p. 152). (And see p. 262.)

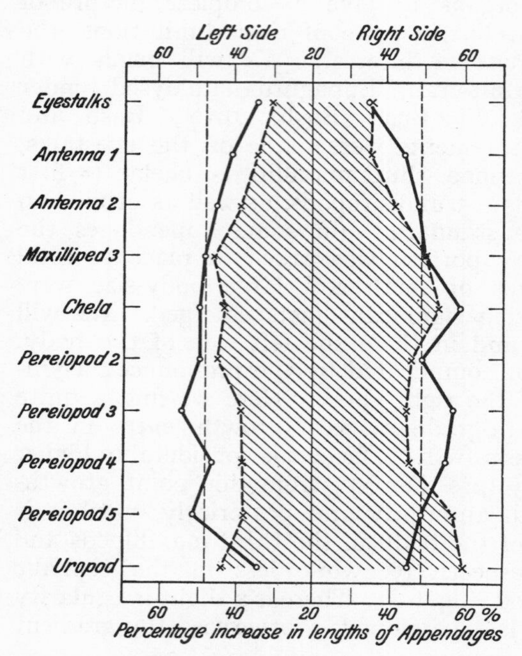

FIG. 61.—Growth profile of the hermit-crab. Growth-gradients along the body of male and female hermit-crabs (*Eupagurus prideauxi*).

The figures give percentage increases of appendage length for a given percentage increase of prothorax length (marked as dotted line).

However, that such a general body-gradient can be locally modified is clearly shown by looking at the curve for the right-hand side of the males, where a second high point is made by the chela.

Further proof of the graded regularity of the changes in growth-intensity is seen when we investigate the asymmetry of the hermit-crab in the two sexes (Bush's Fig. 9). Males and females are approximately symmetrical in the head region (though females appear to be slightly right-handed anteriorly, slightly left-handed rather more posteriorly). The male becomes markedly right-handed in the chela region, and then progressively more left-handed. The same is true for females, but the right-handedness is less in the right-handed region, the left-handedness greater in the left-handed region. In accordance with this greater right-handedness of the males (which is doubtless correlated with the extreme right-handedness of the male chela), the male is still slightly right-handed at the level of the fourth pereiopod, while the female is here symmetrical; i.e. the change from right- to left-handedness is not associated with a particular appendage, but takes place in relation with a graded distribution of growth-intensity.

An extremely interesting case of a large-scale growth-gradient extending through much of the body is provided by the stag-beetles (Lucanidae). It is well known to Coleopterists that large males are characterized, not only by relatively large mandibles, as we have already seen, but also by relatively large heads, prothorax and prothoracic legs, although the relative increase for these is much less than for the highly heterogonic mandibles. So far as I am aware, however, no measurements have been available until those recently made by Mr. Edwards at my suggestion and analysed by me (Edwards and Huxley, unpublished). They refer to *L. cervus*.

The partition-coefficients of various organs, against elytron-length, in males are as follows (between elytron lengths 15 and 21 mm.).

TABLE XIA

Mandible l.	Antenna l.	Head l.	Head br.	Prothorax l.
2·35	1·38	1·62	1·59	1·00

Prothorax br.	Wing l.	First leg l.: Femur	First leg l.: Rest of leg
1·12	1·00	1·09	1·01

Second leg l.: Femur	Second leg l.: Rest of leg	Third leg l.: Femur	Third leg l.: Rest of leg
1·04	0·97	0·95	0·98

There is thus, in general, a decrease in growth-intensity from front of head to hind limb, in other words, a growth-gradient with centre in the mandibles. Heterogony changes from positive to negative between the first and second leg. (It may be noted in passing that the gradient is a little steeper in the femora than in rest-of-leg—evidence of a graded effect within the legs.)

The data for the females are not so full, but indicate a gradient of wholly different shape, much more nearly flat, with very low values for the head appendages, but higher values for the legs, which show positive heterogony. It is worth noting that the distribution of growth-intensity in length and in breadth differs in the two sexes (Fig. 62). In both the intensity of breadth-growth diminishes faster than that of length-growth as we pass forward along the body, but the diminution is greater in females (cf. p. 98).

The best way of comparing the two sexes is to take the ratios of the partition-coefficients of different organs in males and females, and plot these (Fig. 62). It would appear that the main change involved in transforming the female to the male type has been the formation of a very high growth-centre in the mandible region. With this are correlated positively heterogonic effects in the immediately neighbouring regions. But further posteriorly, in the leg region, the gradient is continued into a phase of negative heterogony. It would thus appear that not only may a growth-gradient be steepened, but that marked steepening of one end of a gradient may actually somewhat depress its other

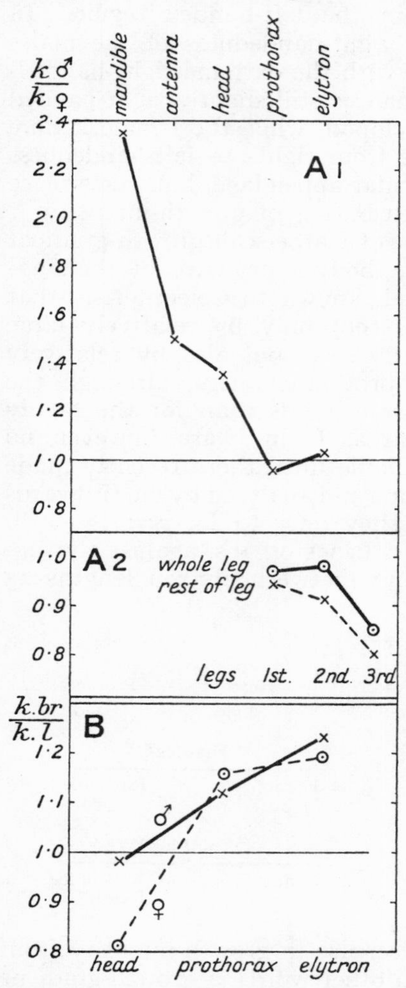

FIG. 62.—Growth-gradients in the stag-beetle, *Lucanus cervus*.

(A) Ratios of the growth-coefficients of the male to those of the female for various organs and regions. For head, prothorax and elytron, the ratios are calculated from the mean between the coefficients for length-growth and for breadth-growth. (B) Ratios between growth-intensity in breadth and growth-intensity in length in 3 regions of the body.

end: the gradient is not only steepened but tilted round a point within its length (cf. Hammond's sheep limbs, p. 88).[1]

Fig. 62 shows the essential facts. It will be seen that besides the parts mentioned, there is also a slight heterogony of the male antennae and prothorax (and see Table XIA); further, that in the regions of the body measured, heterogony in length and breadth is quantitatively different. The most significant point is that the heterogony of the three pairs of legs is graded in an antero-posterior direction. The facts in general support the conception that there exists a growth-gradient with centre in the mandibles, grading down posteriorly[2] (Fig. 63).

Theoretically, two further points appear to me particularly interesting. The first is this, that we can exclude functional hypertrophy as the cause of the correlated increase of the other parts. In fiddler-crabs, for instance, and other crustacea, it might be suggested that the increased weight of the large chela caused extra strain to be put on neighbouring limbs, which then responded by functional hypertrophy. It would be difficult to reconcile this with the correlated *decrease* of the limbs anterior to the highly heterogonic appendage, which will be discussed later in this chapter, but the interpretation is wholly ruled out in a holometabolous insect, in which the male legs have never had to support the weight of the enlarged mandibles before they appear in their definitive,

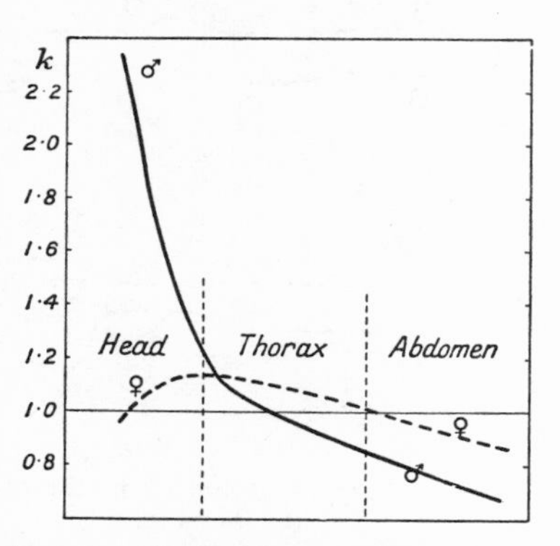

FIG. 63.—Probable growth-gradients in the body of male and female stag-beetles (*Lucanus cervus*).

[1] Champy's figure (1929, p. 198) of the beetle *Oryctes rhadama* indicates a similar gradient with high point anteriorly, but here affecting (*inter alia*) two sexual heterogonic organs, the two 'horns' of the male. The heterogony of the more anterior, cephalic horn is clearly greater than that of the thoracic horn.

[2] Some of the quantitative data for the thorax do not fit in with the idea of a *uniform* gradient. It will be necessary to make additional measurements on other species to clear up this point: as previously indicated, however, there is no necessity to suppose that such gradients are always uniform and not complex in shape.

enlarged form. And further, even if functional hypertrophy of the legs were possible in a beetle, it is impossible to see how it could come into play in regard to the antennae.

The second point is this—that the gradient is concerned with spatial position, not with morphological position. The mandibles are spatially the most anterior portion of the body. Although the antennae are *morphologically* anterior to them, they are *spatially* posterior, and are affected not by a correlated decrease, as appears to be the rule for organs anterior to a growth-centre (p. 122), but by a correlated increase.

Another case in which analysis of existing measurements has permitted us to plot the growth-gradients along the body-axis, giving us what may be called the *growth-profile* of the animal as a whole, is that of the metamorphosing herring, *Clupea harengus* (Huxley, 1931B; data of Ford, partly unpublished, partly in E. Ford, 1930). In this case the growth-gradients are distinctly complex. Their nature will be seen from Fig. 64. Through their operation, the elongated larval herring is converted, while considerably increasing in absolute size, to its post-larval form. Doubtless if measurements had been made vertebra by vertebra, instead of at a few points only along the body, the graded effect would have been more obvious, and interesting results would have emerged as to what happens at the boundaries of regions marked by different growth-coefficients. We may perhaps assume that each region would possess its own growth-gradient. This seems clearly so in regard to the head.

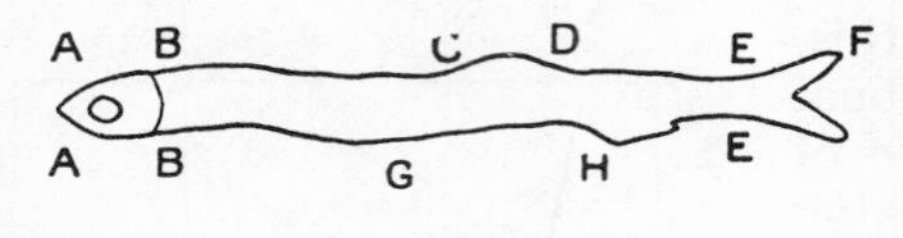

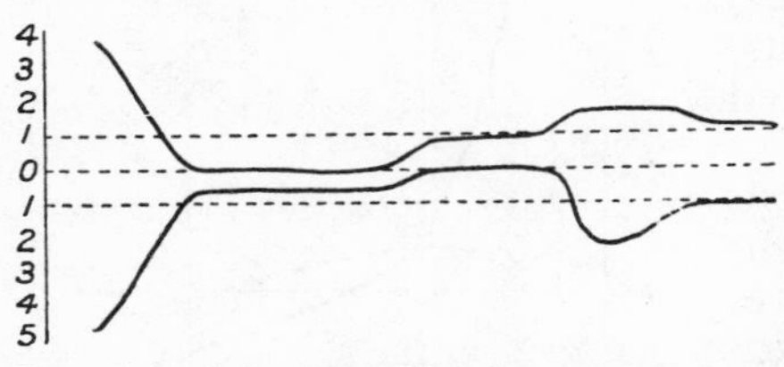

FIG. 64.—Growth-profile of metamorphosing herring.

Above, outline of larval herring at onset of metamorphosis, showing regions measured (after Ford, 1930); below, its growth-profile, dorsal and ventral: the values are growth-coefficients (k), relative to body-length.

Later work by Ford (1931A) indicates that changes similar in principle but different in quantitative detail are at work in the metamorphosis of two other species of the same genus, the pilchard (*Clupea pilchardus*) and the sprat (*C. sprattus*); but growth-profiles have not been constructed for these. In addition the same author (Ford 1931B) gives data on the metamorphosis of the common eel (*Anguilla vulgaris*). Here the tail shows positive heterogony during the process. An analysis of Ford's table (p. 989) shows that when tail length is plotted against rest-of-body length, its growth-coefficient is at first over 1·4, then falls at first rather rapidly, then more slowly to about 1·1. It is, however, probable that it exhibits a growth-gradient with centre of maximum growth anteriorly, so that to get accurate figures for growth-coefficients we should need data on a number of separate regions.

It is clear that regions of different growth-coefficients will often, though not necessarily always, be characterized by special morphological differentiations. Each such region would be characterized both by the possession of its own growth-gradient and its own morphogenetic field (Guyenot). When this is so, we may expect the region to be capable of extension over more or fewer body-segments, and to cover a different serial region of the body-segments in different individuals or species. This would appear to be the case in regard to different regions of the vertebral column (e.g. thoracic, sacral, cf. Bateson 1894). A good example of such a shift during individual development is provided by the Copepods (cf. Calman, 1909, p. 89), in which a broad cephalothoracic region is sharply marked off from a narrow abdominal region. The delimitation of these two regions is not at a constant position in the body, but moves back one segment at each moult, either for two or for three successive moults. It would be interesting to study the growth-relations of the segments which are thus transferred from one region to the other: we may hazard that they would be found to show a sudden alteration of growth-intensity after their morphological changes, so as to fit in with the growth-gradient of the anterior instead of the posterior region.

An interesting example of a general growth-gradient is described by Fauré-Fremiet (1930) for *Zoothamnium alternans*. In this colonial ciliate there are two kinds of zooids, large and small, the latter being in the majority. The multiplication of the small zooids is not uniform; but to use our author's own words, 'decreases according to a kind of gradient in proportion with its removal from the main strain.' It is this growth-gradient which gives the colony its characteristic form. The details are complex, and must be consulted in the original paper; but the existence of a gradient in power of multiplication among the cells of a Protozoon colony is noteworthy.

§ 3. The Two Phases of Growth

Before going further we must consider the difficulty mentioned on a previous page—that it is hard to understand how a regular growth-gradient could produce anything but a series of appendages regularly graded in size, instead of the irregular alternation of large, small and medium-sized appendages which is what we actually find. The answer appears to be that it would inevitably produce such a regular series if it were the only factor concerned in appendage-growth, but that it is not the only factor.

I base this assertion on a study of the figures given by Herrick (1911) for the development of the American lobster, by Giesbrecht (1911) for that of various Stomatopoda, and by Schmalhausen and Stepanova (1926) for that of the appendages of the embryo chick.

In all three cases, there appear to be two successive and quite distinct phases of growth. In the first, the general form of the part is being laid down, and this process is accompanied by very rapid alterations of form, and by marked histological changes; in the second, histological changes are absent or of an entirely secondary nature, and the form-changes are confined to quantitative alterations in the proportions of the definitive structural plan. And it would appear that the regular growth-gradients we have been considering are manifested only (or mainly) in the second of these two phases, some quite other mechanism being at work in the first. Thus not only definitive form-plans, but also marked differences in size, are established in the short first phase, and effects of growth during the second phase are confined to a quantitative modification of the already diversified organization given at the close of the first phase.[1]

The second phase is thus in general of less morphological importance than the first, although occasionally it exhibits differences of growth-potential great enough to effect very

[1] The matter is still further complicated by the so-called law of antero-posterior differentiation of appendages (and by the exceptions to it!); this, however, merely means that the onset of the second phase is not synchronous throughout the body. It is also quite possible that in the first phase, second-phase growth-effects are present, but are masked by the much more radical first-phase effects (see §§ 6, 7).

striking transformations, as in the chelae of Uca, or the abdomen of Pinnotheres. None the less, it is of considerable theoretical importance as revealing the existence of deep-seated and regular growth-gradients which appear to be in the first instance correlated with fundamental properties of the animal body, such as polarity.

Thus the original effects of such gradients upon growth would be, relative to the growth-process itself, secondary; but since all changes of proportions that occupy a considerable time must depend on its agency, in certain cases its growth-affecting properties become modified so as to become of direct importance to the process of growth of parts.

My meaning will be clearer with the aid of an example. There is (Fig. 61) a progressive decrease in relative growth-ratio as we pass forward along the head-appendages of Eupagurus, which obviously results in a slight and continuous change in their proportionate size, relative to the body and to each other, as life goes on; I cannot conceive, however, that these particular changes in relative size are of any biological significance, but am forced to regard them as accidental consequences of the existence of a growth-affecting gradient in the body. The comparatively enormous increase in relative size of the male right chela, however (as of the female swimmerets), does appear to have biological significance; and here therefore we must consider that the form of the gradient has been modified so that its growth-affecting properties become utilized to alter bodily proportions in an adaptive way.

Although clearly much more work must be done before the laws, properties and physiological mechanisms of these two growth-phases are properly understood, general considerations as well as the limited special analysis I have been able to undertake make me feel that their existence is real enough.

If that is so, we may enshrine the distinction in specific terminology, and say that, following the period of chemical predetermination at which the specific fates of different regions of the embryo (or regenerating part) are invisibly determined, there occur two further phases, one of tissue differentiation and the assumption of the definitive general form-plan, and one of subsequent quantitative growth-changes in proportions (and presumably also changes due to functional activity). To the first of these the term *chemo-differentiation* has already been applied (Huxley, 1924c) and adopted by others (Goldschmidt, 1927; Needham, 1931); for the next I propose to use the term *histo-differentiation*, since histological change would here appear to be the most decisive factor ; and for the last, the term *auxano-differentiation*, since quantitative growth-changes are now the most significant.[1]

Histo-differentiation would obviously be at work not only in the first formation of organ-rudiments, but also during any radical metamorphosis. This appears to be the case even when limbs are converted from one structural plan to another during development : e.g. when in a lobster a limb is converted from a biramous swimming appendage to a jaw, or even when the pereiopods lose their exopodites. In this latter case, it is possible that negative heterogony of the auxano-differentiative type is at work, but acts so quickly that between one moult and the next the whole exopodite disappears, but the presumption appears the other way. It would be interesting to study the histology of the different parts of the limb during the instar prior to the exopodite's disappearance.

§ 4. Growth-changes Correlated with High Local Growth-intensity

Another indication of the real existence of these fundamental growth-gradients is afforded by the fact that it appears impossible to effect a marked change in the growth-ratio of one appendage without at the same time effecting slight changes in the growth-ratios of the neighbouring appendages—in other words, that a localized growth-change is not in reality fully localized, but must operate within the framework of the main growth-gradient of the body, and affect its working. The proof of this is afforded by the changes in relative size of neighbouring appendages which are to be found correlated with marked heterogony of a particular appendage. The 'control', by the difference from which the magnitude of the correlated change is deduced, may be provided by contrasting corresponding organs, either those of the opposite side of the body in cases of asymmetrical heterogony of a part, or those of the opposite sex when there is heterogony of an organ in one sex only. As examples of the former we can take the male fiddler-crab (Uca) ; and of the latter, those numerous Crustacea in which one of the pereiopods is enlarged, in the male only, to form a powerful chela.

[1] Most significant, that is, for our present purpose. If we were more interested in the changes directly brought about by function, another term would be needed.

In Uca, it has been known for some time that other parts besides the chelae are asymmetrical in males, notably the carapace and the walking legs, which also are enlarged on the side of the large chela. The only quantitative data on this subject appear to be those of Yerkes (1901), which have been further analysed in this laboratory (Huxley and Callow, unpublished). Unfortunately Yerkes' measurements apply only to large specimens (10–15 mm. carapace-length). Over this range, the large chela is, of course, enormously much larger than the small (at least twenty times as heavy, from my data on *U. pugnax*) : the merus, measured by Yerkes, is about 60 per cent. greater in length. The only other relevant measurements taken by Yerkes are (1) the merus-length of the first walking leg, which is about eight per cent. larger on the large chela side, and (2) the lateral margin of the carapace, which is rather over 5 per cent. larger on this side.

What most concerns us is that for the merus of the first walking legs, the excess of the large-chela side increases distinctly, over the size-range measured, with absolute increase in size of the animal, from below 7·5 per cent. to over 9 per cent. ; and that for the carapace margins, in spite of some irregularity, appears to increase slightly (from below 5 per cent. to 5·5 per cent.), i.e. the asymmetry is progressive.

Work is in progress by Miss Tazelaar in this laboratory on this subject. Her preliminary data indicate clearly that all the walking legs are implicated in the increase of size on the side of the large chela, that the first after the chela (second pereiopod) is most affected, the second next, while the effect on the last two is slight and somewhat irregular ; e.g. six males of mean carapace-length 14·3 mm. gave the following result for the percentage excess of the limb of the large side over that of the small side :

	Pereiopod		
	Second	Third	Fourth
Excess on large side, per cent.	2·6	0·9	0·2

A graded effect is clearly visible.

In Uca, measurements have unfortunately not yet been taken of the appendages lying just anterior to the chelae. This has, however, been done in several cases of sexual difference in heterogony, and in every case we meet with the surprising fact that whereas, as in Uca, marked heterogony of an organ is associated with a slight increase of relative size in the appendages just posterior to the special growth-centre, there is a slight *decrease* of relative size in those immediately anterior to it.

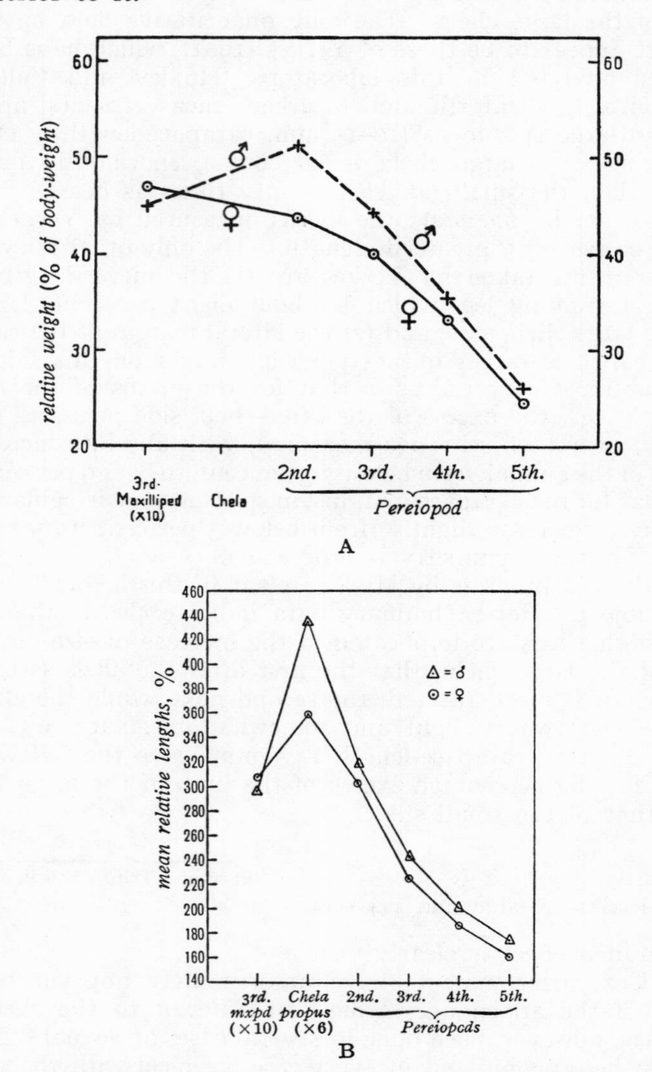

FIG. 65.—Graph of changes in relative size of parts associated with a local region of high growth-intensity (male chela) in spider-crabs, (A) *Maia squinado*, (B) *Inachus dorsettensis*.

This was first discovered in regard to weight-measurements on the spider-crab *Maia squinado* (Huxley, 1927) ; and next confirmed, this time in regard to linear measurements, on another spider-crab, *Inachus dorsettensis* (Shaw, 1928). I then felt that the phenomenon might be due to one of two rather different causes. In both these cases, the organ diminished in size in the male was the third maxilliped, the pereiopod which was enlarged as the male chela being the first of the walking-leg series. It was thus possible that the effect had nothing to do with the main growth-gradient, but was in some way due to the nature of the appendages concerned, maxillipeds for some reason responding differently from pereiopods.

To decide between these two alternatives, it was necessary to find an organism in which some other pereiopod than the first was enlarged to produce a large chela. The true prawns provide a case in which the *second* pereiopod is so enlarged ; but unfortunately in the common British species there is little difference between the sexes in the size of the large chelae. Eventually, through the kindness of Dr. Seymour Sewell, of the Indian Museum, a number of specimens were obtained of the magnificent Indian prawn, *Palaemon carcinus*, a species with marked sexual dimorphism in the chelae, and measurements on these established the fact that the effect is a true positional effect, since in the male, whereas the relative size of third, fourth and fifth pereiopods were increased, that of the first pereiopod as well as of the third maxilliped were decreased.

In general, male crustacea appear to have relatively larger pereiopods than females. But here the difference between relative size of pereiopods in the two sexes is much reduced in the pereiopod anterior to the chela, showing an inhibiting effect of chela-growth. Another method of studying this phenomenon is to take the relative growth-rate of the male and female appendages. When this is done, it may occur, over a given size-range, that the relative growth-rates of the organs anterior to the chela are actually lower in the male than the female (see Figs. 66, 67).

The explanation of this curious effect is for the moment completely obscure, though there can be little doubt that it is connected with the fact that any fundamental growth-gradient along the body-axis must be polarized.

Effects presumably of the same general nature, though not

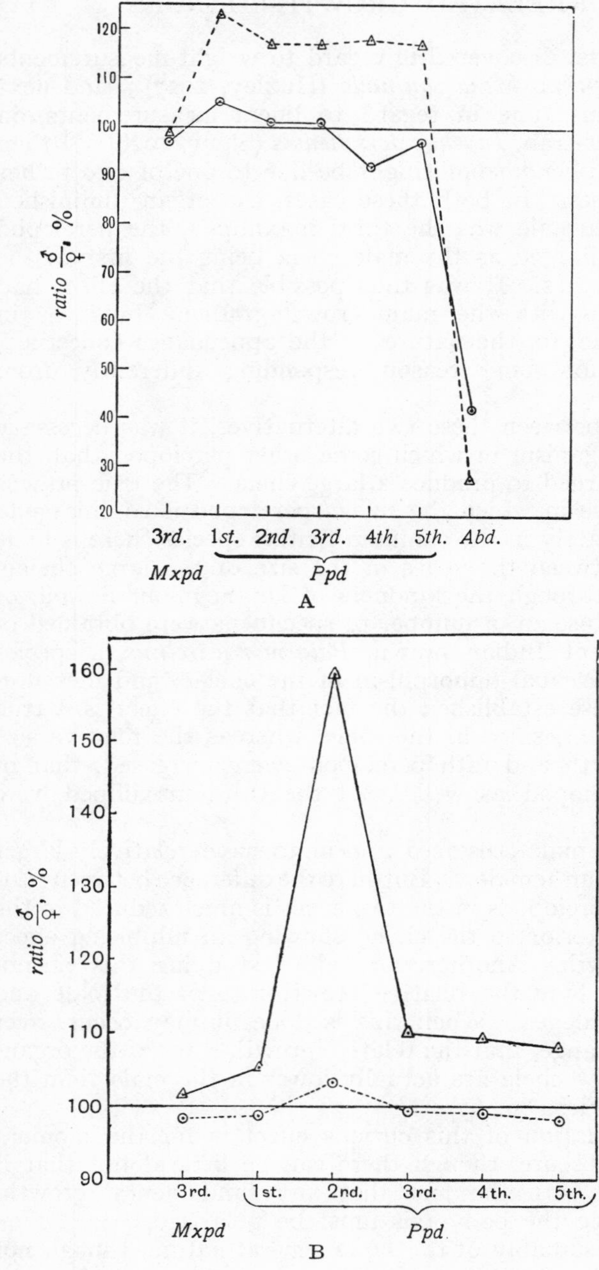

FIG. 66.—Changes in growth-rate of parts posterior and anterior to a local region of high growth-intensity.

(A) *Inachus dorsettensis*; the region of high-growth intensity is the 1st pereiopod.

(B) *Palaemon carcinus*; the region of high-growth intensity is the 2nd pereiopod.

The graphs represent the percentage ratio of male to female appendage-length: ⊙, for small specimens; △, for large specimens.

upon growth, have been recorded by Gabritchevsky (1930) in his experiments on regeneration in spiders. In successive moults, the legs of the species used by him pass through a readily-characterized series of developmental stages. If a leg be amputated, its regeneration is accompanied by alterations in this sequence in the neighbouring legs. Sometimes the normal sequence is accelerated, sometimes retarded; in either case, compensatory effects in the other direction may be subsequently produced. Gabritchevsky's results to date do not permit us to say why either of these two opposed effects may be found; we may anticipate that the cause will prove to lie in the rapidity of regeneration and the time before moulting at which the operation was carried out.

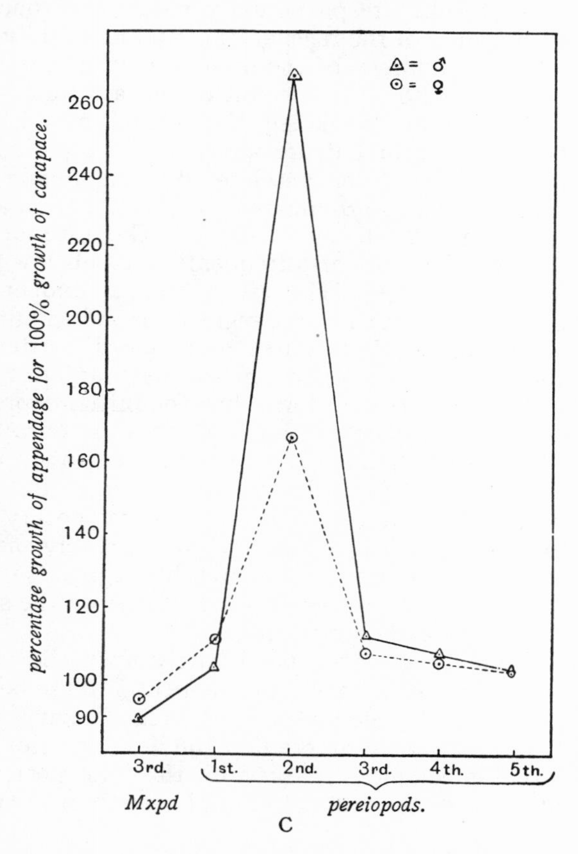

Fig. 67.—Correlated changes of growth-rate in structures immediately anterior and posterior to a local region of high growth-intensity (male chela) in the prawn, *Palaemon carcinus*.

The graph represents the growth-rate (% increase for 100% increase in carapace length) of various appendages in males and females.

Przibram (1917) carried out an elaborate investigation on regeneration of legs in the Mantid *Sphodromantis bioculata.*

His results can be analysed (Huxley, unpublished) to shed some light on our problem, although not so much as if the experiments had been designed for that purpose.

His results are particularly valuable in one respect, as the growth-rate of the regenerating and normal limbs was followed during a number of moult-stages, until the animals became adult (or died). In general it appears that after amputation of the fore or middle leg, the normal growth of the next posterior legs is first depressed during the period of most active regeneration; then accelerated; and finally, as the rate of regeneration approximates to that of normal growth, sinks again, usually below normal. (The rate of growth of the regenerating limb also frequently sinks below normal at about the same time.) The effect on the unoperated leg of the opposite side of the same segment as the amputated leg appears to be similar, but is less pronounced. After amputation of the hind-leg, the effect on the next anterior (middle) legs is of the same general form, but the initial depression is greater, the later acceleration less. (I have no data for the effect of amputation of the middle leg on the growth of the fore-legs.) When, as sometimes happens, no regeneration occurs, the effect of amputation appears to be a temporary acceleration of normal growth in other legs, without any marked depression at all. The effect on limbs in other segments is always identical on both sides; the effect is no greater on the side where regeneration is in progress (Fig. 68).

Thus the effects depend partly upon the rate of growth in the regenerating limb; but there appears to be also a positional effect, the depressant effect of a regenerating (rapidly-growing) limb being more marked on the limb anterior to it, the stimulative effect more marked on the limb posterior to it. This fits in with the facts of normal growth in Palaemon, Maia and Inachus.

Von Ubisch (1915) carried out somewhat similar experiments on the larva of the insect Cloe diptera. Unfortunately his method of recording his results does not permit of a satisfactory analysis of the question which here interests us, namely the effect of regeneration on the normal growth of neighbouring unoperated structures, but only of effects on the rate of regeneration itself. However, some of his results are of relevance. All operations were made directly after one moult, and the final measurements made directly after the next moult. He finds that when two different legs are removed,

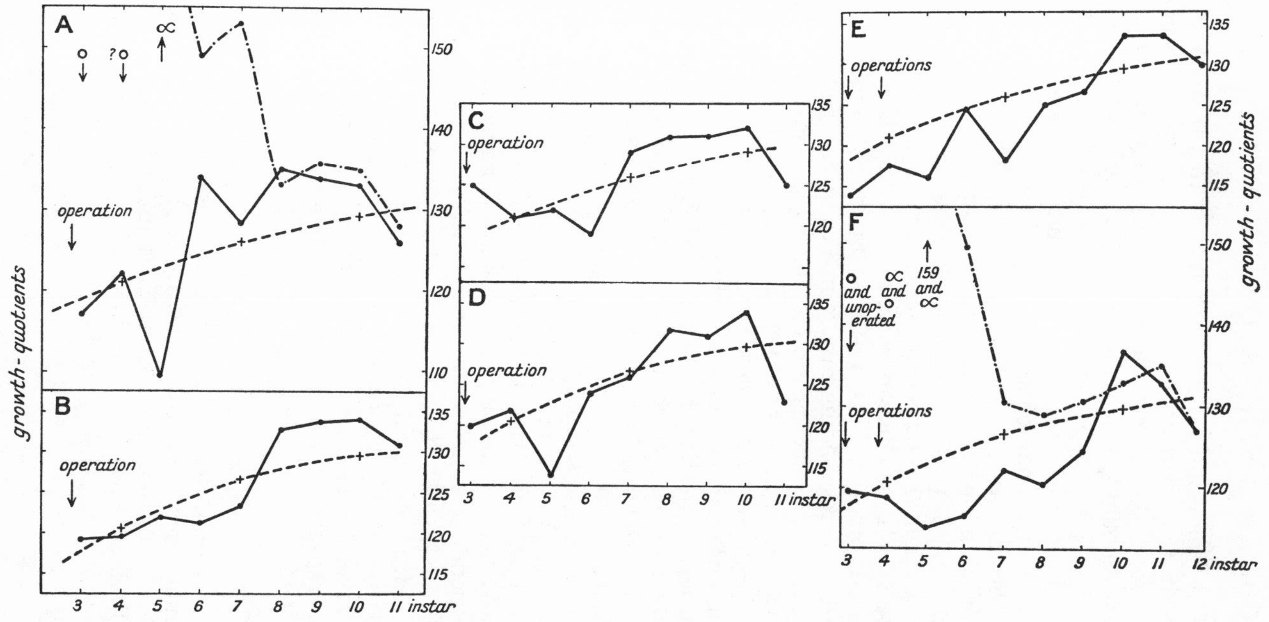

FIG. 68.—Effects of a regenerating limb upon the growth of neighbouring limb, recalculated from the data of Przibram (1917). See text for details.

The growth-rates are given in the form of growth-quotients—i.e. the length of limb at each instar divided by its length for the previous instar. The dotted curves represent the growth-quotients of limbs in unoperated animals (from Przibram's Table B, means (+) taken for moults 4–5, 5–8 and 8–11).

A and B. Effect of amputation of one middle leg on the growth (A) of the middle unoperated leg of the opposite side, (B) of the hind legs; means of 3 animals. The curve for the growth-quotients of the operated leg is also given (—·—); it begins at zero, rises at onset of regeneration to infinity, then falls rapidly.

C and D. Effect of amputation of one fore-leg on the growth of (C) the middle legs, (D) the hind legs; means of 4 animals.

E and F. Effect of the amputation of the hind-leg on the growth of (E) the middle legs, (F) the unoperated hind leg of the opposite side. The curve for the operated leg is also given (—·—); means of 2 animals.

their rate of regeneration is greater when both are on one side of the body than when they are on opposite sides; and the same holds good when three different legs are removed. The mean difference amounts to nearly 5 per cent. of the lower figure.

Further, if two legs are removed, regeneration of a given leg is greater (by nearly 3·5 per cent.) if the other leg is the opposite member of the same pair than if it is a leg of another pair, whether on the same or the opposite side. On the other hand, when only one leg of a pair was amputated, its normal partner on the opposite side was not accelerated in growth, but if anything slightly retarded (mean, 0·4 per cent. retardation—a figure which may however well fall within the limits of experimental error). This result cannot be compared directly with the corresponding results in Przibram's Sphodromantis, since it only concerns growth during one instar.

Going through his tables, however, I have come across one fact which appears to be significant for our problem. When legs belonging either to two or three different segments were amputated, whether on the same or opposite sides, the amount of their regeneration can be compared with the amount made by the same legs when only a single leg is amputated. It is then found that the anterior of the regenerating legs always regenerates relatively less rapidly than the posterior (only one exceptional result in eight series of experiments); in five out of the seven concordant series, the regeneration of the anterior limb is actually below the 'normal' amount for regeneration of the corresponding single limb, and that of the posterior limb equal to or greater than normal. This again points to a depressant effect of rapid growth on anterior structures, and a stimulative effect on posterior ones.

It will, however, be necessary to carry out systematic experiments, using repeated regeneration, before clear-cut results can be obtained.

§ 5. Some Cases of Teratological and Abnormal Growth

In this connexion, the measurements made by Nañagas (1925) on human anencephalic foetuses are of interest. In such monsters, the size and proportions of the trunk and especially of the lower limbs, are practically normal, but, while the cranial region is, of course, markedly underdeveloped, the fore-limbs are hypertrophied (by about 12 per cent.). Furthermore, the hypertrophy is graded (Nañagas' Table I): the upper arm shows an increase over the normal of 24 per cent., the fore-arm of 16 per cent., the hand of 7·1.[1] The length of the middle finger is actually decreased relative to normal. There is thus not only a growth-gradient in the arm, but an indication that, as in male stag-beetles (p. 115), the excessive tilting of one end of the gradient above normal has led to a slight depression of the other end below normal. The fact that the proportions of the segments of the lower limb are not affected is interpreted by Nañagas as meaning that the disturbance to growth occurred very early, before the differentiation of the hind-limb had begun. There appears to be, further, a slight graded effect upon the trunk. The arm circumference is 13·0 per cent., the chest circumference at the nipples 7·5 per cent., at the tenth rib 1·3 per cent. above normal. In any case, the correlation of poor development of head with increased development of the region next behind it is interesting. It may be interpreted as due primarily to the excess growth of the forelimb region; or more probably the primary factor is the failure of the head to develop, which then, in accordance with the views of Child and of Stockard, permits a higher development of regions posterior to it in the body-gradient (Fig. 69, left).

A further teratological study which is also relevant to our purpose is that of Mead (1930), based on a single monster of 34·8 cm. crown-heel length. When the measurements were compared with those computed for a normal embryo of the same crown-heel length, it was found that the head was very much enlarged; likewise the arms, in diameter more than in length. The trunk was enlarged in breadth, more so anteriorly than posteriorly; its anterior portion was markedly reduced in length, its posterior portion slightly increased. And the hind-limbs were slightly increased in length, but considerably reduced in diameter. (The proportions of the segments of the limbs were not disturbed.) This can most simply be interpreted as due to (1) an abnormally high growth-centre anteriorly with growth-gradient extending backwards. But (2) this is complicated by differences in growth-intensity in the different planes of space. From the shoulder region posteriorly there is an abnormal growth-gradient in the breadth-dimension, above normal anteriorly, below normal posteriorly (cf. the arm of Nañagas' specimen, *supra*); and growth in the length-dimension is inversely correlated with growth in the breadth-

[1] This is the statement in Table I, and appears to be correct. On p. 477, however, the value is given as 1·1 per cent., and on p. 485 as 2 per cent.!

dimension, so that the chest is much shortened, the legs slightly lengthened.[1] For some reason, this inverse correlation between breadth- and length-growth does not seem to have affected the head region (Fig. 69, right).

While many details remain obscure, these cases at least provide excellent examples of *graded* abnormalities of growth-intensity. Professor C. R. Stockard has drawn my attention to a similar case, but affecting an isolated organ only. Streeter

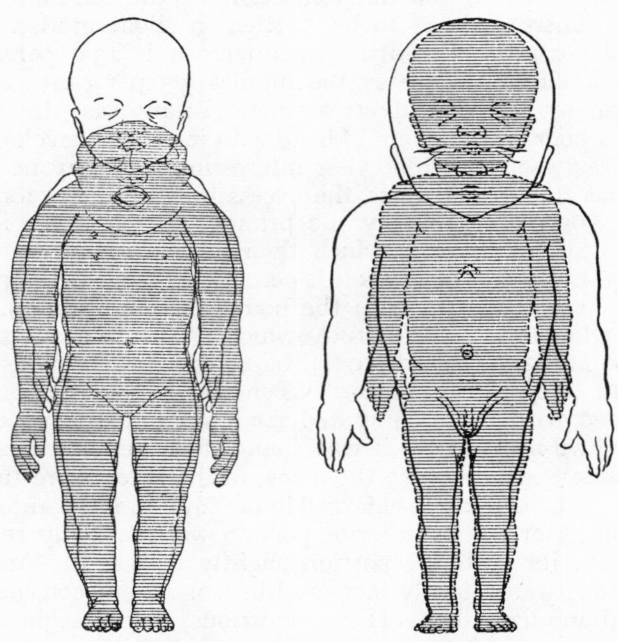

FIG. 69.—Graded-growth effects in two human monsters. (See text for details.)

(1930), in his Plate 12, figures a case of isolated hypertrophy of a single human finger. The point which interests us, however, is that the proportions of the phalanges appear to be abnormal, and to have been affected in a graded way.

While on the subject of abnormal growth, we may refer to some cases of dwarfing and gigantism. Sir Arthur Keith has kindly let me see some MS. notes on this subject. There are

[1] This interpretation is not identical with that of Mead, who prefers to think mainly in terms of the 'law of developmental direction'.

also numerous cases scattered in the *Nouvelle Iconographie de la Salpêtrière*, and Stockard (1931) treats of the question in a comprehensive way. In general, it appears that simple pituitary gigantism is associated with relatively long limbs, but within the limbs there is not much proportionate change, or a slight defect in hand and foot. In acromegaly, however, whether associated or not with gigantism, the relative sizes of hand and foot are increased.

In ateliotic dwarfs, the proportions of the limbs to the trunk and of the limb-segments to each other are not affected. In achondroplastic dwarfs, however, not only are the limbs abnormally short, but the segments are differentially affected; the proximal segments are markedly reduced, the hand and foot scarcely at all. (In mongoloid dwarfs, on the other hand, relatively short fingers and toes are among the most striking differential characters (Davenport and Swingle, 1927).) The hind-limb is usually less abnormal than the fore-limb. In breeds of dogs with markedly reduced limbs, which Stockard (l. c., p. 228, etc.) compares with human achondroplasia, ascribing both to abnormalities in the thyroid, we get a similar modification of the proportionate size of the limb-segments. In the basset-hound he states that the hind-feet are disproportionately enlarged, which would indicate a tilting of the whole gradient so that the proximal region is below, the distal region above the normal level (cf. pp. 89, 115). Unfortunately he gives no precise details.

The relatively large size of limbs in giants can only be due to their increased heterogony during growth; while achondroplastic types afford evidence of a disturbance in the growth-gradient as well as the general growth-coefficient of the limbs. From Stockard's work it is clear that, in dogs, the achondroplastic short type of limb may be inherited separately; it depends upon a single Mendelian gene.[1]

[1] The apparently genetic achondroplasia (chondrodystrophy) found in fowls (Landauer, 1927) differs in its results from that of mammals in various ways. In the first place it does not affect the fore-limb, while the hind-limb and most other parts of the body are markedly affected. Secondly, in the hind-limb the tibia is more affected than the femur. In the pelvis, growth in length is little affected, but many breadth-measurements are enlarged. In a dwarf six-months-old chicken whose dwarfism was apparently of myxoedematous origin, Landar (1929) comes to the conclusion that its deviations from normal proportions (which were negligible in the wing, but marked in the leg, skull and pelvis) can mainly be attributed to a suppression of the later phases of growth.

§ 6. The Law of Antero-posterior Development and its Effect upon Growth

In this connexion, we have the important work of Scammon (references in Scammon and Calkins, 1929) on the growth of the human foetus, from about 5 cm. length to birth. He there finds definite evidence of gradients as regards growth, proceeding antero-posteriorly along the main axis, and centripetally along the limb axes. This main gradient is found for internal organ-systems (gut, vertebral column, etc.) as well as for external form. There are a few exceptions (e.g. the sacral region of the vertebral column) which apparently, like intercalated centres of high growth-ratio in Crustacea, such as the large chela of male Uca, are concerned with special adaptive growth of particular organs.

However, he asserts that practically all parts so far measured grow in linear proportion to crown-heel length, according to the simple formula $y = ax + b$. He further points out that when b is zero, the percentage measurements of the organ, relative to standard length, remain constant throughout the period. If, however, b is positive, the percentage measurements decrease with increase of absolute size, while the converse is the case if b is negative. Since the anterior regions have b positive, while in the posterior regions it is negative, there is the appearance of a growth-gradient. However, this would not be brought about, as in my previous examples, by differences in growth-intensity of the various parts, as measured by constant differential growth-coefficients, but according to whether the organ in question had made much or little growth during the embryonic period (below 5 cm. length). This he would interpret as due to the Law of Developmental Direction, according to which anterior (and proximal) regions are formed first, are soonest through with their origin and histological differentiation, and can embark earlier on their main growth-period. As we may presume that the growth of each separate organ follows the usual rule for the body as a whole, namely that the (compound interest) growth-rate slows off progressively from the beginning of growth, we should expect that there would be a lag between anterior and posterior regions, such as that at any given moment the anterior ones would be at a later and therefore slower phase of their growth than those lying more posteriorly (and see § 7).

It remains to be seen whether this will account for the whole of the difference. Unfortunately, Scammon's final extensive data have only recently come into my hands, and time has not yet been available for their full consideration and further analysis in relation to the ideas set forth in this book.

I incline to the view, after preliminary inspection, that some of Scammon's data would be much better fitted by an expression of the heterogonic type, $y = ax^k$, than by the linear formula he adopts. (It must, of course, be remembered that his linear formula, when $b = 0$, is a special case of the heterogonic formula.) In any case, his analysis is valuable in showing that the time-relations of development during the early embryonic period of histo-differentiation, can exert a marked effect on the percentage changes of parts in the later foetal stage when auxano-differentiation is in progress.

Latimer and Aikman (1931), in a study of the prenatal growth of the cat from total weight 0·3 g. (264 specimens, including 35 newborn), give further interesting data.

The formula for the growth of the weights of various organs (y) against total weight (x) are found to be as follows (all in grams):

Head . .	$y = x^{0.97} - 0.69x$	
Trunk . .	$y = 0.59x - 0.36$	from 1 to 70 g. total wt.
	$y = x^{1.08} - 0.84x + 0.9$	from 70 g. on.
Fore-limbs (2) .	$y = 0.08x$	from 1 to 90 g.
	$y = 0.053x + 2.43$	from 90 g. on.
Hind-limbs .	$y = 0.1x^{1.1} - 0.053$	from 1 to 100 g.
	$y = 0.07x + 3.5$	from 100 g. on.

Unfortunately in their tables the authors only give values calculated on these empirical formulae, and the percentage curves for relative weights in their figures do not appear always to be consonant with the formulae. The percentage curves are reproduced herewith. The constant decrease in relative weight of the head, constant increase of that of the trunk, and increase to a maximum followed by a decrease for the limbs, is clearly brought out (Fig. 70).

For an analogous case on invertebrates of the marked effect which the law of antero-posterior development can exert upon bodily proportions, reference may be made to the measurements of Seymour Sewell (1929) on Copepods. During the later free-living copepodid stages of development, before the adult phase is reached, the number of abdominal segments is increasing owing to the division of the less-differentiated sub-terminal region of the abdomen. Growth is proceeding at

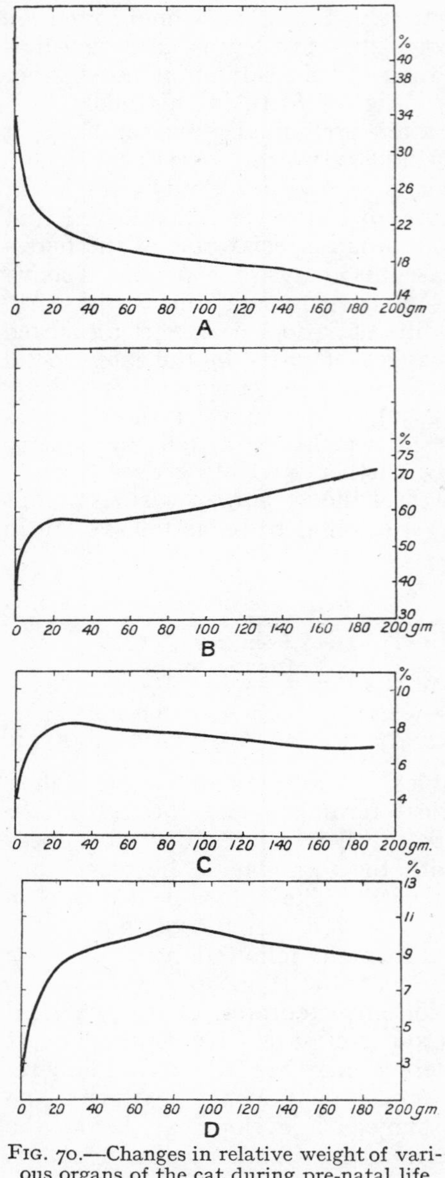

FIG. 70.—Changes in relative weight of various organs of the cat during pre-natal life. (A) Head; (B) trunk; (C) fore-limbs; (D) hind-limbs; all against total weight.

the same time, and in such a way that, as Seymour Sewell, says (l. c., p. 9), 'when a segment of the body divides into two, as for example in the development of the abdomen, the total proportional length of the two daughter-segments is always greater than the proportional length of the parent-segment'. As result, the proportional length of the abdomen increases steadily. But this change in relative size is quite distinct in character from the change in proportional size in different regions of the antennae of the same animals (p. 85), in which the definitive number of segments has been differentiated before the growth-changes occur. Many ontogenies would undoubtedly yield interesting results if analysed quantitatively in the light of the principle of heterogony on the one hand and of that of antero-posterior development on the other. This would apply especially to ontogenies of primitive type showing continuous gradual change, such as those of Trilobites (cf. Raw, 1927).

It is possible that the

facts obtained by Przibram (1917) for the growth of limbs in Sphodromantis are also a consequence of the law of antero-posterior development. His Table B gives the growth-quotients for the middle and hind limbs at each instar—i.e. the ratio of the length at one instar to the length at the instar preceding. If these are averaged for groups of three moult-stages, we obtain the following result :

Moults	Growth-quotients Mid-leg	Hind-leg
2–5	1·211	1·206
5–8	1·260	1·266
8–11	1·293	1·297

i.e. not only is there a steady increase in the growth-quotient during life, but the growth-quotient for the mid-leg begins higher but ends lower than that for the hind-leg. There are, however, considerable irregularities in the values for single instars, and it is possible that this result is not significant.

That the law of antero-posterior development depends upon some fundamental gradient within the body as a whole is indicated by the work of Ruud (1929) who found that the growth of a urodele leg-bud rudiment transplanted to the arm-region was markedly accelerated, that of an arm-bud rudiment transplanted to the leg region retarded.

Regeneration as well as growth may be affected by this law. For instance, Von Ubisch (1923) finds that if three equal V-shaped pieces are cut out of the dorsal fin of Urodele larvae, the regeneration of the anterior piece is almost always less than that of the posterior. This latter result he ascribes to the capacity for regeneration being inversely proportional to tissue-differentiation, and to differentiation proceeding in an antero-posterior direction.

In Cloe, the capacity for regeneration remains unimpaired throughout life, independently of differentiation which is complete in all three limbs, so that we presumably obtain a direct effect of the main axial gradient of the body upon rate of regeneration.

However, that the law of antero-posterior development, as regards growth-effects, need not hold even in mammals is shown by the measurements of whales given by Mackintosh and Wheeler (1929 : see especially pp. 277–95). In these animals the head is, of course, enlarged to carry out the highly specialized straining function of the baleen on the jaws : and accordingly we find that its percentage length relative to total length *increases* steadily from small foetuses to large whales 25 metres long (at extreme sizes it appears to fall again slightly). Similarly, the relative tail-length decreases with absolute size, from the juvenile stage onwards. The results are essentially similar in Blue and Fin whales (*Balaenoptera musculus* and *B. physalis*). Measurements of two dimensions are herewith given for male Blue whales.

TABLE XII

RELATIVE HEAD- AND TAIL-LENGTHS IN MALE BLUE WHALES (*Balaenoptera musculus*) FROM S. GEORGIA (from Mackintosh and Wheeler, 1929

Length, metres	No. of specimens	Relative head-length per cent. (tip of snout to blowhole)	Absolute head-length (calculated) metres	No. of specimens	Relative tail-length per cent. (notch of flukes to end of ventral grooves)
Foetus 1– 2	5	14·45	0·215	5	45·67
Foetus 2– 3	7	14·76	0·369	8	44·06
Foetus 3– 4	3	15·36	0·537	3	43·28
Foetus 4– 5	1	15·16	0·683	1	43·79
17–18	13	15·81	2·77	9	44·95
18–19	14	16·04	2·97	12	44·85
19–20	18	16·58	3·24	13	45·09
20–21	14	17·16	3·52	8	43·81
21–22	16	17·49	3·76	15	43·32
22–23	21	17·72	3·99	18	43·43
23–24	38	18·44	4·33	26	42·91
24–25	56	19·06	4·67	39	41·24
25–26	25	18·72	4·78	19	41·85
26–27	3	19·01	5·04	3	41·98

This is precisely the opposite of what is found by Scammon. Clearly the head is endowed with a specific heterogony which counteracts the effect of its precocious development, and, from the foetal stage on, overrides the effect of the tail's later differentiation.

In passing, it may be mentioned that the measurements given by Mackintosh and Wheeler could furnish a rich mine of information for constructing growth-profiles for both sexes of the two species of whale. Unfortunately, only percentage measurements are given for the means, and without mean absolute measurements for the total length classes. For accurate work it would accordingly be necessary to recalculate the original absolute data in the appendices.

To obtain accuracy I have recalculated the absolute data for head-length on the assumption that the mean total length for Class 1–2 m. is 1·5 m., and so on; the figures appear in Column 4 of the table. A log-log graph from these (for post-natal life) is given in Fig. 71. Up to 25 m. total length, it shows a remarkable approximation to a straight line, giving a growth-partition coefficient of head-length relative to total length of about 1·55. When the figures for foetal life are plotted, they also show a straight line, but indicating a growth-coefficient of only about 1·05. The two lines intersect at about 17 m. total length; if this gives a correct indication, it means that the marked heterogony of the head does not begin until long after birth, which appears to occur at 6–7 m. length, but before sexual maturity, which comes on at 22–23 m. length.

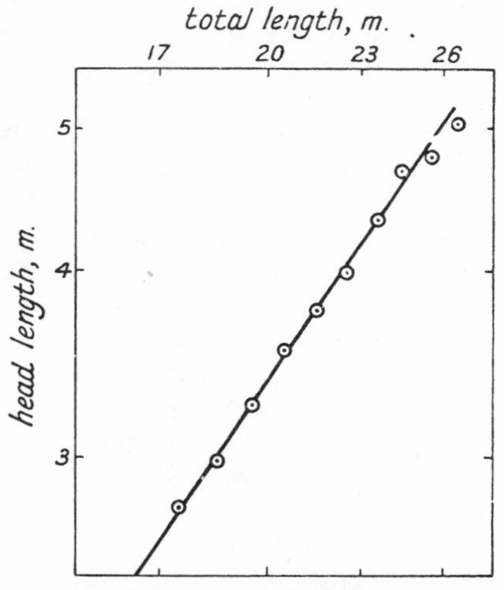

FIG. 71.—Head length against total length in male Blue-whales; logarithmic plotting. $k = 1{\cdot}55$ (except for last point). (*From Data of Mackintosh and Wheeler*, 1929.)

Numerous further exceptions can be found among the limbs of Arthropods. As a simple example, we find that in primitive Copepods both the first appearance and the subsequent growth of the appendages occur in strict antero-posterior sequence; but in many of the more specialized members of the group, while the appearance of the rudiments still takes place in this sequence, the swimming feet of the anterior thoracic segments then grow rapidly, while the more anterior maxillae and maxillipeds remain for some time in a rudimentary bud-like condition (see Calman, 1909, p. 89). Scammon himself mentions some exceptions to the purely linear growth of parts. The work of Schultz (1926, 1930)[1] on other Primates refers to various

[1] Schultz presents his results almost entirely in the form of *percentage values* (ratios) and their changes with age. It would be of considerable interest to undertake an analysis of his absolute data to see whether they conformed to Scammon's linear or to my heterogonic formula.

evolutionary changes in proportion of parts which are difficult to account for solely on variations in the time-relations of early development. (E.g. increase in relative length of arm without a corresponding change for the leg: the fact that the ratio of radius/humerus increases during ontogeny, and increases most in those animals in which it is highest in adult life, which are, further, those with the greatest relative length of the arm). Further, the work of Hammond on sheep previously cited (p. 88) is conclusive proof that growth-gradients as defined by me do exist in vertebrates and may even change their sign, late in development, long after the embryonic period is over; while the disturbances of proportions which occur in conditions due to glandular abnormality (e.g. acromegaly) cannot originate in the embryonic period when there are no functional ductless glands, but must be due to changes in growth-intensity of parts during foetal and post-natal life.

And the analysis of Lapicque (1907) and Dubois (1922) shows that the growth of some organs at least, such as the brain, do not take place according to a linear but to an exponential function of body-size. My tentative conclusion would therefore be that in man as in other forms, the law of developmental direction is of great importance, exerting effects on proportionate size of parts long after it has actually ceased to be at work in the early embryonic period, but that heterogony of parts associated with gradients in actual growth-intensity also operate during the whole of the later period of auxano-differentiation. It must suffice here to point out the similarity between these and the results obtained by Abeloos (l. c.) on Planaria. The existence of an identical type of growth-gradient in two such remote types as a flatworm and a mammal is striking.

It should also be pointed out that the gradients revealed by Scammon's figures are capable of modification and even reversal. In the limbs of man, the gradient in male and female must clearly be quantitatively different, in order to produce the relatively smaller hands and feet of the female. In women, further, the main body-gradient is clearly altered by the accentuation of growth in the pelvic region, apparently associated with diminished growth in the region of the shoulders; and the partition of growth-potential between trunk and extremities appears to be shifted slightly in favour of the trunk in the male as in Crustacea, e.g. Eupagurus (Bush, l. c.) and Gammarus (Kunkel and Robertson, l. c.). A case of reversal of the limb-gradient has already been referred to in

Hammond's sheep (p. 88): here, before birth, the normal effect must have been proceeding; but after birth the distal regions increase least, the proximal regions most, with a graded effect in between.

Thus in the vertebrate body again, we would appear to be dealing with a primary gradient effect, doubtless correlated with Child's physiological or axial gradients, which automatically has an effect upon growth, and secondary modifications of this, imposed to effect growth in biologically advantageous ways. (See also Bray's work, p. 259.)

§ 7. The Mathematical Formulation of Relative Growth in Embryonic Life

Organs developing according to the law of antero-posterior development are special cases of the more general rule that during early development different organs do not originate at the same time. They constitute the most abundant of such cases, and have particular interest, e.g. in relation to gradient theories. But from the point of view of formulating qualitative rules of relative growth, they remain special cases of the more comprehensive rule.

This problem has only been adequately attacked by Schmalhausen (1927A, 1927B, 1930), and in what follows I can do little save summarize his views and to comment briefly upon them.

It is an obvious fact of observation that organs or parts of the body in general grow more rapidly when first formed, and that their absolute growth-rate (when external factors such as temperature are kept constant) diminishes progressively with time. Our previous method of establishing the coefficient of relative growth for an organ by comparing its size with that of some standard representing the measure of the rest of the body at different absolute sizes, is completely valid only on the assumption that the organ and the standard part begin their careers simultaneously, so that the decrease of absolute growth-intensity proceeds *pari passu* in both, and time can therefore be neglected. Even when the origins are not simultaneous, it will, however, usually provide a close approximation when we are dealing with the later stages of growth, for then the difference in time of origin between organ and standard will be negligible in comparison with the time that has since elapsed. This point is brought out by Schmalhausen himself, who gives some theoretical calculations on the subject (1927B, p. 41, etc.).

But during the early stages of development, the effect of different time of origin will be relatively large, and will completely vitiate the method of comparing absolute sizes (l. c., p. 59). What we require to do, if the organ x is first formed n days after the standard part y, is to compare the size of the organ at n, $n + 1$, $n + 2$ . . . days with the size of the standard at 0, 1, 2 . . . days: from these sizes, the growth-coefficient of the organ could be correctly calculated according to our heterogony formula. But to arrive at these values, we require to know the time-relations of early development, which is precisely what we have been able to neglect, with such economy of time and labour, in our previous approach.

Unfortunately, owing firstly to the difficulty of establishing the true time of origin of development, and secondly to variations of developmental rate among individuals, which make different embryos arrive at the same developmental stage at different absolute times, accurate time-relations are hard to establish for embryonic life, and not always serviceable even when established. These difficulties are extreme in the chick, but serious even in mammals, where, e.g., litter-size has a marked effect on foetal size and consequently upon foetal differentiation. For these reasons, Schmalhausen uses an indirect method for estimating true developmental age.

In his previous papers, Schmalhausen was able to show that in the chick and apparently in various other vertebrates, the linear growth of the embryo, measured by the cube root of its weight, $\sqrt[3]{p}$, remained approximately constant throughout embryonic life. Since the specific gravity of the embryo is very close to 1, then if the weight p is taken in milligrams, $\sqrt[3]{p}$ can be taken as giving a value in millimetres; and the constant rate of growth can be expressed in mm. per day. Thus the cube root of the weight of the embryo can be taken as giving a measure of its age.

This approximate constancy of linear growth-rate holds also, according to Schmalhausen, for the separate organs of the body. In all cases, there are considerable oscillations in the value of linear growth per day; and sometimes the value alters progressively during development, so that the method can only be considered an approximate one. None the less, for studies of relative growth, the method appears to be at least as suitable, and certainly much less difficult to arrive at, than accurate time-measurements.

Starting from these assumptions, we arrive at the following

line of argument. We wish to find the *growth-quotient* q of an organ—i.e. the ratio of the growth-rate of the organ during a particular phase of its development to that of the body, or of a standard part representing the body, not during the same period of time, but during the corresponding phase of its development.

Schmalhausen had previously established the following formula for finding what he calls the true growth-rate (C_v) of an organ for a period from time t to time t_1, during which the weight (or volume) of the organ has increased from v to v_1. His formula is

$$C_v = \frac{\log v_1 - \log v}{0{\cdot}4343\,(t_1 - t)} \quad . \quad . \quad . \quad . \quad . \quad (1)$$

Correspondingly the true growth-rate of the body during the same period will be

$$C_w = \frac{\log w_1 - \log w}{0{\cdot}4343\,(t_1 - t)} \quad . \quad . \quad . \quad . \quad . \quad (1a)$$

If organ and body are in the same phase of development during this period, then the growth-quotient q is

$$q = \frac{C_v}{C_w} = \frac{\log v_1 - \log v}{\log w_1 - \log w} \quad . \quad . \quad . \quad . \quad (2)$$

This is simply another method of writing the heterogony formula already arrived at by me, and his q is the same as my k.

For purposes of dealing with embryonic organs, we should take t as t_0, the time at which the organ and the body begin their growth, which in this case we have assumed to be at the same moment. We should then write v and w, v_0 and w_0 respectively. Formula (2) can then be written :

$$\text{Log } v_1 = q \log w_1 + (\log v_0 - q \log w_0) \quad . \quad . \quad (3)$$

Now $\log v_0 - q \log w_0$ will always be the same ; let us call this expression b.

Then $v_1 = bw_1^q$.

As stated above, q is here identical with k in my formula ; but we now have a further analysis of my constant b, which however can only be arrived at if we know the initial size of the organ and the body-standard.

But if the organ and the body begin their growth at different times, then the matter is more complex. We want to compare the growth of the organ with that of the body during corresponding periods of their development. Let t_x be the time which elapses between the origin of the body and that of the organ. Then the length of this period is, by Schmalhausen's method, measured by the linear increase of the body: let this be denoted by L_x. Then if the weight of the organ at times t and t_1 be v and v_1, the linear dimensions of the body, not for the *same time*, but for the *corresponding phase* of its development, will be $(L - L_x)$ and $(L_1 - L_x)$; and the corresponding body-weights (since the linear dimension is derived directly by taking the cube root of the weight) will be as the cubes of these values.

Thus the growth-quotient q for the organ for this period of time can be written

$$q = \frac{\log v_1 - \log v}{3[\log (L_1 - L_x) - \log (L - L_x)]} \quad . \quad . \quad . \quad (4)$$

The corresponding formula for linear measurements of the organ will of course be the same, but with the omission of the 3 in the denominator.[1]

As example Schmalhausen takes the length of the parts of the hind-limb of the developing fowl. Here the development begins proximally, so that e.g. the most distal (4th) phalanx of the 3rd digit begins to develop two days later than the 1st or most basal.

Taking simply the initial and final values for length between the 8th (or 9th) and the 21st day, he arrives at the following result.

	Origin at linear size (L) of embryo mm.	k calculated from Huxley's heterogony formula	q calculated from the linear modification of formula (4) above
Femur	3·77	1·43	1·11
3rd digit, Phalanx 1	5·20	1·65	1·15
3rd digit, " 2	6·87	1·75	1·00
3rd digit, " 3	8·25	1·64	0·84
3rd digit, " 4	8·89	1·85	0·86

[1] If we knew the time-relations precisely, the proper formulation of the growth quotient would be

$$q = \frac{\log v_1 - \log v}{\log (w_1 - w_x) - \log (w - w_x)} \quad . \quad . \quad . \quad . \quad . \quad (5)$$

where w_x is the amount of weight added by the body in the period t_x, between the time of its origin and that of the origin of the organ. Mathematically, it may be pointed out, this is not identical with expression (4); but the latter gives a reasonable approximation.

It is seen that there is in actual fact (q values) a growth-gradient in the 3rd digit dropping distally (with possible slight rise quite terminally again) ; while if we do not take account of the difference in time of origin, the gradient (k values) is quite obscured and the terminal digit comes to have the largest ' growth-coefficient '.

The change in relative size of various organs of the chick during embryonic life is shown in Fig. 72, and the actual growth of some others has been plotted from Schmalhausen's

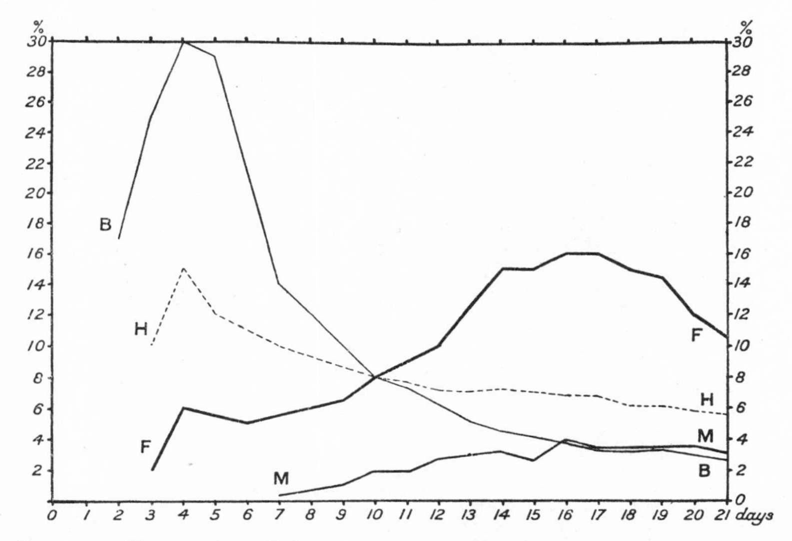

FIG. 72.—Changes in relative size in various organs of the chick during embryonic life.

B, brain ; H, heart ; F, fore-limb ; M, metanephros.
Note the very different shapes of the curves. This depends (*a*) on the time of origin of the organ, (*b*) on its relative growth-rate.

data in Fig. 73. The value of Schmalhausen's method is clearly evident. It means that we cannot discover the true growth-coefficient of an organ during its early stages without precise information as to the time-relations of development.

Somewhat unfortunately from our point of view, Schmalhausen prefers in general not to work with growth-quotients, which are our growth-coefficients corrected for difference in time of origin, but with *growth-constants*. These are obtained by multiplying the growth-rate C_v of an organ (see equation 1) for a given period by the mean age of the organ during that period, since he believes that he has established in his earlier papers the fact that during development, growth-rate sinks in simple inverse ratio to time.

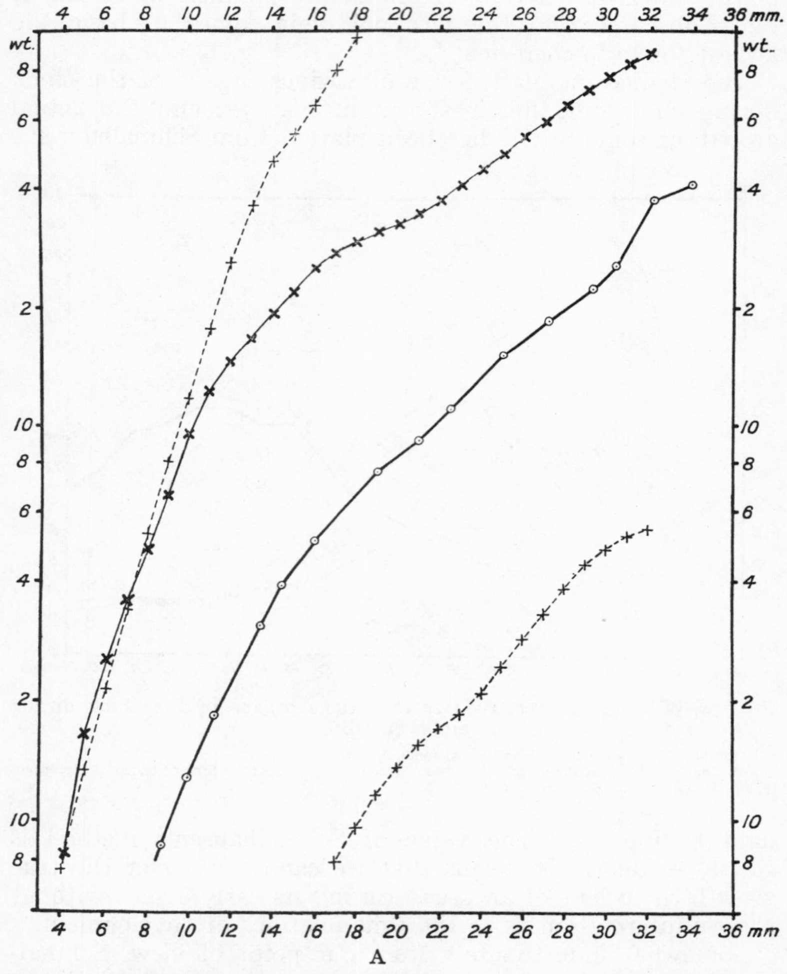

FIG. 73.—Weight-growth of various organs in the embryo chick, plotted on growth-rate, A, against age as determined by increase of $\sqrt[3]{\text{embryo volume}}$ various organs are indicated in the legend. In

In A, × is weight of lens; and + (continued to the right and below) weight of fore-limb. In the portions of development shown the lens increases from 8 × 10⁻⁵ g. to 0·0088 mg.; the fore-limb from 7·5 × 10⁻⁴ g. to 0·54 mg.; the embryo from 0·87 to 41 g.

(*Constructed from the data of Schmalhausen*, 1927A, *Tables* 9, 15, 18; 1927B, *Table* 7.

We need not enter into a discussion of this point here, which concerns the problem not of relative but of absolute growth. In any case, the growth-constants thus arrived at will give us some real measure of growth-rate ; and further, the *relative growth-constant,* which he uses for comparative purposes, is arrived at directly from the growth-quotient above discussed. However, for our purpose we may stick chiefly to the growth-

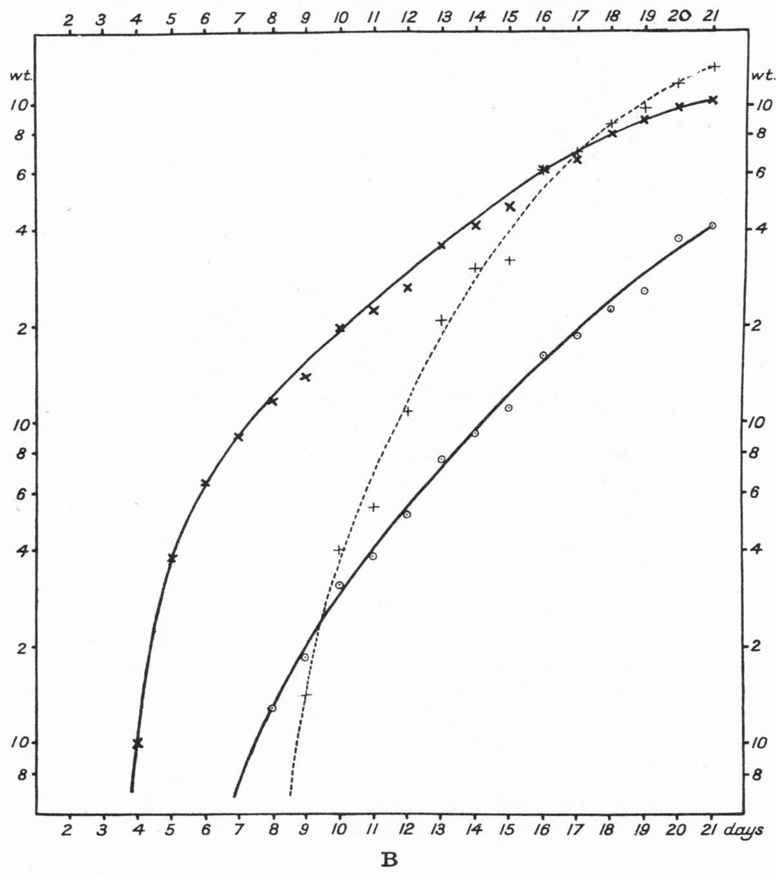

B

arith-log paper so that the slopes of the curves is directly proportional to (see text) ; B, against actual age of embryo. The scales of weight for the both, ⊙ denotes weight of embryo.

In B, + denotes brain-weight (from 0·01 to 1·02 g.) ; + metanephros weight (from 0·0014 to 0·13 g.). The lens and the fore-limb originate nearly together, but the latter grows much faster. The brain starts growth early and then grows more slowly than the whole embryo, the metanephros starts late and throughout grows much faster than the embryo.

quotients, or, as I prefer to call them in accordance with the terminology of this book, the *corrected growth-coefficients*.

These corrected growth-coefficients for various organs of five species of birds are as follows (recalculated from Schmalhausen's Table 9):

	Growth-quotients				Relative mass-factor			
	Brain	Gizzard	Fore-limb	Hind-limb	Brain	Gizzard	Fore-limb	Hind-limb
Chick (*Gallus domesticus*)	0·62	1·20	0·94	1·12	1·27	0·244	0·316	0·371
Duck (*Anas moschata*)	0·67	0·95	0·93	1·19	1·26	0·324	0·316	0·294
House-sparrow (*Passer domesticus*) . .	0·51	1·27	0·94	1·05	1·81	0·369	0·343	0·371
Sand-martin (*Cotyle riparia*) . . .	0·58	0·94	0·79	0·95	1·12	0·491	0·456	0·414
Rook (*Corvus frugilegus*) . . .	0·55	1·14	0·84	1·02	1·97	0·353	0·438	0·386

The heterogony of the brain is always markedly negative, that of the fore-limb slightly negative; that of the hind-limb is always higher than that of the fore-limb; while that of the gizzard is the most variable.

The right-hand part of the table concerns another constant arrived at by Schmalhausen. On the assumption previously arrived at by him that the growth of an organ can be represented by the formula

$$v = (at)^k \quad . \quad . \quad . \quad . \quad . \quad . \quad . \quad (6)$$

where v is the weight or volume of an organ, t is the time elapsed since its origin, and a a constant (if the rate of linear growth of the body or organ is constant, then of course $k = 3$), then a is what Schmalhausen calls the extension factor (Extensitätsfaktor). The constant a can also be calculated for a given interval of time $(t_1 - t)$. A derivative constant is what Schmalhausen calls the mass factor, m. This he takes as a^k. It can be derived from (6) thus:

$$m = a^k = \frac{v}{t^k}, \text{ or } v = mt^k \quad . \quad . \quad . \quad . \quad (7)$$

This constant m characterizes the initial size of the organ-rudiment. Schmalhausen's exposition is here exceedingly obscure, especially as to how he transforms his absolute extension-factor m into what he styles his relative extension-factor r, which alone makes comparison possible between different species. It is clear, however, that if his general argument is correct, the two factors here recorded will serve to give a complete description of the facts concerning the relative growth of an organ from its first inception.

We see, for instance, that the brains of chick and duck start of nearly the same relative size, but that that of the duck then grows relatively more rapidly. The gizzard is laid down of larger size in the duck than in the chick, but then grows much more slowly, so that it ends up considerably smaller. Similarly, the sand-martin, which in the adult is characterized by very small hind-limbs, has large hind-limb rudiments which then proceed to grow very slowly. It is interesting to find that the growth-quotient of the hind-limbs is in all species investigated higher than that of the fore-limbs.

I feel that some of the formulae advanced by Schmalhausen are open to criticism and will need some further corroboration. However, his method for arriving at the true corrected growth-coefficient for embryonic organs is of real value, and if possibly not always fully accurate, is undoubtedly the only way at present available by which we can arrive at a good first approximation. (See also the work of Ford, p. 260.)

§ 8. Conclusion

The chief points in this chapter may be briefly summarized as follows: D'Arcy Thompson's method of employing Cartesian co-ordinates to effect the geometrical transformation of an organism or organ gives evidence of the existence of orderly growth-changes within the body. These may be of complex nature, but can be analysed into a series of *growth-gradients.* Confirmation of this is provided by quantitative analysis of various organisms during their growth. A curious effect is noted by which the presence of a centre of high growth-intensity intercalated in a main growth-gradient is correlated with minor changes in the growth-intensity of neighbouring parts. Those immediately posterior are somewhat increased in size, those immediately anterior appear to be somewhat decreased in size: i.e. the main growth-gradient is deformed in a regular way by the presence of the subsidiary growth-gradient. Finally, it is pointed out that constant growth-coefficients of parts and regular growth-gradients within organs and the body as a whole, such as here described, appear to be operative only during the later phase of growth. During the earlier phase, when histological differentiation is proceeding, quite other quantitative rules apply. To distinguish these two phases of development, the term *histo-differentiation* is proposed for the former, *auxano-differentiation* (Greek *αὐξανεῖν*, to increase) for the latter.

CHAPTER V

GROWTH-CENTRES AND GROWTH-GRADIENTS IN ACCRETIONARY GROWTH

§ 1. The Accretionary Method of Growth

IN the organs we have so far been considering growth is essentially of the compound-interest type. That is to say, the increments of new material produced by growth are alive, and themselves grow and produce new material in their turn, so that growth is a *multiplicative* process. The rate of growth may, and doubtless does, slow down with increasing size and age, but this merely means that the multiplying factor decreases as some regular function of physiological age. The growth remains of compound-interest type, even if the actual rate of compound interest is never constant but progressively decreases.

However, there are many other organs whose method of formation is radically different, so that their growth is essentially of the simple-interest type. In them, the increments of new material produced by growth are turned into non-living material as soon as formed and remain permanently (or until cast off by ecdysis or other means) in the state in which they were laid down. They do not contribute any further new material, so that growth here is not a multiplicative but an *additive* process.

Here again the rate of growth may alter with age, but this only means that the amount of new material added in unit time steadily decreases; and the growth remains of simple-interest type even though the actual rate of simple interest is continually altering.

We may accordingly distinguish these two types of growth as the multiplicative, intussusceptive or compound-interest type on the one hand, the additive, accretionary or simple-interest type on the other.

The most familiar examples of organs growing by the accretionary method are shells such as those of molluscs, brachiopods, or foraminifera; but the horns of antelopes, sheep, oxen, rhinoceroses, etc., as well as the teeth of Vertebrates, also fall into this category. At first sight the forms engendered by this type of growth appear so different from those we have hitherto been considering that we do not even expect to find that the underlying growth-mechanisms have anything in common. The differences, however, depend almost entirely upon the basic difference between any multiplicative and any additive type of growth. It is to my mind one of the most interesting results of these growth-studies that we are able to demonstrate the same fundamental fact of growth-gradients operating to produce these two apparently unrelated types of organic form.

Let me first take the horn of rhinoceroses as example. It has been admirably handled by D'Arcy Thompson in his *Growth and Form.* I here base myself on his lucid analysis, which, like so many other important ways of thinking that enable us to see familiar facts in a new light, seems self-evident once grasped; but I add one or two detailed points, and link it up with the ideas which emerged from the study of multiplicative growth.

The horn of a rhinoceros, then, is produced by intensive production of keratin in special form and abundance over a limited area of the head epidermis. The restriction of horn-producing potency to a limited area is doubtless of the same nature as the other restrictions of potency which occur during early development and sooner or later convert the germ from a plastic construction capable of marked regeneration to a determined construction which we can designate as a chemical mosaic (see Huxley, 1924c). The potency of producing eye, ear, brain or limb becomes similarly restricted and localized in the amphibian and other embryo, and the restriction of horn-potency to a localized area is only another result of this mosaic-producing chemo-differentiation.

On this horn-area, keratin is being produced so as to accumulate at right angles to the surface. In addition, the horn-area itself is enlarging over the surface as the animal grows. Whether, as seems likely, it is enlarging somewhat more rapidly than the surface of the head as a whole cannot be stated with certainty until detailed measurements have been made; but this is immaterial to our present purpose.

§ 2. Logarithmic Spirals as the Result of Growth-gradients

If the rate of keratin-production at any one moment were equal over the whole horn-area, the resultant horn would clearly have the form of a cone, whose precise shape would depend upon the relation between the rate of addition of new material, and the rate of spread of the horn-area over the surface of the head; if the two rates were equal, the cone would be a right-angled one, and so forth.[1] But as a matter of fact, in the common rhinoceros growth is not uniform over the horn-area: it is at its maximum anteriorly, and grades steadily down to the posterior margin. As result, the horn of course curves backwards; and the precise form of the curve is that known as a logarithmic spiral.

The properties of this type of curve have been fully dealt with by numerous authors, and the whole subject ably summarized by D'Arcy Thompson in a series of chapters. I thus need only remind my readers that the most essential characteristics of a structure growing in a logarithmic spiral are that successive increments are all of the same form, though of increasing bulk (gnomonic growth); that the angle which the tangent to the curve makes with the radius vector of the curve remains constant; and that if the spiral grows long enough to form a number of whorls, the ratio, along a given radius, of the breadth of each whorl to that of the whorl succeeding, also remains constant. Further, this logarithmic spiral form must always result in organisms when (*a*) growth-increments are converted into non-living material as soon as produced; and (*b*) there is a constant ratio between the increments at the two ends of the growing structure, with a regular (though not necessarily uniform) gradient of growth-rate between the high and low points. We are thus confronted once more both with the principle of constant differential growth-ratios and with that of growth-gradients. Since, however, these here operate with an additive instead of a multiplicative growth-mechanism, the resultant structure remains of constant (and logarithmic-spiral) form instead of continuously changing its proportions as with a male Uca chela or a female Carcinus abdomen.

But the rhinoceroses teach us a further highly important fact. Some species possess two horns instead of only one.

[1] The extinct Elasmotherium possessed a horn in the shape of a flattened cone, with the diameter of its base greater than its height.

And in these the second and hinder horn is both smaller than the first (and also less curved). This implies that the growth-gradient made visible in the form of the anterior horn is continued across to the second horn-area, causing the growth-intensity to diminish, and therefore resulting in a smaller horn; (and also that the shape of the gradient is not constant, but flattens out, leading to less difference in growth-intensity between the two ends of the horn-area, and consequently to a decreased curvature of the second horn). These facts thus lead to the same important conclusion as did the analysis of the growth of the appendages of Eupagurus—namely, that though intensive growth be restricted to specifically limited areas (there the regions of the limb-buds, here the horn-areas), yet the agency determining the growth-gradients, whatever it may be, is organismal, and extends throughout the body. It can only *express* itself where the potentialities for intensive growth exist, but it is itself continuous (as in a rather different, non-graded way, the hormones are distributed over the entire system, but only exert effects where they meet with tissues specifically adjusted to react to them). Thus we must assume that even in the one-horned rhinoceroses, the growth-gradient is continuous along the head, but can only reveal itself in species where a second specific horn-area is present; and similarly, that in hermit-crabs the growth-gradient controlling the relative growth of appendages is continuous, not merely between the limb-producing areas of successive segments, but even across large regions in which the capacity for limb-production has been entirely lost, as in the anterior part of the male abdomen.

It is worth recalling that we already know of analogous gradients, and to use a more general term under which gradients can be included, *fields*, through the results of experiments on regeneration and grafting. The mere fact that, normally, precisely what is lost in amputation is restored in regeneration points in this direction. The conclusion has been made more probable by the proof given by Schotté, Guyenot and Weiss, proof that in regeneration (of the Amphibian limb) the new tissues are *not*, as was long held, proliferated from the old, but are differentiated from a truly indifferent tissue, and will differentiate normally even if the corresponding tissue has been removed from the basal stump. And finally, it has been clinched by the beautiful experiments of Milojevic, Weiss, Locatelli, Guyenot and Schotté (references in Guyenot and

Ponse, 1930); these have shown that if the regeneration-bud from the tail of a Urodele be removed while still in the indifferent stage and grafted on to the stump of a freshly-amputated limb, it will grow not into tail but into limb; whereas if it has been left a couple of days longer on the tail before being transplanted, it would have been irrevocably determined as tail, and would have become tail even in its new situation. They can only be interpreted as meaning that what has conveniently been called a ' morphogenetic field ' permeates the whole body even of the adult Amphibian. It normally is without effect—in a sense a by-product, we may say, of the construction of the animal; but so soon as indifferent material is placed under its influence, it reveals its presence by the effect which it exerts on that material's differentiation.

Whether growth-gradients and morphogenetic effects on differentiation are both results of one and the same organismal field, or whether two essentially different, separate field-mechanisms are at work, is very difficult to say. Further discussion of this and related points will be deferred to Chapter VI.

Thus the horns of rhinoceroses, when considered from the point of view of relative growth, even without further experimental analysis, reveal interesting and unexpected properties of the animal body. The same point of view, applied to other structures of the same nature, is equally illuminating. The horns of rhinoceroses are median; we should therefore not expect to find a difference of growth between their lateral margins. But as soon as we deal with non-median structures, we should expect, if growth-fields permeate the animal body, to find a difference in growth-intensity not only between anterior and posterior, but also between median and lateral margins. If this is so, the resultant growth will be in what is popularly called spiral form, i.e. not merely curving in a true logarithmic spiral in one plane, but corkscrewing up at right-angles to the first.

This is what actually occurs in almost all horns of sheep, goats and antelopes. Sometimes the lateral growth-difference is very slight, and the ' shear ' at right-angles to the primary plane of the horns' spiral is scarcely perceptible, as in the Sable and other antelopes. At other times it is considerable, and we get the horns of certain sheep corkscrewing out at right angles to the side of the head. It appears that there is no constancy in the sign of the difference, the excess growth-intensity being sometimes on the median, sometimes on the lateral margin. The visible result is that the horns are sometimes coiled clockwise, sometimes counter-clockwise.

Complications, not present in the homogeneous rhinoceros horn, arise in the Cavicorn ruminants owing to the presence of the living bony horn-core within the non-living true horn of keratin. The results of these are analysed in detail by D'Arcy Thompson, but are not relevant to our present purpose.

§ 3. Growth-gradients and the Shells of Molluscs

The most numerous, various and striking of the structures which are based on logarithmic-spiral form and are due to differential accretionary growth are the shells of molluscs. For the moment, we will omit the special case of the bivalves. The problems here are identical with those encountered in the rhinoceros horn, except that the horn is solid and uniform throughout, the shell hollow. In both cases, form depends upon constant differential growth-ratios. These are here of four types : (1) the ratio of growth in length to that in width ; (2) the median growth-ratio ; (3) the lateral growth-ratio ; (4) the ratio of excess growth at specific arbitrary points to that manifested in the major growth-gradients.

(1) *Constant differential ratio of length-growth and width-growth.* In the absence of any other differential growth but that between length-growth and width-growth, the shell (or horn) would assume the form of a cone. The value of this first ratio determines the form of such a cone. In the rhinoceros horn, the physiological mechanisms at work are (*a*) the rate of production of horn-substance, (*b*) the outward spread of the horn-producing area. In the hollow mollusc shell, conditions are quite different, and only one factor is at work, namely, the angle at which the mantle-edge is inclined to the main axis of forward growth. This angle is presumably determined chiefly by the form of the body, and may be modified by functional differences (see Chapter VI). The mantle is thus always laying on material in a direction oblique to the main axis of the shell ; the growth-velocities in length and in breadth are merely components of this single growth-function. A high inclination of the direction of mantle-growth and consequent predominance of the lateral component will naturally produce flattened cones, to which the shell of the common limpet is an approximation. A low inclination, on

the other hand, will produce elongated cones, as in many of the early paleozoic Cephalopods.

(2) *Constant differential ratio as regards growth in the median plane.* If the length-width growth-ratio remains constant, but the absolute magnitudes of both components are greatest at one margin, least at the opposite margin, and are intermediately graded around the two sides of the mantle, our cone will be distorted, and, so long as the growth-ratios concerned remain constant, will grow into a true logarithmic spiral in a single plane. Examples are provided by Nautilus, the great majority of Ammonites, and Dentalium.

Our first growth-ratio will still decide the form of the cone, but now that the cone is distorted this will be measured by the ratio of the shell-diameter at any place to the length of the shell measured from its origin along the curve of the spiral.

The first and second ratio together will decide the tightness with which the spiral is coiled. There are six main possibilities:

(1) As limiting factor, with median growth-ratio = 1·0, or in other words, no growth-gradient from one end of the median plane to the other, a cone results.

(2) When the growth-ratio is low, only slightly above unity, the curve is very slight. In such cases, the mathematical properties of the logarithmic spiral being what they are, a many-whorled structure will never be produced, as the radius of even the second whorl would be of relatively immense extent, and no organism could do more than produce a portion of the first whorl. Such forms are realized in the rhinoceros horn or, among Molluscs, in the shell of Dentalium.

(3) With increasing values of the ratio, the radius of successive whorls rapidly decreases. The next possibility is therefore a shell with more than one whorl, but with no contact between each whorl and the next. This condition is rare, but is realized, e.g. in certain Ammonites.

(4) As the ratio increases further, a specific value will eventually be reached which allows the outer margin of each whorl to be precisely in contact with the inner margin of the whorl following. This condition is not infrequently realized. The precise value of the median growth-ratio needed to produce this result will of course vary with the value of the previous (length-width) growth-ratio. With a narrow elongated cone, a much higher median growth-ratio will be required just to effect contact between whorls than with a broader, less elongated cone (see Fig. 74).

As in each case only one specific combination of these two independent variables will produce such a result, we must suppose that selection has controlled the precise values to secure this result, which obviously secures greater strength than does one in which the whorls do not touch.

(5) The commonest condition, however, which ensures even greater constructional strength than the one preceding, is produced by a further increase of antero-posterior growth-ratio, which has as result the partial overlapping of each old whorl by the whorls formed later. The degree of overlapping will obviously vary with the precise value of the ratio (as well as with other properties of the shell: see below).

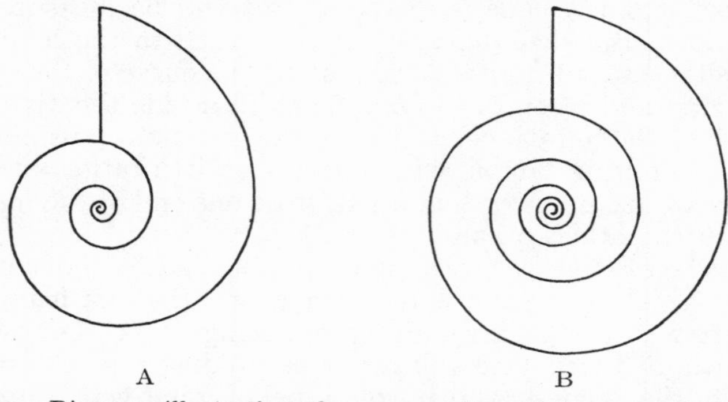

FIG. 74.—Diagram illustrating the co-operation of two growth-ratios in determining the form of the Molluscan shell. The two plane logarithmic-spiral shells both have the outer margin of one whorl just touching the inner margin of the next.

In A the ratio of the distance along a given radius from the centre of the shell to the margin of one whorl to that of the succeeding whorl is 3·0; in B it is 2·0. If there were no lateral growth-ratio (i.e. if the shells were uncoiled), B would be a more elongate cone than A; correspondingly, the lateral growth-ratio in B must be higher than in A to cause contact of successive whorls.

(6) Finally, in some cases the growth-rate of the slower-growing edge of the mantle becomes negligible or even zero, and the growth-ratio accordingly rises towards infinity. In such a case each whorl is completely overlapped by all succeeding whorls. This condition is completely realized in *Nautilus pompilius,* and nearly so in *Nautilus umbilicatus.*

This median growth-gradient may be orientated either way in respect of the main axis of the molluscan body. In Ammonites, for instance, the high point (growth-centre) seems to have been ventral, so that the shell curved upwards over the back. In other cases it is dorsal. In flat-shelled Gastropods (e.g. Planorbis) the orientation is complicated by the fact of torsion.

D'Arcy Thompson (l. c.) has treated the quantitative aspect of this problem at greater length. Unfortunately, he has not analysed the whole process, as would biologically be the ideal method, in terms of two co-operating growth-ratios, but, as regards part of the problem, he has been content to give a purely mathematical description. An analysis entirely in terms of growth-ratios is being undertaken by Professor H. Levy, of the Imperial College of Science; meanwhile it will be useful to give a brief summary of D'Arcy Thompson's treatment of the matter.

The form of a single curve following a logarithmic spiral is given by the expression

$$r = e^{\theta \cot \alpha}$$

where r is the radius of the shell from centre to circumference; θ is the angle of revolution which the spiral has described; and α is the angle between the tangent of the curve and the radius vector of the curve, which remains constant: this is known as the constant angle of the curve. In addition, the ratio of the radii of successive whorls is also always a constant.

The relation between these two constants is given in the following table (abbreviated from D'Arcy Thompson, p. 534):

Ratio of breadth of each whorl to the next preceding	Constant angle of the spiral
1·0	90°
1·5	86° 18′
2·0	83° 42′
3·0	80° 5′
5·0	75° 38′
10·0	69° 53′
50·0	58° 5′
100·0	53° 46′
10,000	34° 19′
1,000,000	24° 28′
100,000,000	16° 52′

In Nautilus, in all ordinary shells, and in all typical spiral shells of Gastropods, the constant angle is rarely below 80°, usually between 80° and 85°, and the ratio of the breadth of successive whorls usually between 3·0 and 1·75. With decreasing values of the constant angle, the spiral flattens out very rapidly. E.g., if the constant angle were 28°, and the first whorl were 1 in. broad, the next whorl would be about 1½ miles broad: this is what happens in shells like Dentalium, which represent but a fraction of the first whorl, and 'never come round', as D'Arcy Thompson puts it.

Now when we are considering, not a single line describing a logarithmic spiral, but a conical shell distorted into this form by growth-forces, the form of the curves described by the inner and outer margin are identical, but the inner margin is retarded in its growth by a constant fraction—in other words, the ratio of their growth-ratios is a constant.[1] The actual retardation can be expressed as the ratio of the length of the inner margin from the centre of the shell, to that of the outer margin, at any point. This figure gives the median growth-ratio we have been discussing (or rather is the reciprocal of it as we have defined it). This value can also be calculated by utilizing the mathematical properties of the logarithmic spiral (D'Arcy Thompson, pp. 541 seq.). We need not go into the calculations, but can confine outselves to the results. The median growth-ratio needed to produce a particular degree of coiling will vary with the constant angle of the spiral.

We will consider only two cases—the median growth-ratio needed to make consecutive whorls just touch, and that needed to produce a shell with spaces between successive whorls, the breadth of each space being a mean proportional between the breadths of the whorls which bound it.

Constant angle α of spiral	Median growth-ratio, of outer to inner border of shell, needed to produce	
	(*a*) Successive whorls just touching	(*b*) Successive whorls separated by a space which is a mean proportional between the breadths of the whorls
89°	1·11	1·05
88°	1·24	1·12
87°	1·39	1·18
86°	1·55	1·23
85°	1·73	1·31
80°	3·03	1·75
75°	4·27	2·32
70°	9·84	3·12
65°	18·2	4·35

[1] Waddington (1929) in various Ammonites finds that this is not strictly true. The ratio is slightly changing all the time, the formula for the spirals being of the form $r + c = e^{\beta\theta}$ instead of $r = e^{\beta\theta}$.

Thus in most Ammonitoid and Gastropod shells with constant angle between 80° and 85°, to produce contact between the whorls the outer border must be growing between 3 and 1·7 times as fast as the inner. For spirals with lower constant angle (i.e. those with high ratio of breadth-growth to length-growth), the median growth-ratio must be much higher. In open-coiled forms, however, the median growth-ratio will be less, and does not have to increase so fast with decrease of constant angle to preserve the same spacing. When the median growth-ratio is very high, we shall get forms whose whorls completely overlap, like Nautilus, if the constant angle is high ; but if the constant angle is low we shall then get types like Haliotis (or most Lamellibranch shells).

(3) *The lateral growth-ratio.* If *in addition* to the preceding two differential growth-ratios, we have also one in a plane inclined (usually, it appears, at right-angles) to the median, the result will be what D'Arcy Thompson somewhat loosely speaks of as a "shear" in the plane spiral, with as result a 'corkscrew' or *turbinate* spiral—i.e. a spiral not confined to one plane. This is prettily shown in the accompanying sketches kindly given me by Miss M. Lebour, of the Plymouth Laboratory, illustrating the origin of this complex spirality in a larval Pteropod. The original shell is a hemispherical cap, produced by growth which is uniform all round. After a certain stage, however, a marked difference appears in the growth at the two ends of the median axis, and a smaller difference in the growth at the two sides ; and the shell at once begins to 'corkscrew'.

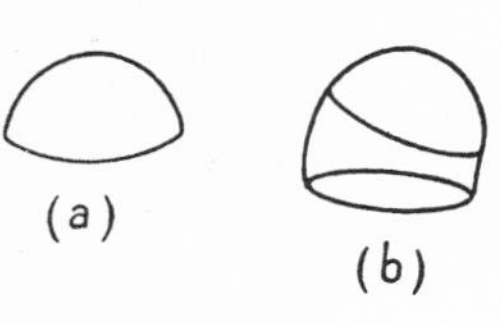

FIG. 75.—Limacina retroversa.

(*a*) shell of larva 1 day old ; the shell is hemispherical ; (*b*) shell of larva 3 days old. Differential growth has begun. (*c*) Larva in its shell, 4 days old. Differential growth has proceeded further, and the spiral shape of the shell is apparent.

In terms of growth-gradients, what happens appears to be as follows (Fig. 76) : If ABCD in (*a*) be the projection of the growing edge of the mantle, with A the region of maximum (+ +), C of minimum (— —) growth, with no lateral differential growth, then the corresponding gradient is shown in (*c*). If, however, a lateral differential is established, the results will be as shown in (*b*) and (*d*). The lateral differential growth-ratio is always smaller than the median (if it were larger, it would of course decide the main spiral, and the other would become the subsidiary differential, concerned with distortion of the main spiral).

Whereas the value of the main or median growth-ratio decides the tightness of the coiling of the main spiral, that of

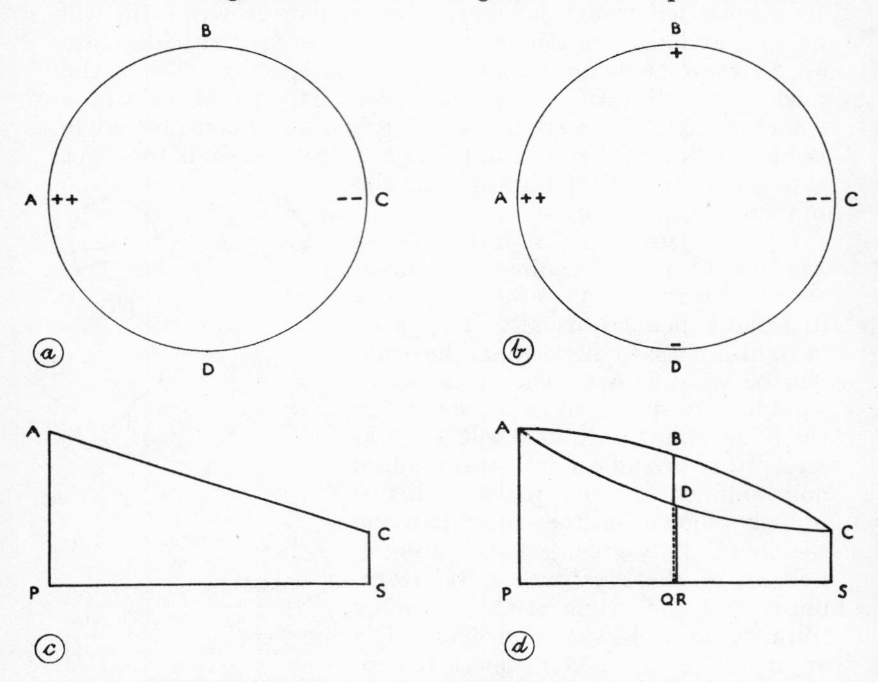

FIG. 76.—Diagram to illustrate the growth-gradients operating to produce the plane and the turbinate spiral shells of Molluscs.

(*a*) and (*b*) Projections of the growing edge of the mantle; (*c*) and (*d*) corresponding elevations, the ordinates representing growth-intensities, the abscissae distance across the shell-opening. (*a*) and (*c*) gradients operating to produce a plane logarithmic spiral shell. There is a centre of maximum growth at A, of minimum growth at C. The growth-gradients between A and C are identical on both sides of the mantle, through B and through D.

(*b*) and (*d*) gradients operating to produce a turbinate spiral. The growth-gradient ABC is of a different shape from ADC; thus a secondary growth-ratio is established between B and D. The growth ratio AP/CS < BQ/DR.

the secondary or lateral growth-ratio decides the degree of distortion of this spiral, the proportionate amount by which it is pushed out of its fundamental plane. When the lateral ratio is unity, the shell is flat, in one plane, like that of Planorbis. When it is low, the shell is low also, like the depressed shells of Helix. When it is higher, the shell becomes more pointed, as in Turritella. Here again, D'Arcy Thompson gives detailed mathematical treatment, although for some reason he has not reduced the degree of 'shear' (distortion of the primary spiral in a plane at right-angles to its own) to terms of differences of growth-rate, as he has done for the degree of coiling (distortion of the primary cone into a spiral).[1]

Typically, of course, the shape of any structure produced by accretionary growth must remain constant so long as the various growth-ratios concerned in its production remain constant. But as a matter of fact, the various ratios often alter with age or size, in some cases suddenly, in other cases progressively. Thus certain Ammonites have their oldest portions uncoiled, while the earlier-formed part of the shell is of typical form—an example of sudden alteration. In others, the ratio of the diameters of successive whorls does not remain constant as in the true logarithmic spiral, but increases progressively. This gradual change, due to a progressive change in the shape of a fundamental growth-gradient, is not infrequent, and may possibly prove to be associated with senescence. Among Gastropods, the tapering of the oldest part of the shell in Pupa, Clausilia, etc., in place of the continued expansion to be expected if the growth-ratios remain constant, is another case, and there are numerous other examples.

(4) Finally, there may be excess or defect of growth-ratio at special points, not in connexion with the main growth-gradient. This, of course, has its analogy in multiplicative growth, e.g. in the development of markedly heterogonic appendages which break the main growth-gradient of the body, as in the right chelae of male hermit-crabs.

The most obvious examples of such growth are found in lamellibranch shells. The 'ears' of the shells of scallops and other species of Pecten are an excellent case. The general growth-gradient is of usual Molluscan type, with high point directly opposite the hinge, and a uniform and symmetrical double gradient extending thence round both sides of the shell. Just before reaching the hinge, however, the growth-ratio, after sinking very low, increases rapidly and then abruptly descends to zero, thus generating the 'ears' of the shell.

[1] Interesting numerical details concerning various structures of logarithmic-spiral construction may be found in Petersen, 1921.

A rather different example is provided by the razor-shells (Solen). In these, the normal symmetrical gradient of the lamellibranch shell, with growth-centre opposite the hinge, is completely distorted by the development of a second growth-centre at the morphologically posterior (siphonal) margin of the shell. This occurs, of course, in numerous other forms, making the shell asymmetrical along the antero-posterior axis; but in Solen this 'secondary' growth-centre has become more important than the phylogenetically primary one, and its markedly higher growth-ratio converts the shell into the well-known elongated blade.[1] The margin of the shell at this end is truncated, but the opposite margin is rounded, the shell thus consisting of two markedly different but homologous halves, the one conforming to the normal lamellibranch shape, the other pulled out to ten or even twelve times the diameter of the former. Still further examples, again of a somewhat different type, are seen in the spines which beset the shell of such forms as the Spiny Cockle (*Cardium aculeatum*) or various Gastropods. These represent localized centres of excess mantle-activity, which, however, are only active periodically. Permanently active fluctuations in growth-activity along the gradient are revealed in such forms which have a crenellated margin to the shell (e.g. Tridacna). It is not known whether the presence of such special centres of high growth-ratio is correlated, as in Crustacea, etc., with slight excess or defect of growth in adjacent regions.

As in multiplicative growth, these subsidiary regions of special growth-activity are themselves constructed on the usual plan, of growth-gradients—single or double—culminating in a high point or growth-centre.

The bivalve shell (Lamellibranch and Brachiopod) demands a few words to itself. It owes its form to the existence of two separate gradients of accretionary growth, each forming a shell of logarithmic-spiral form. Typically, the two gradients are equal but of opposite sign, each symmetrical about a line drawn from the hinge to the ventral margin; the growth-ratio at the hinge is so close to zero as to be negligible, while the component of growth in shell-width is relatively so large that the spiral is a very high-angled one, and never forms even one complete whorl.

[1] When distortion of this type occurs, it is usually the posterior half which is enlarged; but there are a number of genera which show the opposite tendency (e.g. Donax).

This last feature is sometimes carried to an extreme, as in the lower valve of Pecten shells, where all growth is in the direction of width, with a perfectly flat shell as resultant. Or, as in other forms which habitually lie on one valve, the original direction of the gradient is reversed, and the curvature of the two valves becomes similar in sign, instead of opposed (e.g. Productus, occasionally Anomia).

Still more complexity of detail is shown in many Pteropod shells; these, as well as the shells of Foraminifera, which achieve logarithmic-spiral form by a somewhat different method (the addition of whole chambers instead of the mere prolongation of a single shell), have been well analysed by D'Arcy Thompson, and need not detain us here. Mention should also be made of the interesting paper of Sporn (1926), who applies a different set of mathematical ideas to the analysis of growth in molluscan shells. These have been related to growth problems in a more general way by Smirnov and Zhelochovtsev (1931).

§ 4. CONCLUSION

We may sum up the most important points of the present chapter as follows: Accretionary growth, in which the new material deposited is not itself capable of further growth, gives rise to structures whose general appearance is radically different from those produced by ordinary intussusceptive or multiplicative growth. But the differences turn out to be due only to this difference in the fate of the new material added by growth, as result of which the fundamental law of accretionary growth is one of simple interest, that of multiplicative growth one of compound interest. In other respects, the relative growth obtaining in the two kinds of structures is similar. In both we find growth-centres and growth-gradients, not only major gradients extending through major regions or the whole body, but also minor gradients superposed upon and locally overriding the main gradients.

The prevalence of the logarithmic-spiral form in nature is due to the fact that a uniform single growth-gradient, combined with the method of accretionary growth, *must* produce a structure in the form of a logarithmic spiral. Departures from the strict logarithmic-spiral form are due to irregularities in the growth-gradients, or to changes, either sudden or progressive, in one or other of the growth-ratios concerned.

The precise form of the shell or other accretionary structure depends upon the numerical values of the one, two or three main growth-ratios concerned in the production of conical, true logarithmic-spiral, or sheared (turbinate) logarithmic-spiral form respectively.

Thus the essential growth-mechanism underlying the auxo-differentiation stage of development of a single appendage (chela, vertebrate limb), a series of appendages (hermit-crab, stag-beetle), a region of the body (brachyuran abdomen), a rhinoceros horn, a vertebrate tooth, or a molluscan shell, are all of the same fundamental nature. In every case we find constant differential growth-ratios, and these are arranged in growth-gradients culminating in growth-centres. If we wish to think analytically about organic form and proportions, we must think in terms of constant differential growth-ratios organized in the form of growth-gradients. And if we are ever to solve the problem of growth physiologically, we should do well to concentrate on discovering the biochemical basis for growth-centres and the physiological reasons for the graded distribution of growth-potential on either side of these centres.

CHAPTER VI

HETEROGONY, GROWTH-GRADIENTS AND PHYSIOLOGY

§ 1. Normal Proportions as Result of a Partition-equilibrium

THIS chapter cannot but be an unsatisfactory one, for the simple reason that we know so extremely little about the physiological or biochemical processes underlying growth in general and growth-gradients in particular. All that can here be attempted is to bring together some of the scattered facts and indications which are in any way connected with the problem.

In the first place, as we have seen in the first chapter, there are strong grounds for believing that the normal growth-ratio of a heterogonic organ is in some way determined as result of a balance between its size and that of the body, equilibrium being attained when the formula $y = bx^k$ is satisfied. Whenever $y < bx^k$, the growth-ratio of y (the heterogonic organ) is accelerated ($k_1 > k$), and becomes normal once more ($= k$) when the organ reaches the size demanded by the original formula. The organ can become smaller than its size for growth-equilibrium in a number of ways. The most obvious is by amputation of the organ, in types where regeneration is possible. Its regenerative growth-ratio is then much more rapid than its normal growth-ratio would have been, and gradually slackens down until it becomes normal with the attainment of proper relative size by the organ. This is only a special case, for all organs, whether heterogonic or not, appear to exhibit the same behaviour during regeneration. Przibram (1917) has in mantids given a beautiful analysis of the way in which the excess growth-ratio falls away to normal as regeneration proceeds. The close approximation of his results (p. 51) to the laws governing the flow of E.M.F. between regions of differing electric potential is a further justification for the provisional use at least of the term 'growth-potential'.

One peculiar fact demands notice. We have already seen that the chela of the Gulfweed crab, *Portunus sayi*, affords a good example of constant differential growth-ratio. Zeleny (1905, analysed in Huxley, 1931B), however, also carried out regeneration experiments; and these are especially suited to

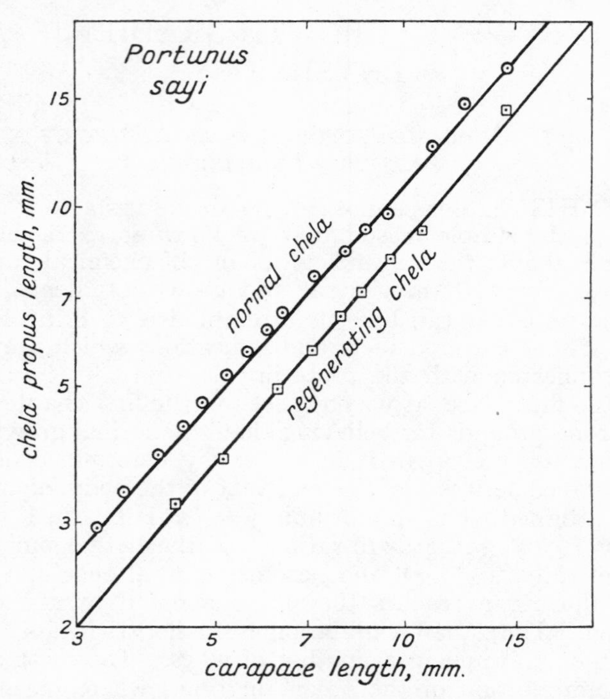

FIG. 77.—Graph to show (upper curve) simple heterogony of the large claw in the Gulfweed Crab, *Portunus sayi*, and (lower curve) the relation of the amount regenerated during one instar to normal size of the claw (logarithmic plotting).

(*Constructed from the data of Zeleny*, 1905.)

our purpose since (*a*) he found that moult-period, not time, was the essential factor affecting amount of growth in regeneration, and (*b*) he always amputated a limb immediately after one moult, and measured it immediately after the next.

When his figures, both for the size of normal claws and for

those that had thus been regenerating for one instar, are plotted on a double logarithmic grid against body-size, an interesting result is obtained. Both sets of points approximate nicely to straight lines; but the line for the regenerates is inclined at a slightly higher angle than that for the normal claws—$k = 1{\cdot}20$ instead of $1{\cdot}15$. In other words, during the first instar after operation, large crabs regenerate a slightly greater fraction of their claws than do small ones, and the absolute amount of increase is multiplied by a constant factor for each unit of multiplicative increase in absolute size. The amount of this factor is, of course, obtained by dividing the 'growth-ratio' determined from the slope of the line for the regenerating claws by the actual growth-ratio as found for the normal claws. This paradoxical result may possibly be accounted for on the principle, which holds in many cases, that the rate of regeneration increases with the amount removed.

We have already seen (p. 52) that when a heterogonic organ is grafted on to a body relatively too large for it, its growth-rate is decreased, so that here too it tends towards its proper relative size. Thus here also the idea of a partition-coefficient, representing an equilibrium between the amount of material in the organ and in the body, is supported.

§ 2. The Initial Determination and Physiological Basis of Growth-gradients

An important question is that of the physiological and biochemical bases of heterogony and of growth-gradients. Some experiments of Morgan (1932A) on male fiddler-crabs give some indications on this problem. His results may be summarized as follows. Young male fiddler-crabs, at a very early instar of their post-larval existence, produce two claws of male type. If one of these be amputated, the other is very shortly afterwards fixed as permanently male-type, and will regenerate male-type if later amputated. The amputated claw regenerates of female type and is from thenceforth fixed in this condition. Normally, it is the loss of one claw during the first few instars which determines the right- or left-handedness of the males.

If both claws are cut off during this early symmetrical stage when both are of male-type, both regenerate of female type, and remain so permanently. Occasionally, however, a male is found in nature which has lost neither of its claws during youth, and has grown to a considerable size while still the possessor of two symmetrical male-type claws.[1]

When either of these is amputated, it regenerates male-type. Thus we have the remarkable fact that while in early youth the amputation of both male-type claws is followed by a loss of all the male-type potentialities, and the amputation of one by a loss of male-type potentialities on that side of the body, this does not hold when the symmetrical double-male-clawed stage has lasted to a considerably later period of life.

One possibility that suggests itself is as follows. We know that during the phase of chemo-differentiation, prospective potencies are sharply localized. It is reasonable to suppose that the potency for growing into a large instead of into a small chela is localized in the claw's growth-centre, viz. the propus. Whatever the chemical substances responsible, they are then wholly removed by amputation, just as those responsible for limb-differentiation in Amphibia are wholly removed if a particular disc of material be removed during embryonic life. But we have seen that in the male-type chelae of Uca there is a growth-gradient. If, as a result of this during growth, the substance determining heterogony and masculine type should spread proximally to the breaking-joint, then amputation should now permit the regeneration of a male-type chela.[2]

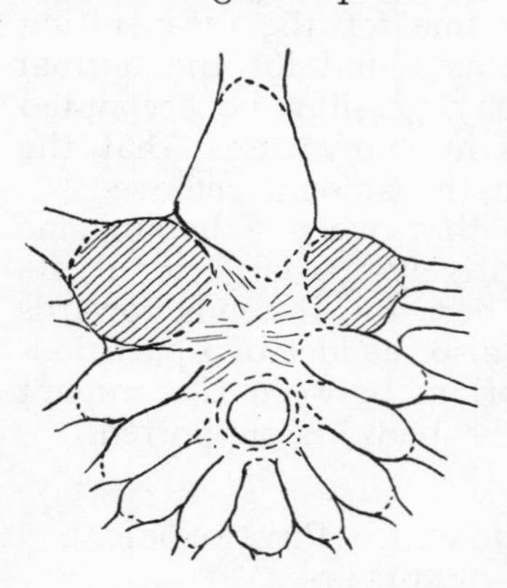

FIG. 78.—Sketch to show asymmetry in the thoracic portion of the central nervous system of the male fiddler-crab, *Uca pugnax*. The shaded regions are the ganglia supplying the chelae.

[1] In the only specimen of this type which I came across, the two male claws together weighed more than the mean for a normal single male-type claw for that body-weight, but considerably less than the sum of two normal single claws. The range of individual variation being considerable, however, one should have a number of specimens before attempting to generalize.

[2] In this connexion, reference should also be made to the results obtained by Haseman (1907A and B) on the direction of differentiation in segmenting Crustacean appendages. He finds that some regenerate basipetally, others centripetally; further, there is sometimes (e.g. in many antennae), but not always, a particular segment which produces

Another alternative is to suppose that the nervous system is implicated. This is suggested by unpublished observations of F. N. Ratcliffe on the nervous system of male Uca. He finds a marked asymmetry in the thoracic ganglionic mass, naturally largest in the ganglion of the chelar segment but visible in other segments as well. There are two classes of cells in the nuclei associated with the ganglia, large and small. The large are usually of different sizes but the same in number on the two sides of the body. But the small, at least in the ventral nucleus, appear to be equally distributed in number in early stages and unequally distributed (larger number on the side of the large chela) in later stages. Ratcliffe tentatively suggests that this condition is brought about by the permanent transfer to the side of the large chela of a certain number of originally median cells which could be transferred to either side of the body. If this suggestion be substantiated (admittedly it needs further work for its verification), it would imply that the normal fixing of the male-type potentiality would be finally due to a certain number of 'neutral' nerve-cells being transferred to that side, while the capacity of both claws to regenerate of male type in the doubly male-clawed older males would be due to a loss, with age, of the capacity of the median cells to transfer themselves from one side to the other.

This suggestion would obviate the need of postulating the proximal spread of a specific growth-promoting capacity in the limb. It is, however, perfectly compatible with the idea of an initial chemo-differentiative localization of high growth-potential in the male chela.

In this connexion also the observations of Perkins (1929) are interesting. In crabs (Carcinus, Cancer) and lobsters (Homarus) he finds a gradient in the body as regards the content of glutathione, sulphydryl and other reducing compounds known to be associated with growth. And this runs parallel with the actual growth-relations of the various appendages and regions of the body. He further advances a biochemical hypothesis to explain the existence of growth-gradients; but as this is highly speculative, and as I do not pretend to

the other segments by repeated fission. In certain cases (1907B) he was able to show that the normal directions of regenerative differentiation could be reversed by special conditions. It is probable that these facts are to be in some way related with those of growth-gradients, but for the moment the connexion remains obscure.

specialized biochemical knowledge, I will merely refer the reader to his article (see Fig. 79).

It is noteworthy that in other respects he has found an association of sulphydryl with growth-potential: e.g. the decline with age both of sulphydryl content and growth-rate (found in Carcinus, Pandalus, Sacculina embryos and Periplaneta). The coincidence of a gradient in this important 'key' metabolic agent with observed growth-gradients is

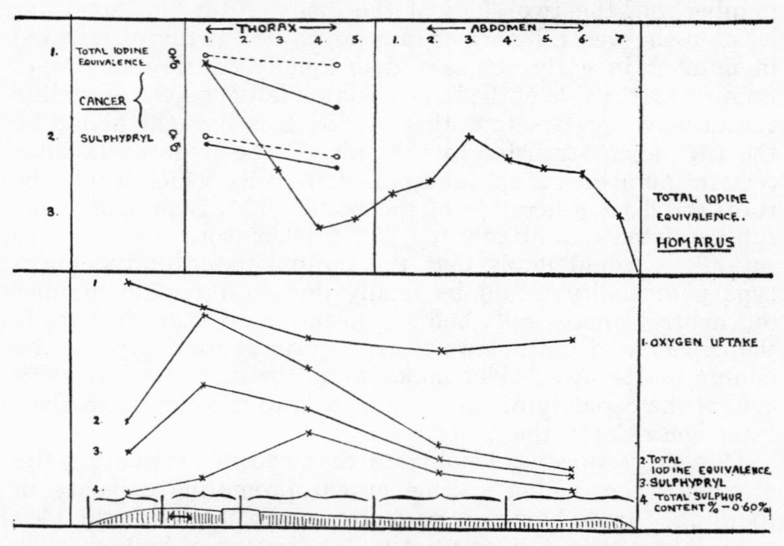

FIG. 79.—Gradients in content of various sulphur compounds and in oxygen uptake in the Crustacea, Cancer and Homarus (above), and the earthworm (below).

obviously a fact of considerable interest. Interestingly enough, the gradient in sulphydryl content does *not* coincide with the gradient in oxygen metabolism (in earthworms), so that Perkins concludes that the total oxygen uptake, being concerned more with katabolic than anabolic processes, is not a good measure of any gradients primarily concerned with growth, a fact which clearly has important bearings upon Child's work on the metabolic basis of his axial gradients.

§ 3. Other Gradient Theories

This brings us face to face with the relation between the growth-gradients here described and other types of gradient-effect. The existence of such effects has been casually recorded on a number of occasions, and their importance for morphogenesis has been emphasized by such workers as Boveri, von Ubisch, D'Arcy Thompson, and notably Child,[1] who has systematized the theory more thoroughly than other workers. The more general conception of the morphogenetic field, of which I believe gradient phenomena to be a particular case, has been analysed from somewhat different points of view by Weiss (1926), Guyenot (see Guyenot and Ponse, 1930), Hirsch (1931) and Bertalanffy (1928). This is not the place to enter into a general discussion of the subject, and I propose merely to refer to a few relevant facts and ideas.

In the first place, two sets of essentially morphological facts concerning gradients have long been known and recognized. The first is generally subsumed under the title of the law of antero-posterior development. It points out that during development differentiation begins anteriorly and gradually spreads posteriorly. Often the development of the head is far advanced when that of the hinder end is not yet begun. In some forms the undifferentiated posterior region may persist throughout life; or, as in many crustacea, it may persist through a considerable phase of free-swimming existence although lost in the adult. In addition, there exists in bilaterally symmetrical animals a similar gradient in time of development between dorsal and ventral surface. In Vertebrates the region which leads the way is the dorsal mid-line, in Invertebrates in general the ventral mid-line. Subsidiary graded effects of similar nature also occur within the appendages.

Secondly, a gradient also almost invariably occurs within the ovum, along the main axis. This may be revealed in the stratification of yolk or other materials, or in the greater rate of segmentation at the animal pole, or in both ways. It is the merit of Child that he has linked up these two sets of facts in one general physiological theory. He has further shown that the physiological gradient effects of which these

[1] I prefer not to use the term metabolic gradients, also sometimes used by Child. Child has not conclusively demonstrated that his gradients are fundamentally metabolic in character; but he has demonstrated that, as regards morphogenesis, gradient-systems do exist and are operative. The most important facts about these morphogenetic gradient-systems are (*a*) that they are *field systems* in which all the parts are interdependent within one plastic system, and (*b*) that in some respects at least they are *quantitatively graded*. Personally I would prefer the phrase *morphogenetic gradient-fields*, but there is no need at the moment to complicate terminology thus. For an excellent discussion, see Needham, 1931, p. 582 seq.

are particular morphological expressions may continue throughout life. He and his pupils have most thoroughly demonstrated this for hydroid polyps and planarian worms. By this work, certain important empirical laws have been established, notably the fact that the first region to differentiate in regeneration normally acts as a 'dominant' region which has a morphogenetic effect on the regions which differentiate later. This has now been brought into line with the facts concerning Spemann's 'organizer' in Amphibian development (see Santos, 1929). Further, such experiments as Stockard's celebrated production of cyclopia in Fundulus, cannot be interpreted except in terms of axial gradients. Many of Child's empirical facts have been independently conformed by Abeloos (1930), and the existence of a dominant region with morphogenetic effect by various workers, of whom Berrill (1931) is the latest. (See also the work of Buchanan, p. 260.)

In regard especially to hydroids and planarians, Child has been able to show that the physiological gradients constitute a true field system, e.g in regenerating fragments of Planaria the gradient can be either steepened (e.g. by optimum temperature) or flattened (e.g. by cold or by narcotics): and when this is done, it is found that the dominant region whose activity has been depressed induces smaller dependent organs, at a smaller distance than normal from itself, while the converse holds when it has been heightened. Analogous experiments have been performed on developing eggs, by using stimulatory or depressant drugs, or by applying temperature-gradients.

The continuance of the gradient throughout life is shown in many forms by their graded capacity for regeneration, e.g. the head-frequency often decreases steadily in an anteroposterior direction. There exists also a graded susceptibility to poisons along the main axis. In forms which bud or divide by transverse fission, the distance between the dominant regions of old and new zooids appears to be determined by the extent and steepness of the gradient. The production of axial heteromorphosis in regeneration, such as biaxial heads or tails, can also be satisfactorily interpreted in terms of the gradient hypothesis. Child has attempted to explain these facts in relation to differences in metabolic rate, but the proof cannot yet be said to be conclusive. However, whether or no this metabolic interpretation be correct, a set of important empirical principles remain—notably that physiological gradients exist in early stages of development, that they may persist throughout life, that their slope and extent are determined with reference to a first-formed dominant region which also has a morphogenetic or organizing effect, and that the parts and organs involved in an active gradient system are bound up together in a single physiological field, so that alterations in one part will necessitate correlative alterations elsewhere.

In connexion with the persistence of gradient effects throughout life, we may refer to some further examples not cited by Child. Apart from the continuous growth-gradients in single organs, in body-regions and in the body as a whole which we have discussed in this book, we have also discontinuous growth-effects which we can only understand on the basis of continuous underlying gradients (p. 152). And we have also gradients affecting the rate of growth or regeneration of epidermal structures such as feathers and hair (p. 100).

Clausen (1929), by grafting methods, has shown the existence in the tadpole tail of a gradient in susceptibility to the autolysing agencies which operate at metamorphosis: skin and muscle grafts from the anterior regions of the tail when transplanted to the back undergo more rapid histolysis during metamorphosis than do similar grafts from more posterior situations. This is a significant fact, for it cannot well have any particular functional or adaptive significance in relation to the metamorphic process, and the difference, like that between the two horns of the rhinoceros, merely acts as an indicator for the existence of some fundamental inherent property of the organism.

To take quite another example, Alverdes has demonstrated the existence of graded peristaltic activity in regions of the mammalian gut, showing that physiological gradients may exist even where no morphological differentiation is visible.

A gradient as regards regeneration is also seen in the results of von Ubisch. From his data (1915) on regeneration of limbs in the insect larva Cloe diptera, we can calculate that the length of the femur of regenerated limbs after one moult-interval is 44·2 per cent. of the length of the normal femur for the fore-limb, 40·6 per cent. for the middle limb, and 36·7 per cent. for the hind limb. This differs in sign from that for growth in Sphodromantis (p. 135), but the data are not wholly comparable.

In a later paper (1922) he refers to the fact that Przibram (1919) and Krizenecki (1917) obtained a gradient of opposite sign for the regeneration of the limbs of mantids and mealworms and thinks that his work on Cloe might have given different results if extended over a longer period of regeneration. But he adduces new and important results from grafting experiments on earthworms, showing that a difference in 'differentiation potential', as when a young head is grafted on an old body, gives much better results than the converse experiment or than even autoplastic grafting of young heads on young bodies. From this he deduces the existence of a 'differentiation gradient' of importance in morphogenesis and regeneration.

Sinnott (1930) has elicited a curious fact in regard to the variability of cell-size in tissues of the petiole of maple-leaves (Acer). He finds that there exists a gradient between surface and exterior, variability being least in the size of epidermal cells, and increasing cell-layer by cell-layer towards the centre of the petiole.

We have also the well-known work of Lund (e.g. 1923, 1928) who finds in various organisms a system of gradients in bio-electric potential persisting throughout life, and has shown that organic polarity and morphogenesis can be controlled by electrical means. A correlation between electrical and metabolic gradients has been shown by Purdy and Sheard (p. 260). And on the chemical side, the recent paper of Watanabe (1931) may be consulted; he finds a gradient in amount of oxidizable substance in earthworms, running parallel with the gradients postulated by Child for this organism.

It would doubtless be possible to multiply examples; but this deliberately heterogeneous list will serve to emphasize the wide range of gradient phenomena which may exist in the adult animal body.

It next falls to discuss the relations between the growth-gradients with which we have been concerned, and other gradient-systems, notably the axial gradients, as I shall term them.

Here we are on speculative ground; but there are certain indications which make us suspect some real connexion between the two. In the first place, there is the co-existence of a growth-gradient and an axial gradient in Planaria (see Abeloos, 1928). It would be extremely interesting to follow this up in greater detail and especially to see whether changes in steepness of the axial gradient were quantitatively associated with changes in the growth-gradient. In this case, it is worth recalling, size is the chief index of physiological age, and growth-partition is almost entirely a matter of size, whether in fed specimens which are increasing, or in starved specimens which are diminishing in size.

It is important to note that in Planarians the morphogenetic effects associated with the gradient (induction of pharynx and other organs by the dominant region) take place early, but do not interfere with the persistence of physiological effects of the gradient (e.g. regenerative capacity, growth-intensity) throughout life. In a not dissimilar way, the realization of normal morphological differentiation of a limb or tail in Triton does not interfere with the persistence of the morphogenetic potency of the surrounding area to produce a new or additional differentiation of the same type throughout life (Guyenot and Ponse, 1930).

The gradient, though persisting throughout life, might be altered in shape. It is more natural on various grounds we would expect that it was more likely to be flattened than steepened with age. If it were flattened, we should expect that the potency of differentiation at a given body-level would be altered to a potency originally characteristic of a more posterior body-level. This is what appears actually to occur in serially heteromorphic regeneration in crustacea and insects. The regenerated heteromorphic appendage almost invariably is of a type which normally belongs to a more posterior segment—e.g. antenna regenerated in place of eye-stalk (Palaemon), or fore-leg in place of antenna (stick-insects.)[1]

For these and various reasons we may regard it as probable that the primary axial gradient of the egg and early embryo will normally persist, although doubtless often in somewhat altered form, in later periods and probably throughout life, even when we have no ready means, such as antero-posterior differentiation, heteromorphic regeneration, or head-frequency in regeneration, of deducing its existence.

Further, if it does persist, we may again regard it as probable, especially in view of the facts in Planaria, that it will exert some influence upon growth. This influence may be direct or indirect, but in any case would be graded in its effect.

Further, to reverse the approach, we may as a matter of speculation conclude that it is probable, when gradients specifically concerned with growth, like those in the male chelae or female abdomena of crustacea, are found, that these

[1] Recent work by Przibram (*Akad. Wiss.*, Vienna, 9. vii. 1931) and Suster (ibid.) confirm this view. In Dixippus, antennae amputated in the 1st instar regenerate as antennae, while in later instars they regenerate leg-like organs. In Sphodromantis, the regenerate forms an antenna at 25°, but a leg-like organ at lower temperatures. In both cases the leg-like organ is produced when the axial gradient may be presumed to be flattened.

have elements in common with the axial gradients of Child—viz. a dominant region (here the growth-centre) which exerts a graded effect on neighbouring regions, so that the whole system is a field-system.

It is already clear that in Child's scheme, place will have to be found for qualitative differences among gradients. The primary morphogenetic gradient in, e.g., a worm must differ qualitatively from the activity-gradient concerned with the addition of new segments throughout life. The primary animal-vegetative axial gradient in an Amphibian egg must differ qualitatively from the latter field-system of which the organizer (dorsal lip) is the ' dominant region '. And specific growth-gradients will constitute another main type.

This section has been admittedly very speculative, but as tentative conclusion we may suggest that gradient-systems, all perhaps of essentially similar nature, are concerned with primary differentiation, the time-relations of early development, certain physiological properties of parts of the adult organism, certain regenerative capacities, and with growth-intensity. These gradient-systems will all obviously in the long run be ' metabolic ', but may be specialized in qualitatively different ways according to the type of activity which is graded within them.

The primary gradient of the egg and early embryo may be expected to persist throughout life and to have a minor effect on the graded distribution of growth-promoting substances—i.e. some growth-gradients will be secondary effects of the primary axial gradient. But in addition we may expect that gradients concerned specifically with growth-intensity may come into existence supplementary to and largely independently of the primary axial gradient, but will then exert their indirect effect upon such proportions of the primary gradient as still persist.[1]

§ 4. Heterogony and Hormones [2]

We must also consider the relation of hormones to heterogony. Here we must at the outset remind ourselves of an important point—that the action of a hormone always demands

[1] The important paper of Smirnov and Zhelochovtsev (1931) has appeared too late to receive the discussion it merits. It contains a detailed mathematical analysis of relative growth in the leaves of the Nasturtium (*Tropaeolum major*) under conditions of normal and reduced illumination and brings the results into relation with a general conception of a ' gradient-field ' of growth. See also Werner (p. 258).

[2] See also the work of Robb (p. 257).

two specificities—the specificity of the hormone, and the specificity of the tissue which reacts to the hormone. For instance, there may or may not be a hormone concerned in the growth of the fiddler-crab's large chela. Should there be one, however, its action would in this case be subordinate to the action of the specific capacity for heterogonic growth possessed by the male type but not by the female type of chela, whereas in a case like that of the fowl's comb, the tissue-specificity is apparently the same (or almost so) in both sexes, and the sexual differences in comb-size are brought about by the specificity of the two sex-hormones.

So far as we know, there exist no sex-hormones in insects. Accordingly, in this group any secondary sexual heterogonic organs will depend for their development entirely upon their inherent growth-capacities, which differ in the tissues of the two sexes according to the cellular metabolism induced by one or other sex-chromosome complex. In vertebrates, however, the reverse is usually the case: the tissue-capacity is the same or highly similar in both sexes, and the sexual differences are due to differences in sex-hormones. A good deal of work has been done on the growth of the fowl's comb by Pézard, Benoit, Lipschütz and others (references in Goldschmidt, 1923). It appears that at the onset of sexual maturity in males the growth of the comb becomes highly heterogonic, and approximates to a constant differential growth-ratio. The growth of the comb in females is also heterogonic, but mildly so; while in castrates it is isogonic. The marked comb-changes associated with the onset of a laying period in a pullet may be associated with a change in sex-hormones, or quite possibly with a change in general metabolism. Benoit (1927B) has shown that the growth-coefficient of the comb in growing fowls differs according to the season at which they are hatched. In those hatched in March–June, the coefficient is high early, then decreases markedly in late summer, to resume its high level in October or November; in those hatched after June, there is no slackening, but the initial coefficient is lower. Benoit suggests that all the phenomena are due to seasonal variations in testis activity. It is probable that besides the sex-hormones, many other factors, such as nutrition, influence the growth-partition coefficient of the comb.

Castration in adult males is followed by a regression in comb-size; this takes place according to a well-defined mathematical formula. As set forth by Pézard (1921) the comb's regression-curve is parabolic, being represented by the formula

$$L = l + \tfrac{1}{2}C(\theta + t)^2$$

where L is the length of the comb after the lapse of time t from the onset of regression, l the final length at the end of regression, θ the total time taken for regression, and C a constant (varying from individual to individual). This is of some interest, as body-size does not enter into the formula at all. The converse curve, of comb-growth produced by injection of male sex-hormones in capons has been recently determined by Blyth, Dodds and Gallimore (1931). The authors do not discuss this aspect of their work, but plotting their data shows that the curves for re-growth are quite different from Pézard's curves for regression. They show a well-marked point of inflexion and are often very regular. They could be represented by an expression of the form $x = A + Bx^2 - Cx^3$, which is equivalent to saying that Robertson's autocatalytic growth-formula would apply to them. It is interesting to find this difference between the positive and negative aspects of the same growth-process.

Grafts of female comb or wattles on to male hosts become larger than similar grafts on hosts of their own sex (Kozelka, 1930). The response of the comb to the sex-hormone is brought about by a specific mucoid layer in the dermis (Hardesty, 1931) : i.e. this is the true heterogonic tissue.

A great deal has been written as to the 'all-or-none' law of the action of the sex-hormones on comb-growth, some writers, like Pézard, maintaining that it holds, others, like Benoit, opposing the idea. It would appear that Benoit is correct, but that the range over which the action of the sex-hormones is proportional to its amount is very limited. Above this point, the maximum reactivity of the comb has been reached, and increase of testis-size is not followed by further increase of comb-size.

That the matter need not be so simple as this, however, is shown by experiments on mammals. Collip (1930), by injecting a particular fraction of placenta-extract into rats, has produced in both sexes accessory sexual organs (seminal vesicles, vagina, etc.) far exceeding in absolute and relative size anything normally found in the species ; on the other hand, injection of anterior pituitary (in females) causes an acceleration of maturity but without disproportionate size of the accessory organs relative to the gonad (see references in Parkes, 1929, pp. 158–9). Thus it would seem that pre-pituitary extract causes acceleration of the growth of the gonad and accessory organs, but without inducing an increase in the final maximum size of either, or a disproportionate development of accessory organs relative to gonads; while injection of placenta-extract leaves the gonad of normal relative size, but apparently induces a supernormal production of sex-hormone, this in turn resulting in supernormal hypertrophy of accessory sexual organs. We may expect to find similar complications elsewhere as regards the relation of hormone-producing organ and reactive organ.

Champy (1924) has given us an interesting experiment on the dorsal crest of male newts (Triton). This is a male secondary sexual organ, appearing at the onset of the breeding season, and exhibiting marked heterogony. Champy finds that it shows a differential response to starvation, being reduced relatively faster than other organs.[1] This may be due to the crest consisting of material which is readily drawn upon in starvation, or may point to some more general law of the reversibility of heterogonic growth, according to which a heterogonic organ would always tend to approach the size appropriate to its partition-equilibrium whatever the bulk of the body, and whether that bulk was being increased by normal growth or reduced by starvation. That something of this latter sort may occur is shown by the well-known fact that planarian worms reduced in size by starvation revert to juvenile proportions (Child, 1915). This relation has recently been worked out quantitatively by Abeloos (l. c.), who finds not only that the change of proportions during normal growth is truly heterogonic and approximates to that obtainable by constant differential growth-ratio, but that it is almost exactly quantitatively reversed during reduction due to starvation. In any case, the example of Triton clearly demonstrates the co-operation of nutritive and hormonic factors in determining the size of an organ (Fig. 80).

Non-sexual organs may, of course, also show heterogony, and their heterogony may also depend upon hormones. This is best seen as regards the growth of limbs in Anuran metamorphosis, which depends upon thyroid hormone, and is, up

[1] Unfortunately no weight-measurements of crest and other soft parts were made; but Champy's illustrations appear in general to bear out his assertion.

to a considerable dosage, above which no further increase of reaction is obtainable, a function of the amount of hormone administered. Champy (1922), as a result of investigations on thyroid-fed frog tadpoles, concludes that thyroid, during

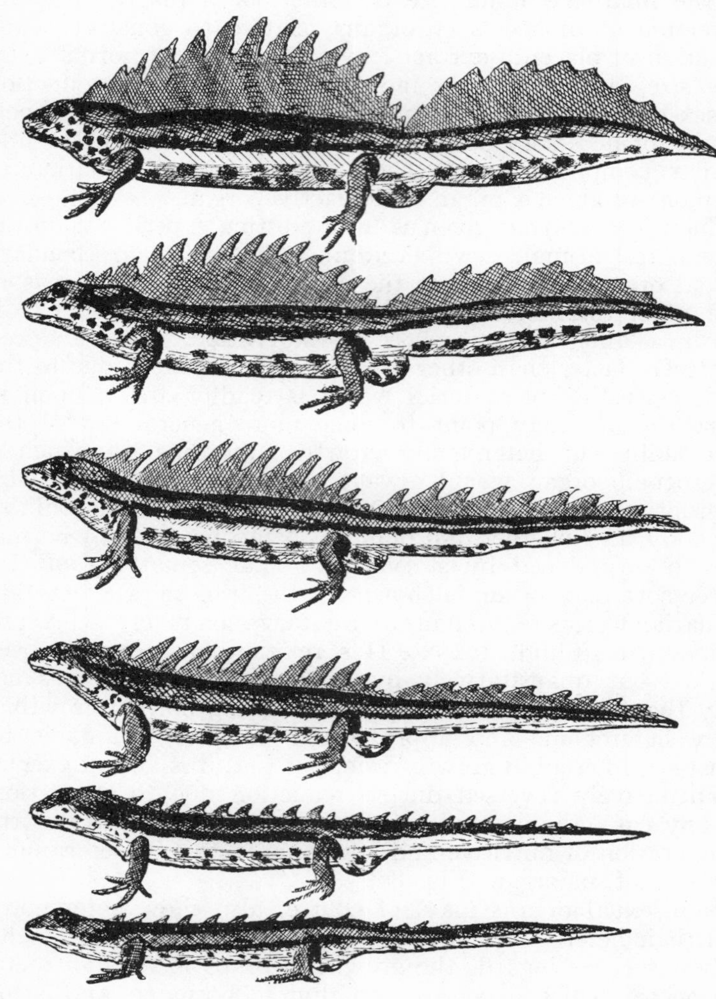

FIG. 80.—To show disproportionate reduction by starvation of a heterogonic organ (male dorsal crest) in a male newt, *Triton cristatus*. The drawings are to scale, and were made at 0, 5, 9, 16, 24, 32, and 45 days from the beginning of starvation.

the few days between its administration and metamorphosis, causes a progressive geometric increase in the number of dividing cells. This would imply a progressive *increase* in growth-ratio during the period. Further researches on this interesting point are needed.

In any case, the legs of Anura are of interest to us in several respects. First of all, they emphasize once more the importance of tissue-specificity. While their growth responds very sensitively to thyroid, the larval Urodele limb is wholly insensitive to the same hormone. Next, they show us that even the threshold of reactivity to one and the same hormone may vary from tissue to tissue. Anuran limb-buds appear to be sensitive to any dose of thyroid, starting from zero. But the equally specific degenerative response of the tail-tissues does not begin until a considerable concentration of thyroid-hormone is reached in the blood. Thirdly, they show us a clear-cut case of the growth-coefficient of an organ varying within wide limits with the dosage of a hormone. Even the normal growth of the Anuran tadpole's limbs is slightly heterogonic, as is shown by extirpating the thyroid in the embryo, when growth is much lower, and apparently isogonic (Allen).

Fourthly, they show in diagrammatic form the interrelation of the factors of growth-coefficient and available time as regards the problem of relative size of organ. Normally, the slight leakage of thyroid hormone into the blood during larval life produces a mild limb-heterogony. By the time the metamorphic crisis occurs (apparently due to a specific change in the pituitary which causes the thyroid to liberate most or all of its secretion suddenly into the blood), the hind-legs, originally mere buds, have had time to increase in relative size until longer than the trunk, although the trunk itself has increased perhaps two- or three-fold in linear size.

If, on the other hand, a moderately strong dose of thyroid be administered to small or small-medium tadpoles, with limb-buds hemispherical or conical but not yet fully differentiated, then although the growth-ratio of the limb-bud is raised far above normal, the accelerated growth can only operate for a few days before metamorphosis supervenes, with the result that the transformed froglet has relatively very small limbs. It would be of great interest to see whether these proportions were later regulated to or towards the normal, but the experiment has not so far been tried. As suggested in Chapter II, this factor of the amount of time available for heterogonic growth appears to be of importance in holometabolous insects in general, and in the dimorphism of male Forficula in particular. (See Fig. 81.)

Finally, anuran limbs well illustrate the dependence of differentiation upon growth. It is often asserted that in Amphibia the thyroid hormone 'favours differentiation but inhibits growth'. This is inaccurate and misleading. Slight thyroid activity, as in the normal tadpole, is not incompatible with total weight-increase. Excess thyroid causes loss in total weight, but this is an effect on balance, many organs losing weight, others, like the limbs and skeleton, gaining weight. As Champy (1925) has clearly shown, the effect of thyroid on some tissues is to halt their growth (gut) or even to cause their atrophy (tail, gills), on others is neutral, and on still others is to increase their growth (limbs). Further, the differentiation of the limbs is not a specific effect of the thyroid hormone, but a secondary effect of their growth in size. Differentiation of limb-segments, digits, etc., occurs at certain limb-sizes. And it will do so even when no thyroid hormone is present, as is shown by Allen's thyroidectomized tadpoles (Allen, 1918, 1919). These grew to a giant size (for tadpoles). As result, their limbs, though growing isogonically, attained an absolute size comparable to that reached by the limbs of normal tadpoles a few weeks before metamorphosis; and they showed a comparable degree of differentiation. This dependence of type of differentiation upon absolute size of organ is frequently to be met with (cf. in prawns with different heterogony of the chelae in both sexes, the resemblance of the proportions of heterogonic male and female chelae of the same absolute size, but of very different ages and attached to bodies of very different absolute size: see Chapter III).[1] In a later chapter we shall see that it has important taxonomic and evolutionary consequences.

FIG. 81. — Disproportionately small limbs caused by precociously induced metamorphosis in the common frog. The snout-anus length of the froglet was 8 mm.

The work of Hutt (1929) shows an interesting effect of the male sex-hormone upon the proportion of limb bones in

[1] It is however not invariable. Male and female chelae of Maia, Uca, etc., grow according to quite different growth-gradients (Chapter III).

fowls. The annexed table, modified from his Table 5, shows the chief results.

TABLE XIIA

PERCENTAGE CHANGE IN SIZE IN PARTS OF THE LIMBS OF MALE FOWLS INDUCED BY CASTRATION

	Humerus	Radius, Ulna	Carpo-metacarpus	Phalanges of digit 3 (mean)
Fore-limb .	+ 2·6	+ 2·25	+ 2·2	− 0·1 per cent
Hind-limb .	+ 3·0	+ 3·4	+ 3·9	+ 4·7 per cent
	Femur	Tibio-tarsus	Tarso-metatarsus	Phalanges of digit 3 (mean)

It will be seen that castration causes in general an increase in the size of the limb-bones. But whereas in the hind-limb the increase itself increases as we pass distally, in the fore-limb it is graded in the reverse sense, leading to an actual decrease in the terminal segment. The existence of the gradients is interesting, but the explanation of their opposite sign in fore and hind limbs is at present quite obscure.

Hammett (1929B) has published a summary of our knowledge of the effect of thyroid upon growth. In the first place, the effect upon total growth is an affair of dosage. In intact young mammals (and probably many other vertebrates) slight excess of thyroid causes an increase of growth in weight, while heavier doses, by a differential encouragement of katabolism, reduce it. In thyroid-deficient animals (whose growth is usually retarded), much greater doses will, of course, still permit increased growth.[1]

Our chief interest, however, concerns the *differential* effect of thyroid activity upon bodily proportions, and here Hammett himself has made elaborate experiments upon albino rats. Groups of these were thyroidectomized at 23, 30, 50, 65, 75 and 100 days respectively, and their organ sizes and weights determined and compared with those of unoperated controls at 150 days. When the increments made by the various organs measured in the operated animals are calculated as percentage of the increments made in the same space of time by the same organs of the controls, some important facts emerge. First, in every case the effect of thyroidectomy in retarding growth (not the absolute effect, but the *relative* effect,

[1] Thyroidectomy in some animals (anuran tadpoles, axolotls, etc.) is not accompanied by any change in growth-rate. In anuran larvae, excess thyroid is usually accompanied by a decreased growth (incipient metamorphosis). Even here, however, very minute doses accelerate growth during the early part of the pre-metamorphic period.

as measured in the way described) increases with the age of the animal. To take but a few examples, the eye, one of the organs least affected by thyroidectomy, in males operated at twenty-three days, by 150 days had made nearly 100 per cent. of the growth of the eyes of the controls in the same period. For those operated at 100 days, however, it made less than 80 per cent. of the controls' growth between 100 and 150 days.

In regard to total body-weight, males operated at twenty-three days showed about 60 per cent. of the increment of the controls, while for those operated at 100 days, the percentage was below 30 per cent. And as regards kidneys, males operated at twenty-three days showed under 50 per cent. of the controls' increment, those operated at sixty-five days showed hardly any increase at all, while those operated at 100 days, had before 150 days *lost* in absolute kidney-weight an amount equivalent to some 50 per cent. of the increment made by the controls in the same period. Thus the sensitivity of growth to thyroid-deficiency increases with age.

Secondly, most of the organs of the body fall into distinct groups as regards differential sensitivity to thyroidectomy. The eye-balls, central nervous system, length-growth both of body and tail (doubtless determined by growth of the axial skeleton), and both length- and weight-growth of humerus and of femur are relatively resistant, being retarded less than the weight of the body as a whole. On the contrary, the adrenals, spleen, kidneys, liver, heart, submaxillary glands, pancreas and to a slighter extent the lungs, make less increment than general body-weight (Fig. 82).

Hammett points out that these results are all consistent with the idea that the influence of thyroidectomy is greater on growth by increase of cell-size than on growth by cell-multiplication; greater on that fraction of the metabolism concerned with function (work)—e.g. secretion, muscular activity, etc.—than on that concerned with growth; and greater on labile than on stable chemical compounds. The last point is a correlate of the obvious fact that in conditions of subnormal nutrition (as in thyroidectomy) labile materials, such as the contents of glandular tissues, are more readily drawn upon than stable or inert substances, such as the salt-deposits in the skeleton, the lipoids of the nervous system, or the humours of the eye. The second is a correlate of the fact that in malnutrition, growth gives way to the mainten-

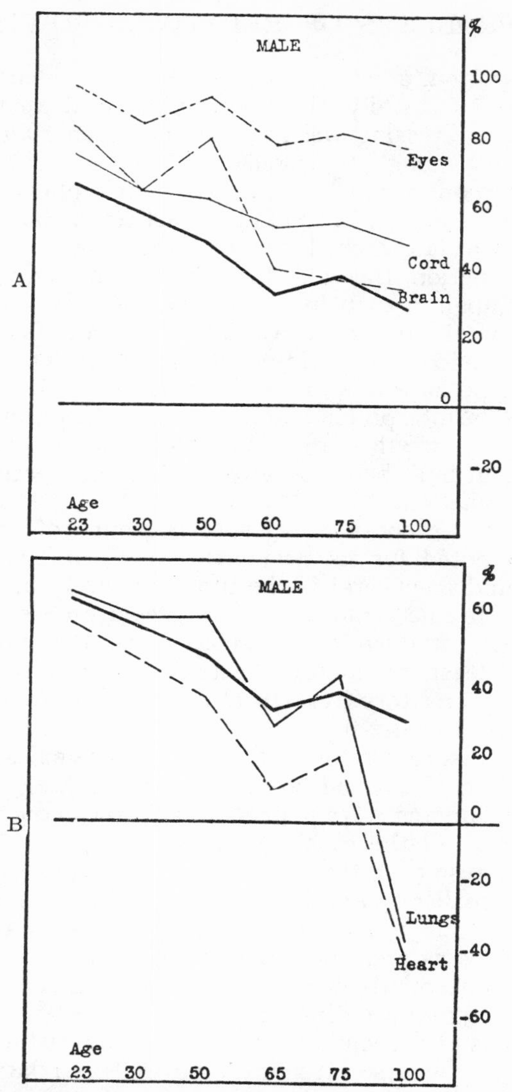

FIG. 82.—Differential effect of thyroid removal upon the growth of various organs in the male Albino rat.

The curves indicate the amount of growth made by an organ, represented as a percentage of the amount of growth made by the same organ in unoperated controls. For each organ there are 6 points, representing operations carried out at ages from 23 to 100 days; all the 6 series were allowed to grow until 150 days of age before final organ-size was determined. In all graphs the heavy line represents the curve for body-weight in thyroidectomized animals. In all cases the effect, relative to the control, is greater in animals operated later.

(A) Organs of resistant type: eyes, spinal cord, brain. All are less reduced than body-weight.

(B) Organs of sensitive type: heart, lungs. The heart in particular is more reduced than body-weight. In animals operated at 100 days of age, both heart and lungs actually lose weight (negative increment).

ance of function, as shown by the fact that young mammals can live fairly healthily when kept at constant weight or even, for a time, decreasing weight, and that planarians, etc., can keep healthy though enormously reduced in bulk by starvation. In organs where, to use Hammett's phrase, the *work-growth ratio* is high, as in glands, heart, etc., since a decrease of metabolism falls more heavily on the growth-function than the work-function, therefore the growth-function will be more seriously impaired than in organs such as C.N.S. or skeleton, where the work-growth ratio is low. And the first point, the greater influence of thyroid-deficiency on growth by cell-size increase, is perhaps due to a specific action of thyroid, though it too could be interpreted in terms of a greater effect of thyroid-depression on growth than on function or work, the cell which is growing in bulk being usually already differentiated for its definitive function.

Hammett also considers a ' special group ' of organs. The thymus is noted for its sensitivity to all unfavourable conditions ; and in general is the most adversely affected of all organs by thyroid removal. The testes are resistant because continuously producing new cells, while the ovaries, after puberty at least, are highly affected. The pituitary is known to show compensatory hypertrophy on thyroid removal, and in consequence *increases* its growth, notably in males (Fig. 83).

There is also a specific sex difference, females showing a greater effect of thyroidectomy as regards body-weight and almost all organs. And finally, puberty accentuates the sensitivity of all the sensitive organs.

The net result of thyroidectomy is the production of an absolutely smaller 150-day animal, but one which is abnormally slender, and has relatively larger eyes, testes and seminal vesicles, nervous and skeletal systems than a control of the same age, but relatively smaller viscera and glands ; further, the normal sexual size-difference is increased. When compared with animals of the same body-weight, the resistant groups of organs are, of course, absolutely as well as relatively larger (Hammett, 1929A).

Hammett further states that there is no correlation between the degree of sensitivity to thyroidectomy and the normal growth-rate (growth-coefficient) of organs. However, he has for the most part compared qualitatively different organs, in which the work-growth ratio, the cell-multiplication : cell-size-increase ratio, and the ratio of labile to stabile materials may readily obscure differences due to inherent growth-potential. To detect the latter, it would be necessary to compare the effects on different members or regions of a single organ system —e.g. different bones—which have different growth-ratios. This interesting study at least shows the complexity of the factors by which 'normal proportions' are determined.

Here we may cite the paper of Nevalonnyi and Podhradsky (1930), which does indicate a differential result of *excess* thyroid (and of excess thymus) on various parts of the skeleton of fowls. Unfortunately the thyroid results are based on only two experimental and four control animals, so that confirmation is required. It would appear, however, that on the whole there is a tendency for excess thyroid (in this dosage) to produce relatively thicker bones; further, to encourage the total growth of the femur, shoulder-girdle and hip-girdle, but to decrease the length-growth at least of the metacarpals and metatarsals; the radius, ulna and tibia occupying an intermediate position. This is evidence, so far as it goes, of a graded centrifugal effect. We may also note that Hammett found femur less sensitive to thyroid defect than humerus, and the Czech authors found it respond more by excess growth to thyroid excess.

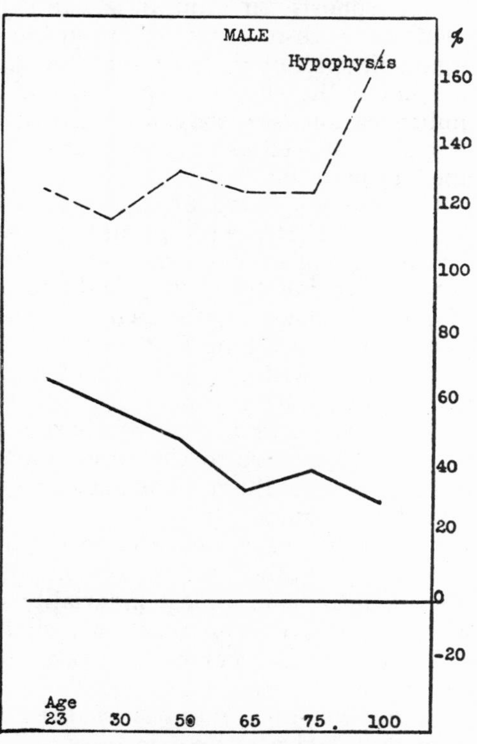

FIG. 83.—Effect of thyroidectomy on the growth of the hypophysis in the male Albino rat.

The graph is constructed as described in Fig. 82. The hypophysis is the only organ which increases more in thyroidectomized than in Control animals.

There is one point in Hammett's discussion which needs further consideration. He maintains that all the growth-effects of thyroidectomy are what we may call non-specific, due to its differential effects on different general kinds of materials or metabolic activities. That this need not always be so, however, is conclusively shown by the facts of Amphibian metamorphosis ; in Anuran larvae thyroidectomy has a much greater growth-retarding effect on the limbs than on any other organ, in spite of the fact that the growth of these, normally and under the effect of excess thyroid, is mainly due to cell-multiplication (see above). There is, in fact, the possibility of specific as well as of general differential effects of hormones upon growth.

This has been well brought out by Keith (1923), who has pointed out that the pre-pituitary affects growth differentially, acting most of all upon the extremities and on the parts connected with jaw-function. And similar specific sensitivity is, of course, abundantly shown in regard to the sex-hormones (see above). Cushing (1912) points out that hyperpituitarism is associated with relatively stout digits, hypopituitarism with relatively slender ones.

Although there are no sex-hormones in insects, reference may be made here to Champy's interesting discussion (1929, pp. 204 seqq.) of the relation between heterogony and secondary sexual characters. In various genera of beetles, in which a horn or other excrescence is normally present in males only, it occurs in both sexes in certain species. When this is so, the organ appears almost invariably to be highly heterogonic when restricted to one sex, only slightly heterogonic or even isogonic when present in both sexes (e.g. the beetle *Phanaeus lancifer* as against other species of the genus). In *Enema infundibulum*, and other beetles, the female possesses a cephalic horn, but lacks a thoracic horn : in the male, the latter is highly heterogonic, the former only slightly so.

The same phenomenon is found as regards the ' tails ' of the swallow-tail butterfly, *Papilio dardanus*. In the few sub-species where the sexes are similar, the ' tail ' is isogonic ; in the rest, it is heterogonic. He further states that the same phenomenon occurs in Chameleons, Sheep, Antelopes, and Deer (Reindeer, as against the other forms). Here, however, quantitative data are lacking, and the statement needs confirmation.

Champy correctly stresses the frequent relation between

heterogony of a character and its restriction to one sex. It will be clear, however, from our previous discussion that heterogony is a quite general phenomenon, and that it is merely very obvious in many secondary sexual characters because these are frequently of an exaggerated nature, demanding high growth-ratios for their development.

§ 5. Heterochely and Relative Growth-rates

Some suggestive facts concerning differential growth emerge from the experiments on heterochelous crustacea, which have been summarized and interestingly discussed by Przibram (1930). As is well known, in forms like Alphaeus, while amputation of the small or nipper claw is followed simply by the regeneration of a new nipper, with no change in the crusher, amputation of the crusher is followed by a reversal of the type of claw, the old nipper growing into the new crusher, the claw which regenerates in place of the old crusher growing into the new nipper. (This, it may be recalled, is closely parallel with the results of Zeleny (1905) on the polychaete Hydroides. This sedentary worm has paired opercula, of which one is large and functional, the other rudimentary. Amputation of the large operculum causes the rudiment to grow into a functional organ, while the amputated organ regenerates of rudimentary type.) In certain other heterochelous crustacea, such as the lobster (Homarus), it was long supposed that amputation of the crusher was not followed by this reversal. But later work has now shown that this only applies to later stages : if the operation is performed on quite young animals, reversal does occur. And it now seems to be a general rule that reversal will occur at small sizes in all heterochelous forms, although the size-range over which it occurs differs considerably from form to form. In some cases no reversal is possible after quite a young stage. In others, like Alpheus, it can occur throughout life. In both Homarus, in which the crusher (like the large chela of male Uca) may be either on the right- or left-hand side at random, and in the crabs Portunus and Eriphia, in which the crusher is normally always on the right, reversal can only take place before *visible* differentiation of claws has occurred. The result of amputation of future crusher in such cases is that the other claw (as in young Uca, where, however, both chelae are similar, of male type) becomes the crusher, while the amputated claw regenerates as a nipper.

Further, in practically all species, even in those where reversal will take place after visible heterochely has appeared, it will be either slowed down or totally prevented after the attainment of a certain quite small body-size. Above a cephalo-thorax length of about 1 cm., if the crusher be amputated, both chelae appear of nipper type at the next moult following amputation of the crusher; and in most such cases the regenerating claw usually turns into the crusher again after one or more further moults. (The only exceptions are hermit-crabs, in which asymmetry of the whole body, and consequently of the chelae, sets in during the larval development, and appears to be invariably determined by the start of post-larval life.)

Przibram ingeniously suggests that these facts are due to the ratio between rate of regenerative and of normal growth at the time of operation. Rate of regenerative growth is always high. We have little information as to its decline with total size, but the previously cited experiments of Zeleny on *Portunus sayi* (l. c.) show that there need be no falling-off. Probably rate of regenerative growth remains about constant where regeneration is possible at all. Normal growth, on the other hand, falls off very markedly with absolute size-increase. Thus the growth-ratio of regenerating crusher to normally-growing nipper will *increase* with increase of absolute size. There is, further, some evidence which suggests that a smaller organ (such as the nipper) can in the absence of competition from a corresponding larger organ (such as the crusher) draw on the blood for a greater amount of nutriment.

Przibram accordingly suggests that when the growth-ratio of regenerating crusher to normal nipper is relatively low, the impetus given to the nipper's growth by removal of the crusher can reach the threshold of growth-coefficient needed to produce a crusher before the regenerate can catch up, upon which the regenerate is then partially inhibited and must remain as a nipper; but when this growth-ratio is high, the regenerate can differentiate as a crusher at the next instar, because the amount of excess growth achieved by the nipper is relatively so low. An intermediate condition is seen when at the next instar both chelae regenerate as nippers, though the original nipper is still the larger; the nipper has not yet got sufficient impetus to differentiate into a crusher at the first instar; and although the growth-ratio of the regenerate has not been sufficiently high to reach crusher-type at this instar, its growth-ratio is still considerably higher than the old nipper, and at the next moult it will almost inevitably overhaul its partner (Fig. 84). This hypothesis fits the facts very prettily; it only remains to confirm it quantitatively by appropriate experiments. It should, however, be pointed out that in any case it only explains why one of two alternatives is realized on a given side of the body in each particular case: it gives us little light on the question of the nature and determination of the two alternative types themselves. The difference between crusher and nipper is doubtless in part one of absolute growth-coefficient; but as we have earlier seen, there are also differences between them in the spatial distribution of growth-potential, and qualitative differences in teeth, etc. The changing growth-ratio between regenerating and normal claw acts primarily as a 'realization factor', to use a German term, just as does the female sex-hormone of birds in effecting the difference in colour and form between male and female feathers. The determination of the difference in growth-gradients and in teeth is not in the least explained by this.

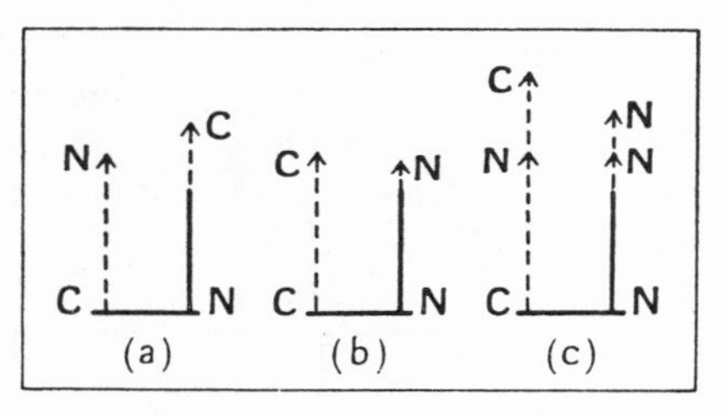

FIG. 84.—Diagram to illustrate Przibram's hypothesis as to reversal of chelae in heterochelous Crustacea.

C, crusher; N, nipper claw.
In each case the crusher has been removed, and its regeneration rate is assumed to remain constant in spite of age. Dotted lines indicate growth. The height of the lines is not supposed to indicate absolute size, but is adjusted so that greater height indicates capacity to develop into a crusher and to inhibit the other claw, causing it to become a nipper.
(*a*) In young animals the nipper grows relatively fast and can thus differentiate into a crusher at the next moult, causing the regenerating crusher to become a nipper; (*b*) in old animals the nipper grows relatively slowly; the regenerating claw can thus differentiate into a crusher again at the next moult, and the nipper remains a nipper; (*c*) at an intermediate stage, neither claw gets sufficient advantage to develop into a crusher by the first moult; but at the succeeding moult the old crusher will become a crusher again.

§ 6. SPECIFIC GROWTH-INTENSITIES AND THEIR INTERACTION

A series of highly important experiments bearing on our problems has been carried out by Harrison (1924, 1929), and later by Twitty and Schwind (1931), in grafting organs from one species of urodele larva to another which has a different normal growth-rate. The species used were *Amblystoma tigrinum*, which has a high growth-rate, and *A. punctatum* which, even when fed as fully as possible, grows at a considerably lower rate in the same conditions; and they then grafted limb-disks and eye-rudiments from one species to the other. The main important finding is that the organ retained their specific growth-rates even on the bodies of their new hosts, provided that conditions were equalized by keeping both species at the same nutritive level (which was most simply accomplished by giving both the maximum amount of food they would take). Under these conditions, the transplanted organ attained the same size which it would have had if left in the body of its own species; *tigrinum* organs grew just as fast as normal though on the slow-growing bodies of *punctatum*, *punctatum* organs just as slow as normal though on the fast-growing *tigrinum* bodies (Fig. 85). On the other hand, this inherent growth-

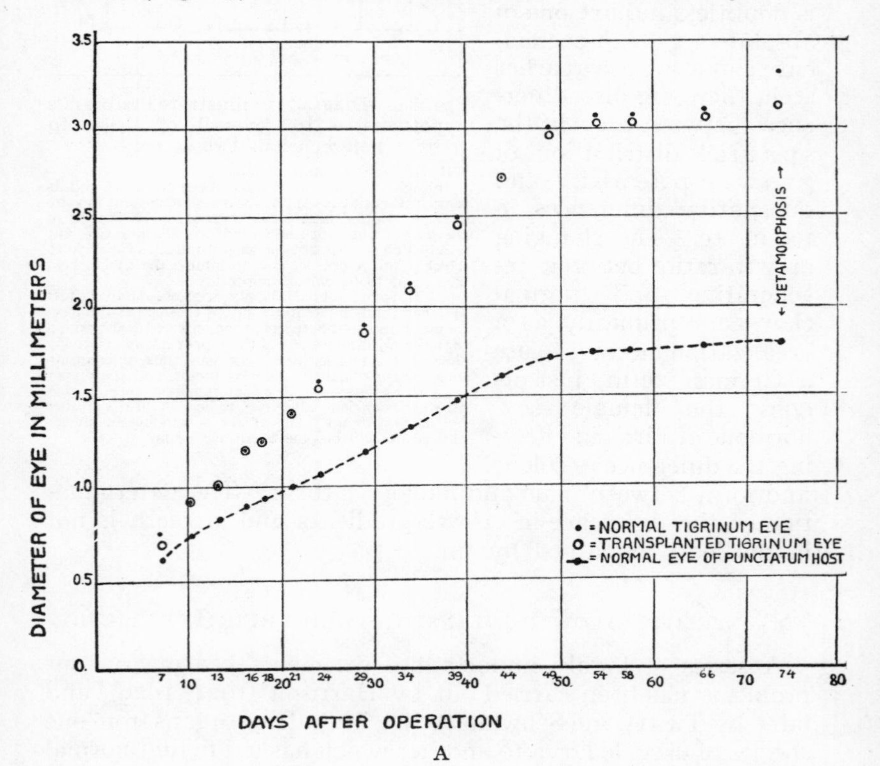

A

Fig. 85.—Graph showing the growth of normal

(A) Growth of *tigrinum* eye grafted on to *punctatum*, compared with that of a normal *tigrinum* eye, and that of the normal *punctatum* eye of the host.

capacity of the organ was related to the growth of the body (I am now interpreting the data in terms of the views put forward in this book) according to the law of constant coefficient of growth-partition (Figs. 31, 86). In other words, the amount of growth made by the organ was a function not only of its own higher or lower growth-potential, but also of the amount of growth made by the body, and this was equally true both for normal and transplanted organs. For instance, at first it appeared that *tigrinum* (rapid-growing) organs grew even faster when transplanted than when left in place, and Harrison put forward an elaborate hypothesis to account for this. It later turned out, however, that this was due to the *punctatum* hosts having been at an optimum nutritive level, while the *tigrinum* controls were not being given quite as much food as

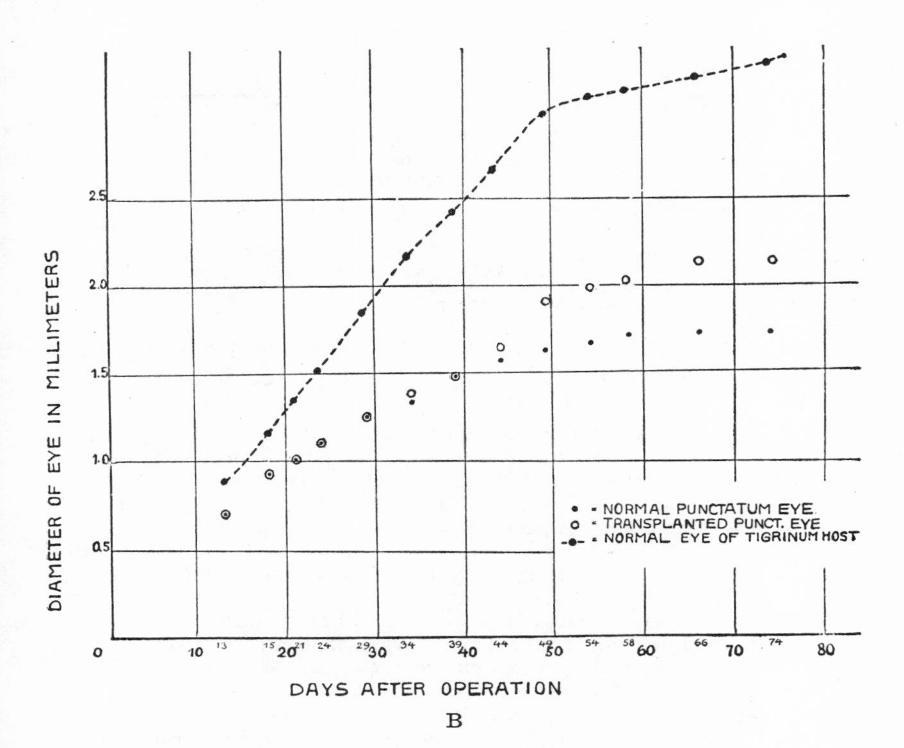

B

and grafted eyes of two species of Amblystoma.

(B) Reciprocal experiment. Growth of *punctatum* eye grafted on to *tigrinum*, compared with that of a normal *punctatum* eye, and that of the normal *tigrinum* host eye.

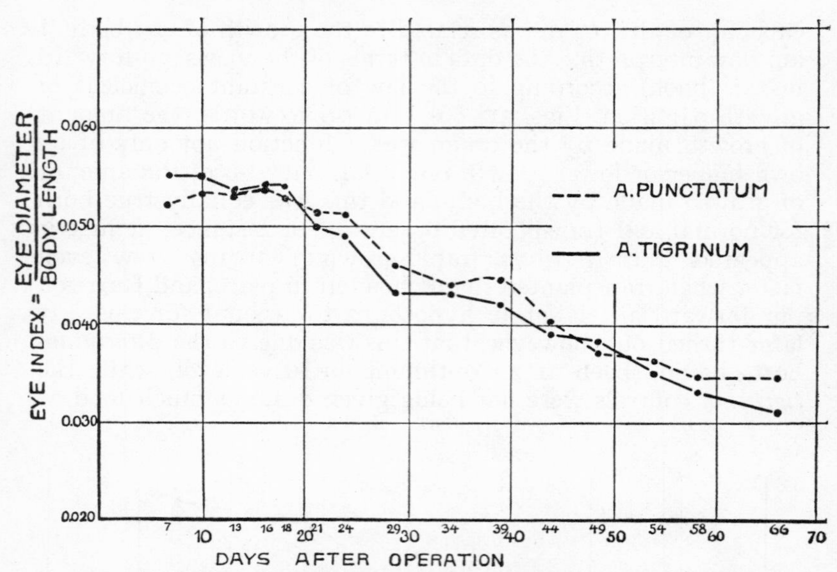

FIG. 86.—Similarity of relative growth of eye of two species of Amblystoma with different absolute growth-rates.

The curves are derived from the means from 5 specimens of each species, which were used in eye-grafting experiments. They represent the ratio between the diameter of the normal unoperated eye and the total length.

The ratios decrease (negative heterogony) but are closely similar for the two species in spite of the fact that during the period the mean-length of the *tigrinum* larvae had increased from about 14 to 110 mm., while that of the *punctatum* larve had only increased from about 13 to nearly 55 mm. The operation was carried out in the embryonic period.

they could eat.[1] By reversing the nutritive situation, the amount of growth made by the grafted limbs could be reduced

[1] The matter is actually not quite as simple as this. From the detailed results of Twitty and Schwind (1931) it appears that the correspondence between grafted and control limbs only remains perfect, even in maximally fed specimens, up to a certain time in larval life. After this, there is a slight decrease in the size of the large limbs grafted on to the small host, and a more considerable increase of the small limbs grafted on to the large host (Fig. 85)—an effect the reverse of that first posited by Harrison. It would thus appear that the size of the host gradually comes to influence the inherent growth-capacity of the graft, although this influence is always slight compared to the difference in the inherent growth-capacity of the organs of the two species.

Further, Detwiler (1930) and Severinghaus (1930) claim that the host may in some cases exert some effect upon the inherent growth-rate of the graft in early stages before feeding begins. But whatever the extent of these minor effects, the specific growth-rate of the organ is the *main* factor in determining its relative growth-rate when transplanted.

below that made by the controls left in place. Thus we clearly see that the relative size of any organ is due to two variables —its inherent growth-capacity, and a regulatory partition-coefficient which allots material as between organ and rest-of-body.[1] We could have deduced this from our studies on the chelae of Uca, etc., but this work clinches the matter decisively by means of experimental tests.

This work of Harrison's has also shed important light on growth-gradients. In addition to grafting whole eyes, he made separate inter-specific grafts (*a*) of optic vesicle and (*b*) of lens-producing epithelium (Harrison, 1929B). Eyes were thus produced which contained rapid-growing optic cup and slow-growing lens, and others with slow-growing optic cup and fast-growing lens; and either sort in both kinds of host. In every case, there was a mutual influence of the components upon each other. The growth of the eyes as a whole was of intermediate rate, that of the rapid component was slowed down, that of the slow component speeded up.[2]

Fig. 87, constructed from Harrison's data, shows the results quantitatively. The association of a fast-growing optic vesicle with a slow-growing lens retards the growth of the optic vesicle by 22 per cent. in a *punctatum* body (No. 2 in Fig. 87: ratio 1·26 instead of 1·61); and by 17 per cent. in a *tigrinum* body (No. 4: ratio 0·83 instead of 1·0); similarly, it accelerates the growth of the lens by 13 per cent. in a *punctatum* body (No. 9) and by 6 per cent. in a *tigrinum* body (No. 11). The reverse

[1] It is interesting to note that Twitty and Schwind (l. c.) in their latest paper, published after this chapter was first written, also adopt the phrase *partition-coefficient of growth*. They, however, think of the partition-coefficient as a *percentage* ratio, and therefore, since the ratio eye-diameter : body-length decreases considerably during larval life, find it necessary to postulate a progressive change in the coefficient. If our conception be adopted of a partition-coefficient which is an *exponent* of body-size, this difficulty disappears, and the coefficient can be considered to remain constant throughout growth.

[2] That such interaction of neighbouring parts as regards their growth-intensity need not always occur is shown by the recent work of Schwind (1931), who made heteroplastic grafts of parts of the shoulder-girdle between the same two species. He found that the presence of a grafted coracoid region growing at a different rate from the host's organs had no effect on the growth of the host's scapular region. Thus we have here an example of mosaic growth-rates, as opposed to mutual regulation, which is perhaps comparable to the mosaic differentiation of some developing eggs as opposed to the regulatory capacities of others.

association, of a slow-growing optic vesicle with a fast-growing lens, accelerates the growth of the optic vesicle in a *tigrinum* body by 25 per cent. (No. 5), and by 22 per cent. in a *punctatum* body (No. 3). And it retards the growth of the lens by 20 per cent. in a *punctatum* body (No. 9), by 21 per cent. in a *tigrinum* body (No. 10). The regulation is a gradual one, as shown by Harrison's Fig. 37.

Further, Harrison gives conclusive proofs that the regulation is due to chemical and not in any appreciable degree to merely mechanical causes. He also finds some regulation of a similar sort in other tissues in the proximity of the eyes.

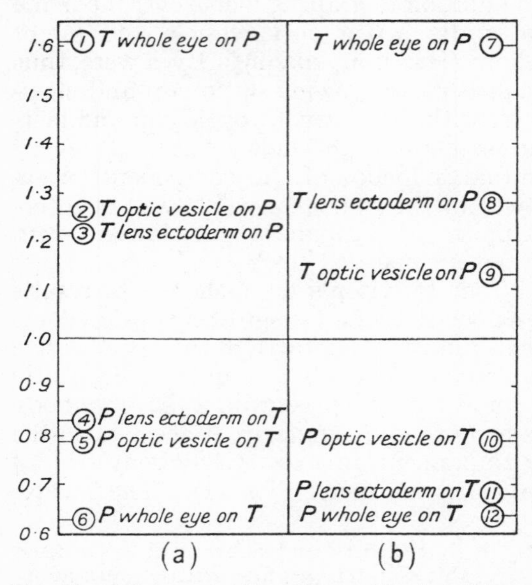

FIG. 87.—Diagram to show the results of grafting whole eyes and their parts between *Amblystoma tigrinum* (T) and *punctatum* (P).

The ordinates give the ratios of the linear dimensions of (*a*) the optic bulb and (*b*) the lens on the side receiving the graft to those of the intact eye of the other side (see text for details).

(*Constructed from Table II in Harrison,* 1929B.)

We could interpret these results in various ways. Perhaps both rapid-growing and slow-growing organs produce specific substances which interact with each other. Perhaps there is only one type of substance, a positive or growth-promoting substance; in such case, when there is a deficiency of this in, e.g., the slow-growing lens, there will be a greater difference in concentration of this substance between optic vesicle and lens, therefore more diffusion of the substance left in the optic vesicle and accordingly less rapid growth of the opt c vesicle. This latter suggestion would be in harmony not only with the facts as to growth-centres and growth-gradients here set forth, but also with recent work on plant growth, e.g. by Dolk, who finds a specific growth-hormone with quantitative action. On this, however, only further experiment can decide; but the results are clearly of importance in that they show experimentally that, as we have deduced from mere measurements, rapidly-growing organs may exert an influence upon the growth of neighbouring organs.

§ 7. The Influence of External Conditions

We next have to consider the effect upon growth-gradients and heterogonic organs in general of the natural or experimental alteration of conditions.

The heteroplastic experiments with *Amblystoma* eyes which we have just been discussing provide us with some facts bearing on this problem. In Chapter II reference was made to Twitty's experiments in grafting on to *punctatum* host-larvae, *tigrinum* eyes of the same size but less physiological age than the host's eyes.[1] He found that when the hosts were starved, their body-length (five animals, over an average of thirty-five days) decreased by 18 per cent., their own eyes remained unchanged in diameter, and the grafted eyes increased very slightly, by 2·5 per cent. When, however, they were fed just more than enough to maintain their size, their body-length (four animals over an average of thirty-one days) increased by 7 per cent., their own eyes by practically the same amount (6·5 per cent.), but the grafted eyes by 40·5 per cent. One of these specimens showed an increase of 50 per cent. in the diameter of the grafted eye, but without any change in body-length. It is thus clear that competition for food plays an important part in the regulation of relative size of parts. The rapidly growing eye, whether rapidly growing by virtue of its youth or its higher specific growth-intensity, can draw disproportionately upon the supply of food available (for further discussion of this general problem, see Jackson, 1925, and Huxley, 1921). It is clear that much valuable work would be possible in determining the quantitative growth-relations, in varying nutritive conditions, of heteroplastic grafts of the same and of different ages, and of homoplastic grafts of different age from their host. (Fig. 31.)

We have already considered the effect of poor nutritive conditions upon the forceps of male Forficula (Chapter II) and upon the dorsal crest of male newts (Chapter VI). Perkins (1929) has analysed the effect of the parasite Sacculina upon

[1] A similar regulation of rate of development was found by Choi (1931), who grafted lateral half-larvae of two species of frog together.

the growth-gradient in Carcinus. He finds that the effect of the parasite is primarily to reduce general growth of the crab

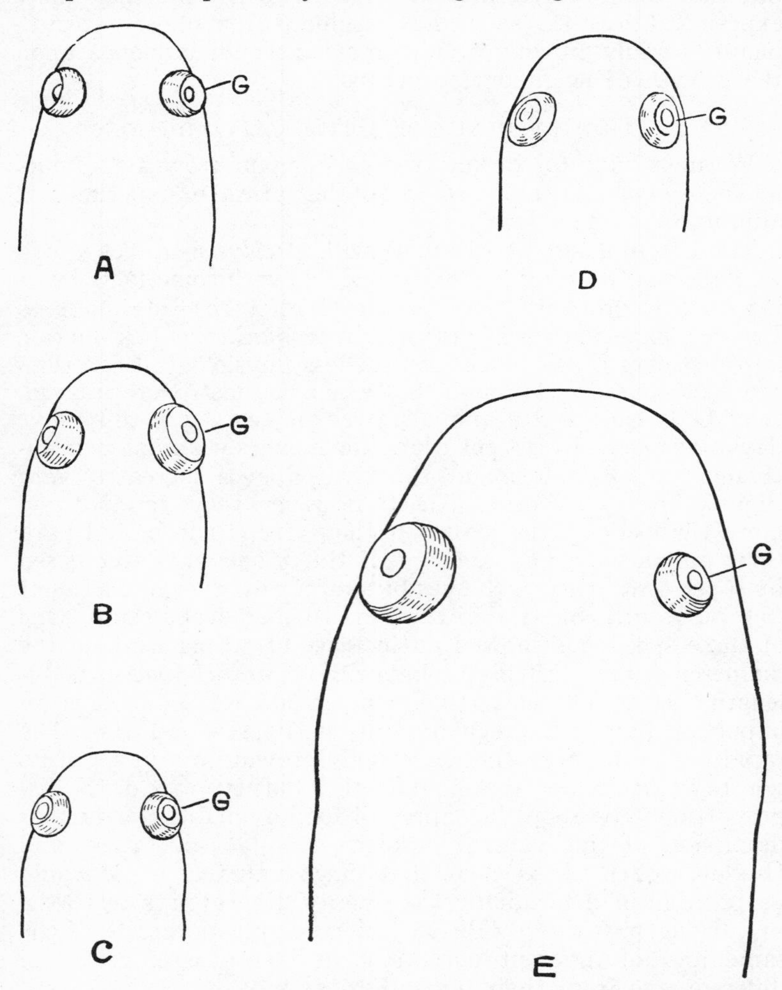

FIG. 88.—Regulation of growth of eyes of one species of Amblystoma grafted to individuals of another species and of different age. All × 6.

G, grafted eyes.
(A) A specimen of *A. punctatum* with an eye from a younger specimen of the larger species, *A. tigrinum*.
(B) The same 40 days later, after feeding on maintenance diet; the grafted eye has enlarged.
(C) A similar specimen to (B), but starved since the operation; the grafted eye has not grown.
(D) *A. tigrinum*, with an eye from an older *A. punctatum*.
(E) The same specimen 45 days later, after normal feeding. The grafted eye has lagged in its growth.

by acting as a new and a very active 'growth-centre' which, of course, enters into competition for food-material with the organs of the crab's own body. Secondly, however, he finds that this effect is not uniform on all the organs. It is greatest in those of most rapid inherent growth-potential (highest growth-coefficient). Thirdly, he believes that normally excess of growth of one part is associated with a 'drainage effect' upon the growth of other parts, and that when the growth of the organs with high growth-coefficient is inhibited, as in Sacculina, there is a releasing effect on those of low growth-coefficient leading to an actual increase in their relative size. It is by this means that he would account for the well-known fact of the increase in the abdomen-width (approximation to female type) in sacculinized male Crustacea.

This third conclusion of his does not appear to be well founded. It is in disagreement with the correlation we have found between the presence of a highly heterogonic organ (e.g. chela) and excess growth in the appendages posterior to it; and with the important fact discovered by G. Smith (1906A), to which sufficient attention has not been paid, that parasitized female crabs, even if Perkins be right in supposing that their general growth is checked, show *accelerated* development of the female secondary sexual characters of their abdomen. This last point, together with the fact that in males which have recovered from the parasite, the gonad develops ova, seems to me conclusive proof that the effect of the parasite here is in major part a specifically sexual one, and not merely a quantitative growth-effect. Be this as it may, Perkins has shown that there does exist such a quantitative growth-effect of Sacculina upon its host, and that in regard even to appendages which show no obvious secondary sexual differences, it exerts a differential and graded influence; and this opens the door to all kinds of future work.[1]

Another influence upon differential growth is shown by the work of Thiel (1926) upon the Lamellibranch mollusc Sphaerium. He shows that in clean water with high oxygen content, the form of the shell becomes more rounded than in the dirty water of its normal habitat. This, however,

[1] In this connexion the specific and differential morphogenetic effect of the parasitic worm Mermis upon its ant hosts should be noted, although for the moment its relation to the growth-gradients of the ant-body cannot be clearly envisaged. See Wheeler (1928), Vandel (1930).

appears to be mainly or solely a mechanical effect due to an alteration in the direction rather than the amount of growth of the mantle-edge, and to be determined by an increase in the number of embryos contained in the gills.

Similarly, Huntsman (1921) found that light not only markedly retarded the growth of mussels (*Mytilus*), but also affected their shape. Although the depth of the shell relative to length was not altered, the relative breadth was increased from 42 to 46 per cent ; as result, the ratio of weight to the product of $(l \times d \times br)$ of course increased. The light appeared to inhibit the growth of the edge of the mantle, which is responsible for additions in length and in depth. Relative breadth, on the other hand, again depends on the angle at which the mantle-edge is growing.

Entz (1927), in an exhaustive study on Peridinians, has shown that the relative length of the horns is markedly altered by salinity, while the size of the body is little affected. In different external conditions, the antapical horn may vary in the ratio of 4 to 1 ; the post-equatorial horn in the ratio of 5 or 6 to 1 ; and the other two horns from zero to a considerable development. As Entz puts it, the relative size of the horns renders small differences of salinity visible in an obvious and exaggerated form. (See also Gajewski, p. 259.)

What again appears to be a comparatively simple influence is that revealed in what Przibram (1925) calls the 'Tail-thermometer' of rats and mice—namely, the fact that the relative tail-length of these rodents (pp. 22, 40) increases with increase of the external temperature in which they grow up. The relationship is an approximately linear one. If relative tail-length be expressed as *trunk length* per cent, the value of this index of tail-length (for young Albino rats) decreases by 0·75 for each rise of 1° C. in external temperature. He has also shown that this is correlated with an actual increase of internal temperature (0·2° C. for each 5° of external increase) ; and further, that this is really the important agency in producing the effect. We may therefore conclude that the effect is due in part to a differential temperature-coefficient for tail-growth. On the other hand, the fact that the tail (and other organs such as ears and limbs, which also show a similar effect with temperature) have a relatively large surface, and therefore a lower temperature than the rest of the body (cf. Crew's and Moore's proof of this fact for the mammalian scrotum), doubtless enters into the picture. Higher internal temperatures, especially with higher external temperatures, would reduce the temperature-difference between tail and body, and would therefore accelerate the tail's *relative* growth.[1]

Information perhaps more important, and certainly more directly correlated with Perkins's results, is afforded by our knowledge of the differential effect upon bodily proportions produced by malnutrition, by hormones, and by teratological embryonic agencies.

Jackson (1925) has published interesting data concerning the effect of malnutrition upon differential growth in rodents. An important fact is that in young animals held at constant body-weight, certain organs, notably the skeleton, continue growth, and continue on the whole in the same differential way as in the normally-fed animal—a pretty demonstration of the existence of inherent growth-potentials and their variation in different parts of the same organ-system.

The actual details of the process vary according to the age at which the animals are subjected to the treatment. (In acute inanition leading to actual weight-loss, the changes, as expected, are different in degree.)

In albino rats kept at maintenance from three weeks of age, the tail actually grows in absolute length, and considerably in relative length. The absolute head-weight is slightly increased. When the experiment started at birth, the increase in head-weight was much more marked. The brain and the testes are other organs which show increase of weight when growing animals are kept at constant weight, especially in the new-born (Fig. 89). Full details are given by Jackson.

Hammond (1928, 1930, 1931) has also some interesting remarks on the effect of undernutrition at various periods upon relative growth in mammals (sheep, rabbit, etc.). In an adult, the loss of weight occurs first mainly from fat, then mainly from muscle, last of all the bones and the more ' vital ' organs, such as heart, lungs, etc.

In a young growing animal the vital organs appear to have first call on nutritive material: during moderate undernutrition they thus continue their growth more or less unaffected, while that of muscles (and fat) is markedly inhibited: since muscles show positive heterogony in juvenile life, the animal will retain juvenile proportions (and will, of course, not be so valuable as a carcass).

In general, when organs have their period of maximum absolute growth at different times (from whatever physio-

[1] See also Abe (1931), *Endokrinol.*, **9**, and Przibram, ibid., 110, for effect of internal temperature upon relative cell-size in rats.

logical cause), ample nutrition at the period of maximum growth, followed or preceded by undernutrition at other periods, will produce an animal with the organ of unusually large relative size ; e.g., in rabbits, the ears have an early period of maximum growth : accordingly maximal feeding when young, when followed by low feeding later, will produce a rabbit with ears above the normal proportionate size.

Hammond further points out that since in meat animals the maximum percentage of flesh is only produced by a high level of feeding throughout most of the growth-period, animals cannot show their full potentialities as meat-producers if kept in unfavourable nutritive conditions. When practising genetic selection for meat qualities, it is therefore necessary to provide optimum conditions for individual development in order to be able to pick out extreme variants. Conversely, if a breed is required for a poor environment, selection should be practised with sub-optimal nutrition, of the type to be expected in this environment.

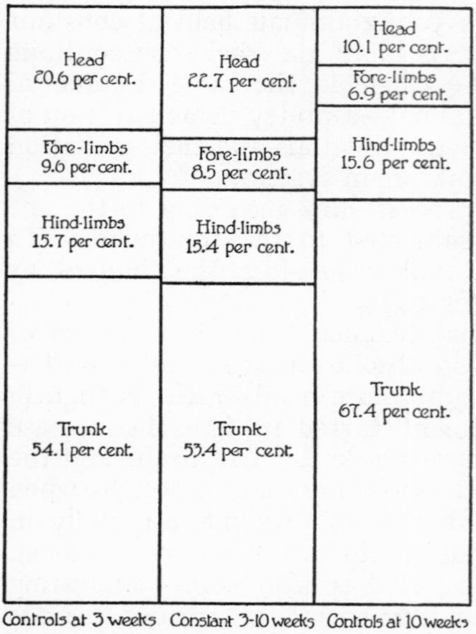

Head 20.6 per cent.
Fore-limbs 9.6 per cent.
Hind-limbs 15.7 per cent.
Trunk 54.1 per cent.
Head 22.7 per cent.
Fore-limbs 8.5 per cent.
Hind-limbs 15.4 per cent.
Trunk 53.4 per cent.
Head 10.1 per cent.
Fore-limbs 6.9 per cent.
Hind-limbs 15.6 per cent.
Trunk 67.4 per cent.
Controls at 3 weeks Constant 3-10 weeks Controls at 10 weeks

FIG. 89.—Diagram showing changes produced in relative weight of various parts of the body in young rats held at constant weight for 7 weeks by under-feeding (centre) as compared with controls at the beginning (left) and end (right) of the experiment.

In regard to plants, numerous examples are known of external conditions affecting proportions. We may cite two papers of Pearsall and Hanby (1925 and 1926). In the former they showed that in Potamogeton leaves excess calcium causes a predominance of growth in breadth, excess potassium a predominance of growth in length. This was due partly to the greater relative length of the individual cells in the Ca cultures, partly to the arrangement of cells in the young leaf-initials. In the latter paper they found that reduced hydrostatic pressure causes palmate leaves to become not only smaller, but more dissected and with relatively smaller basal lobes. However, as growth in plants clearly follows very different laws from growth in animals, we will not attempt to enter into any detail.

§ 8. Rhythmical Irregularities of Growth-ratio

Finally, we must just mention a curious but possibly important fact, namely the tendency of structures to grow by abnormally high growth first in one dimension of space, then in another, so that the growth-ratio for breadth on length does not remain constant, but fluctuates more or less rhythmically round a mean. Przibram (1902) has pointed this out in regard to the carapace of crabs. In these he followed individuals through several moults, so he could be certain of the fluctuation in proportions.

I. W. Wilder (1924) finds a similar alternation, of periods of filling out and periods of rapid length-growth with drop in the body-build index, in the larval growth of the salamander Eurycea bilineata, and the same phenomenon is stressed for human children by various authors, notably Bean (1924), Harris (1931 A and B), and with all the apparatus of biometrical method by Berkson (1929)[1]. Thiel (1926), as the result of careful measurements on the skull of the bivalve mollusc Sphaerium, finds an alternation of periods of increased intensity of growth in height and decreased intensity of growth in breadth with periods where the reverse relation holds. Coghill (1928) finds evidences of the same sort of thing for definite centres within the embryonic nervous system of amphibia. (He also finds evidence that this periodicity alternates in closely adjacent centres.) And various authors have emphasized that the same phenomenon occurs in regard to the post-natal length- and breadth-growth of the human cranium. Thus the phenomenon would appear to be general. This again emphasizes the need for large bodies of observations on proportionate growth; but its physiological basis is at present quite unexplained.

This chapter, as I pointed out in the introduction to it, has inevitably been both discursive and inconclusive. However, the facts and ideas set forth in it may serve to point the way to more crucial and more radical methods of analysing the problem.

[1] See also the recent paper of Davenport (1931), *Proc. Amer. Philos. Soc.*, **70**, 381.

CHAPTER VII

BEARINGS OF THE STUDY OF RELATIVE GROWTH ON OTHER BRANCHES OF BIOLOGY

IT will be clear from preceding chapters that the study of relative growth has important bearings upon many other branches of biology. In the present chapter I will attempt to summarize a few of these as succinctly as possible. We will begin with its bearing upon systematics.

§ 1. Heterogony and Taxonomy: Sub-species and Taxonomic Forms

Systematists are agreed that mere size-differences may have no taxonomic significance, since they are often the direct effects of environmental conditions. But they usually attach much greater importance to differences in the *percentage* size of parts. Our studies of heterogony, however, make it obvious that such differences may have precisely as little taxonomic significance as those in absolute size, for wherever an organ is heterogonic, differences in absolute total size of body will bring about differences in relative size of the part.

When such problems crop up, they require analysis along the following lines:

(1) Is there a difference in total absolute size between the varieties considered?

(2) If so, is the difference due (*a*) wholly to environmental differences, (*b*) wholly to genetic differences, or (*c*) to a combination of the two?

(3) Is there a difference in the relative size of any parts or organs?

(4) If so, is there also a difference in total absolute size?

(*a*) If not, heterogony is not at work, and the difference is genetic and is of taxonomic value.

(*b*) If yes, then do the relations of relative size of part to absolute size of whole obey the simple heterogony law?

(i) If not, then some at least of the difference in relative size (percentage proportions) is *prima facie* of independent genetic origin, and therefore of taxonomic significance.

(ii) If yes, then *prima facie* the differences in proportions are merely secondary effects of the difference in absolute size, and therefore not of taxonomic significance.[1]

(5) While measurement and simple mathematical analysis will thus give important *prima facie* evidence, it is of course desirable to apply experimental tests wherever possible to clinch the matter.

A few examples will illustrate my meaning.

(1) The Red Deer (Cervus elaphus) of Scotland rarely exceed 125 kg. in (clean) body-weight, and rarely show more than twelve points on their antlers. In some localities, the usual run of body-weight is only 75–90 kg., with six to eight points. In some districts of Europe, however (e.g. the Carpathians), stags run to double the maximum Scottish body-weight, with twenty, twenty-five or more points, and there appear to be no continental districts in which the mean weight or point-number is as low as in Scotland. Further, the relative size (weight) of the antlers is clearly much higher in the larger continental stags (Huxley, 1931A).

It would be natural to consider the low size and antler development of the Scottish deer as constituting criteria of a true geographical variety or sub-species, as has also been done with some island races, which are also small in both these respects.

The problem is complicated, however, by the fact that in the peat-bogs of Scotland are found the skeletons of deer rivalling the biggest existing European specimens in size and point-number; and that these date back only a few thousand years (Ritchie, 1920); Fig. 90. It would seem difficult to imagine that true evolutionary (genetic) change reducing the body-weight by half could have occurred in such a short space of time. It is true that persistent killing for trophies of the biggest stags, with the best heads, would have had a genetic effect, but this could only have been intensively operative for a very few centuries at most; and in any case the same process should have been at work on the Continent. Ritchie himself

[1] See Champy (1929, p. 239) where the same problem is discussed along somewhat similar lines.

(l. c.) ascribes the change to the altered environment produced by the cutting down of the forests in Scotland, and the giving over of many of the original haunts of the deer to agriculture. The Red Deer is by nature a forest species; driven from the lowlands, and when deprived of the shelter and succulent browsing provided by forests, it grows stunted. This, however, takes no account of the difference in relative size of antlers between Scotch and e.g. Carpathian strains. Here, however, analysis of the available weight-measurements

FIG. 90.—Differences in absolute body-size and relative antler-size of prehistoric (left) and modern (right) Scottish Red Deer.

(Huxley, l. c.) reveals, as we have seen (p. 42), that in spite of large individual variations, the relative weight of antlers increases regularly with absolute weight of body (Figs. 25, 29). The antlers thus constitute a heterogonic organ, and we should *expect* stunting to be followed by a diminution in relative antler-size. Furthermore, the number of points on the antler is also found to be a function of absolute antler-size (Huxley, 1926, 1931A). Here again there is considerable individual variation, but the means show a regular curve. It would appear to be an automatic consequence of increase of antler-bulk that the rapidly growing bone-material should branch more profusely: the same phenomenon is found, e.g., in the antler-like mandibles of stag-beetles (Lucanidae) and the horns of other beetles (see Fig. 93).

We thus arrive at presumptive, *prima facie* evidence for the truth of Ritchie's view. Finally comes experiment, and settles the question. During the last hundred years, a number of Red deer from Scotland (and elsewhere) have been imported into New Zealand and there liberated. The regions where they were liberated were profusely forest-covered; and the stags of Scottish strain in this new environment attained body-weights of 200 kg. and over, with relatively large antlers showing twenty and more points: in other words, they bridged most of the differences between the Scottish and Carpathian strains in one generation. Still further evidence is provided by later events. Lacking natural enemies, the herds of deer in New Zealand became too numerous, and began to destroy the forests; with overcrowding and less favourable food conditions, 'degeneration' set in, and now in many regions the mean and the maximum body-weight and point-number are far below what they were a few decades back. See Thomson (1922) and Huxley (1931A).

We are thus perfectly justified in concluding (1) that the bulk of the observable differences in size and proportions between the existing Scottish strain of Red deer and (*a*) the existing Carpathian strain and (*b*) the sub-fossil Scottish type are non-genetic and of no taxonomic significance. (2) However, we must admit that killing of the finest males is likely to have had some genetic effect; this will have been exerted on both strains, but may have been more intensive in Scotland. In any case, its primary effect will have been to reduce the mean size of both races, the effect on relative antler-size and point-number being consequential. It should further have had a 'dysgenic' selective effect, in that among stags of equal body-weight, more of those with genetic tendency to relatively larger antlers will on the average have been killed off. (3) There may also be true genetic differences between the Scottish and Carpathian strains. But (4) the differences due to (2) and (3) are relatively small, and the whole problem must be studied afresh before we can be sure what they are, and indeed, as regards those under (3), whether they exist at all. As corollary, we may take it that differences in proportional antler-size in dwarf island races of deer are probably only secondary effects of differences in absolute size, and that when an island race is smaller, its small size is probably due in the main to environmental stunting. This is supported by the findings of Antonius (1920), who shows that in the dwarf race of deer on Sardinia, the length of the face relative to the cranium is reduced. Since the face is positively heterogonic, this again is to be expected if the animal is stunted. The stunting has produced a strain which, because dwarfed, has permanently juvenile proportions. See also Klatt (1913) on dog skulls.[1]

I have spent some time on this case, as its analysis is fairly complete. We may presume that similar differences in relative antler-size found in geographical races of Roe-deer are secondary, though here the differences in absolute size (which on the whole grades upwards from west to east across Europe) are themselves more likely to be genetic, since the type of habitat of the Roe-deer is approximately the same over its whole range.

It is equally clear that many other percentage measurements —e.g. of chela-size or abdomen-size in crabs—will in themselves be of no systematic value. What *are* taxonomically distinctive in such organs are their growth-constants—the values of the growth-ratio and of the partition-constant in the heterogony formula, and the absolute body-sizes at which heterogony (if it is not continuous or uniform) begins, ends or alters. These numerical values, if properly established, would be true taxonomic characters; since, however, they are far less convenient to ascertain (though probably of much greater biological importance) than the usual type of specific difference, which is selected for its obviousness and ease of recognition, they are not likely to be much employed.

A peculiarly interesting example, constituting in a sense a test-case, is provided by the Lucanid beetle Cyclommatus tarandus, with markedly heterogonic male mandibles. This has been studied by Dudich (1923), whose measurements and conclusions have been further analysed by me (1927, 1931c). Coleopterists have distinguished a number of forms of the male of this stag-beetle—five main types, or if we include sub-types, no fewer than seven, all based on structural characteristics of the mandible. (See also Griffini, 1919, *Natura*, **10**, 13.)

They are as follows (I omit the sub-types):

[1] A slightly different case of retention of juvenile proportions is given by Drennan (1932) who discusses the skull of the pre-Bushman race of South Africa. This appears more akin to paedomorphosis (p. 239).

(1) *Prionodont.* Mandible beset with inwardly-directed teeth throughout its length (2 sub-forms).

(2) *Amphiodont.* A toothless gap on the mandible, separating 'prebasal' and 'sub-apical' teeth.

(3) *Telodont.* Prebasal teeth lacking.

(4) *Mesodont.* With a new, single 'submedian' tooth, on the ventral side (2 sub-forms).

(5) *Mesamphiodont.* Numerous teeth in homologous position to sub-median tooth of (4); and also a set of teeth in position of pre-basals of (2). (Fig. 91.)

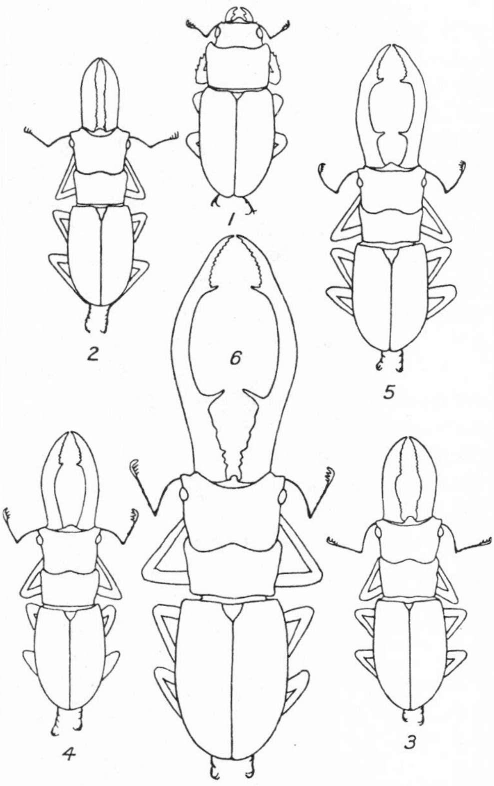

FIG. 91.—(1) Female and (2—6) five different-sized males of the stag-beetle, *Cyclommatus tarandus*, to scale, to show the change in form and relative size of the male mandible with increase of absolute body-size.

Male forms: 2, prionodont; 3, amphiodont; 4, telodont; 5, mesodont; 6, mesamphiodont.

These types grade into each other imperceptibly, and the distinctions are admittedly arbitrary: e.g. a specimen is classified as Amphiodont if a gap is present which is considered larger than the normal gap between two teeth, Prionodont if it is considered not to exceed this size. Furthermore, there is a correlation between tooth-type, mandible-size, and body-size, the five types, in the order listed above, succeeding each other with increase of size; and even within each 'form' there is a trend, with increasing size, away from the type of the previous form and towards that of the one succeeding.

It might be at once concluded, one would think, that all

the 'forms' are purely arbitrary growth-types, the heterogonic increase of mandible-size with total size being accompanied, as often, by changes in tooth-characters and other morphological alterations. But Dudich states that he is forced to regard them as in some way 'real' and of taxonomic importance, for the following reasons: (1) Because the several 'forms' are transgressive as regards body-length, the smallest of form 3, for instance, being smaller than the largest of form 2. (2) Because, although the frequency-curve for female body-length is unimodal and approximates to normal type, that for the body-length of the males is irregular, multimodal, and skew. He attaches more importance to (1).

Analysis (Huxley, 1931c) of his figures by body-size (178 specimens grouped into 28 classes) and a plotting of the resultant means showed that when all the 'forms' were lumped together, a clear approximation to the simple heterogony formula (constant growth-partition coefficient of male mandible) was found (with the deviation from it at highest absolute sizes which we previously established as usual in holometabolous insects: see Chapter II). When however the 'forms' were taken separately, and each subdivided into size-classes, the resultant plot resembled a series of overlapping tiles on a roof, thus giving graphic proof of Dudich's contention that they are transgressive as regards body-size (Figs. 35, 92).

The type of overlapping, however, at once gives the key to the situation. (1) What we have plotted are the means. (2) But we should expect a definite range of variation in the relative size of mandible for given body-size, whether this be due to variations in growth-partition constant (k) or fractional constant (b). (3) Over the range of size where two 'forms' overlap (as elsewhere, of course) we should accordingly expect to find some with mandible-size well above the mean for that particular body-size, others with mandible-size well below that mean. (4) But it is a general rule that increasing absolute size of many organs (other Lucanid mandibles—see especially the paper by Griffini, 1912—deer antlers, etc.) is correlated with morphological changes (Fig. 93). (5) We should therefore expect that, over each region of the size-range where overlap occurs between two 'forms', individuals which vary in the direction of high mandible-size should on the whole have the tooth-characters of the form above, those which are low variants as regards mandible-size should on the whole have the tooth-characters of the lower form. (6) The trans-

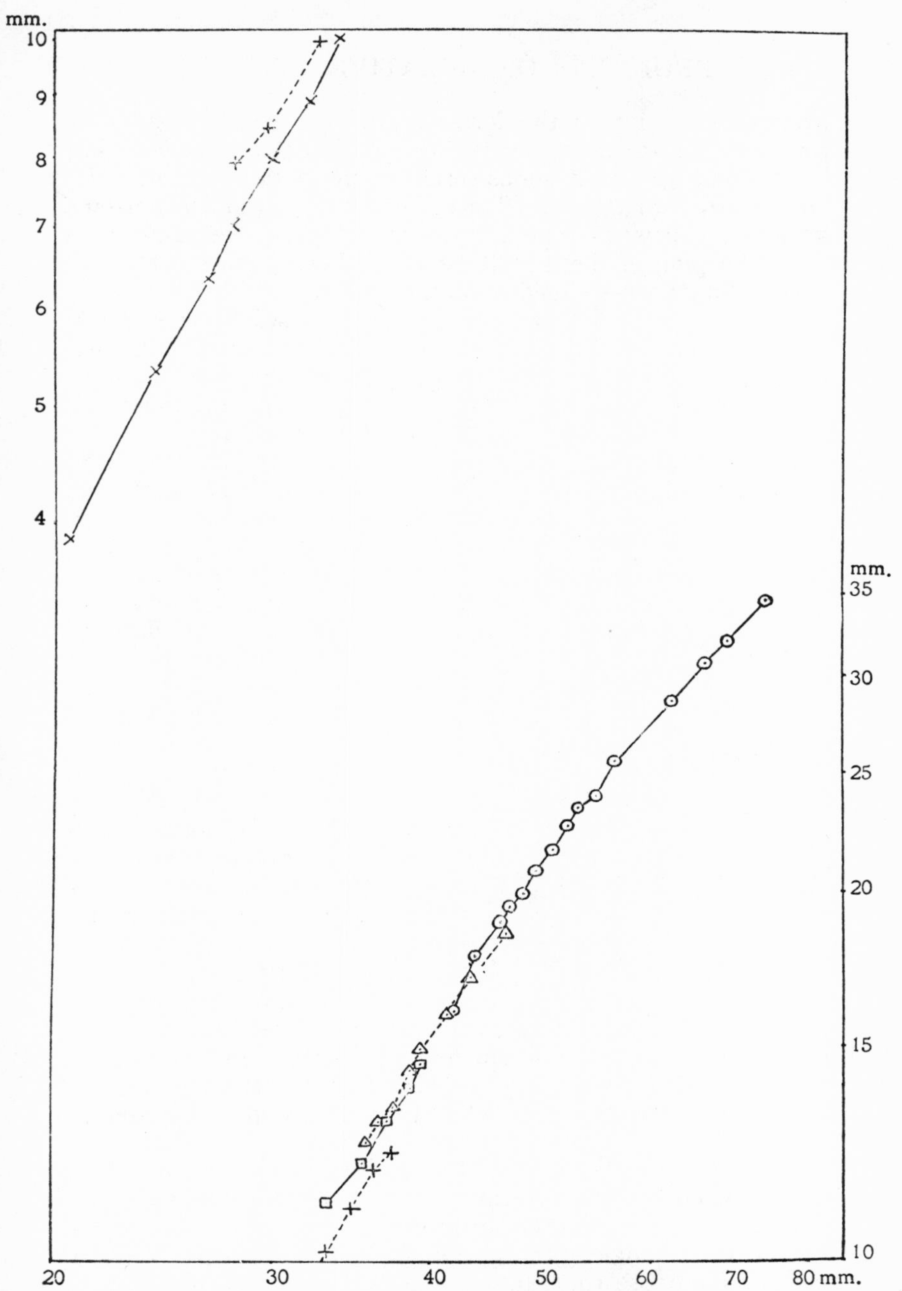

FIG. 92.—Graph to show relative growth of the mandible in the different male forms of the stag-beetle, *Cyclommatus tarandus*.

Mean mandible-length (mm.) has been plotted logarithmically against mean body-length (mm.) for various size-classes of each of the five male forms (see Fig. 91): ×, prionodont; +, amphiodont; ⊡, telodont; △, mesodont; ⊙, mesamphiodont.

The various curves overlap each other in a regular way (see text).

gressive variation of the forms as regards size has therefore no 'real' or taxonomic meaning, but is an automatic result of splitting up a continuous series on the basis of purely arbitrary characteristics. (7) Thus, if you were to plot the extreme variants, you would find the means for the series taken as a whole keeping in the middle between the extremes, while the curves for the class-means of the 'forms' taken separately would each begin close to the extreme plus limit of variation, and end near the extreme minus limit.

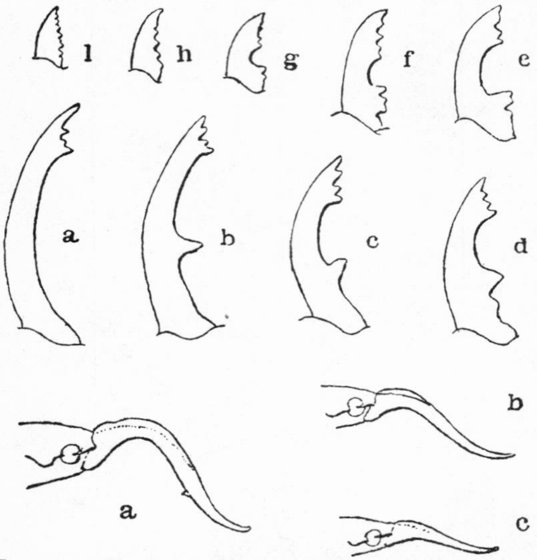

FIG. 93.—To illustrate change of form of mandible with increase of its absolute size in male stag-beetles.

Below, a, b, c, side view of the head in the *minor*, *media* and *major* forms of *Psalidoremus inclinatus*. Above, a—l, mandibles in the genus *Odontolabis*.

I have treated this case also in some detail, since it provides an excellent example of the dangers inherent in mere labelling: although the several forms are actually quite continuous one with the other as regards their diagnostic characters, the mere fact of having separated them, though on arbitrary grounds, made it possible to group the facts so as to make the arbitrary separation appear based upon some biological reality; and it is only further analysis based upon a study of the laws of relative growth and its variation which enables us to detect the fallacy.

§ 2. Heterogony in Groups higher than the Species

There is no need to imagine that the heterogony mechanisms and other laws of differential growth are confined within the boundaries of single species. On the contrary, there is every reason to suppose that, like the activities of ductless glands, these growth-mechanisms may operate in very similar fashion throughout considerable groups of animals. What we have just been saying about systematics has an immediate bearing upon certain evolutionary problems. For we should then expect to find heterogonic organs varying in their relative size with absolute size of body in different-sized species of the same genus, or different-sized genera of the same family, just as in different-sized individuals of the single species. Our supposition is confirmed: this does occur. It is not universal or inevitable, but it does occur frequently.

The rule was independently discovered by Lameere (1904; see also 1915) and Geoffrey Smith (1906B), and numerous further instances of it have been put on record by Champy (1924

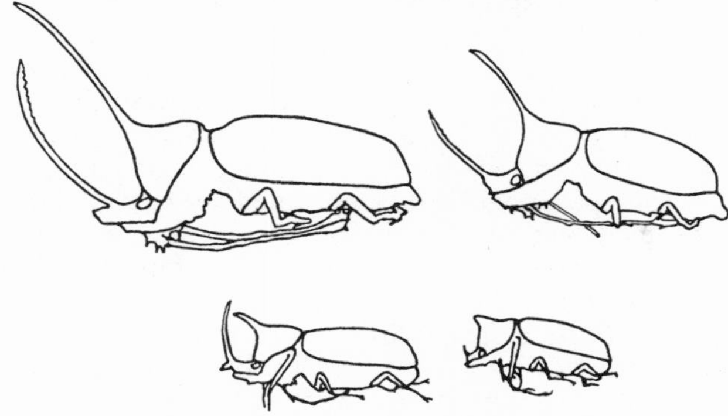

FIG. 94.—Phenomenon of Lameere and Geoffrey Smith in species of the Dynastid beetle Golofa.

Below, two specimens of *G. porteri*.
Above, right, *G. cæcum*; left, *G. imperialis*.

and 1929): Lucanidae (1924, pp. 142, 152); Dynastidae (ibid., pp. 152–4); and see Fig. 94.

In the most striking cases, the heterogonic organs in the males of the large species are relatively enormous, while in those of the small species they are scarcely more developed than in females (e.g. Fig. 94; the beetle Golofa porteri against G. imperialis, Champy, 1929, p. 230). This at first sight appears as an example of orthogenesis; but if orthogenesis be taken in its strict sense, of a determinate evolutionary change in the germ-plasm, this is not so. All that we have is a type of growth-mechanism, which, if not modified, will yield certain predictable results, as regards relative organ-size, with any given absolute body-size. If a new high or low extreme of body-size, not previously attained in the evolution of the group, is reached by a given species, the growth-mechanism will see to it that new potentialities of relative organ-size are realized. But though the proportions of the organ are new, the same single growth-mechanism is at work. There has been no change in its hereditary basis, determinate or otherwise; merely new results, potential in it from the first, have been realized.

If we wish to search for evolutionary parallels, we find that a similar situation has occurred in regard to many time-relations in body-processes of warm-blooded vertebrates, which take place at temperatures well above the lethal temperatures for most cold-blooded vertebrates. Accordingly such processes occur at speeds which are never realized in cold-blooded forms. The speed of such processes in the cold-blooded forms varies with temperature in a regular way. The maximum speed in the cold-blooded forms is limited by the temperature which they can tolerate without dying; but the actual, and much higher, speeds in warm-blooded forms approximate closely to the speeds which are obtained by extrapolating the curve obtained for the cold-blooded forms to the blood-temperature of the warm-blooded. An example of this is seen in the lapse of time between onset of convulsions due to insulin injection in a frog, which dies at about 30° C., and in a rabbit, which dies at 37° C. (Huxley and Fulton, 1924).

The evolutionary importance of the facts lies in this: that whenever we find the rule of Lameere and Smith holding true for a series of separate forms, we are justified in concluding that the *relative size* of horn, mandible, or other heterogonic organ is automatically determined as a secondary result of a single common growth-mechanism, and *therefore is not of adaptive significance.* This provides us with a large new list of non-adaptive specific and generic characters.

The operation of the rule, however, is not constant: it is, in fact, merely a rule, with numerous exceptions. There are a number of cases in which related species of very dissimilar absolute size show secondary sexual characters of the same relative size. Champy (1924, p. 155) lists a number of these, and Dr. Arrow, of the British Museum (Natural History), has informed me that there are numerous other examples to be found in Coleoptera. This should, however, not surprise us. The fact that the growth-ratio of the male chela of Uca changes during the life-history, or that the onset and end of heterogony may vary for the same organ in different types, indicates that such growth-mechanisms *can* be quite markedly modified should biological need arise.

We meet with a precisely similar state of affairs in regard to other biological rules. For instance, there is an undoubted tendency for secondary sexual characters acquired by one sex to be transferred, in whole or in part, to the other sex, even if of no biological value in the other sex. But this general tendency is constantly being modified or overruled by other agencies, e.g. the need for protective coloration in certain female birds frequently inhibits the transference of male display characters to the females, and the tendency can only manifest itself, among females of these groups, in those species which are not in need of protection as regards colour, because they nest in holes (e.g. tits, red-breast).

Doubtless, as well as such total exceptions to Lameere's rule, we should also find many minor modifications, revealed by organs whose relative size was not precisely what should be expected on an embracing formula, though in part a function of absolute body-size. But to detect these we should have to make quantitative studies of the variation of relative organ-size (*a*) within the individuals of the several species, (*b*) among the species of the genus ; and this has not yet been attempted.

The undertaking of this task in insects and crustacea should be all the more interesting, since in regard to relative brain-size in mammals, Lapicque (1907, 1922) and Dubois (1914, 1922) and Klatt (1921) have established the remarkable fact that different formulae apply to the intra-specific and inter-specific variation respectively. Variation of body-size within the species (e.g. in different breeds of dogs or rabbits) has less effect on absolute size (though more on relative brain-size) than does variation of body-size from one species to another. In both cases the formula $y = bx^k$ is followed, but whereas the value of the growth-partition coefficient (k) of the brain is 0·56 when different species are considered, it is only 0·22 for different-sized individuals of the same species. In this particular case, the facts are doubtless to be correlated with the artificial nature of the intra-specific selection that has been at work. Man has been concerned to produce large or small breeds, irrespective of their general adaptation or their efficiency in a state of nature. The changes in absolute brain-size produced by such changes in body-size are presumably the minimum to be expected as automatic or secondary result of the change in body-size ; while those found as between wild species would represent the best possible physiological adjustment between brain-size and body-size. We should also expect that the genetic variations in body-size *within* a wild species would also presumably be largely independent of those affecting absolute brain-size ; but so far as I know, data on this point are lacking. (See also p. 224 ; heart.)

Whether any such discrepancy between intra- and inter-specific formulae would be found in regard, e.g., to the horns or mandibles of male beetles remains to be seen. If it did occur, it might well be expected to be of opposite nature, the larger species having relatively smaller horns than would be expected from the extrapolation of the intra-specific curve of smaller species. But this, as I have already said, also remains to be determined.

§ 3. Heterogony and Evolution

A further evolutionary implication of these facts is this : that the existence of an organ with high growth-ratio tends to limit the extreme size attainable by the type in question during evolution, for with very large absolute size, the relative size of the organ will tend to become so huge that it becomes unwieldy or even deleterious. Doubtless the organ could be held in check by a modification of its growth-ratio such as we have been envisaging. But for increase of absolute size, two concomitant processes of mutation would then have to occur—mutations favouring decreased growth-ratio of the organ, as well as those favouring increased absolute body-size, so that the evolutionary problem is thereby complicated.

It can be calculated that a male Uca with body weighing 1 kg. would, if its claw's growth-ratio had remained unaltered, boast a large chela weighing some 10 kg.—which, as Euclid says, is absurd : and it is perhaps no coincidence that the largest fiddler-crabs attain sizes far below those of many other Brachyura, and even far lower than those of other land or semi-land crabs (Ocypoda, Birgus, etc.) ; see p. 32. The excessive antlers of the Irish 'elk', Megaceros, may very likely be accounted for on similar grounds. An organ which is verging on the deleterious may rapidly become actually deleterious if conditions change.

A further important modification of the rule has been pointed out by Champy (1924, p. 156 seq.), viz. that it applies only to the *size* of the organ. The details of *form* may be radically modified, and yet the general size conform to the rule.

This implies that we have a fundamental growth-mechanism controlling size and location of organ, but that its form is capable of modification by numerous independent genes. We

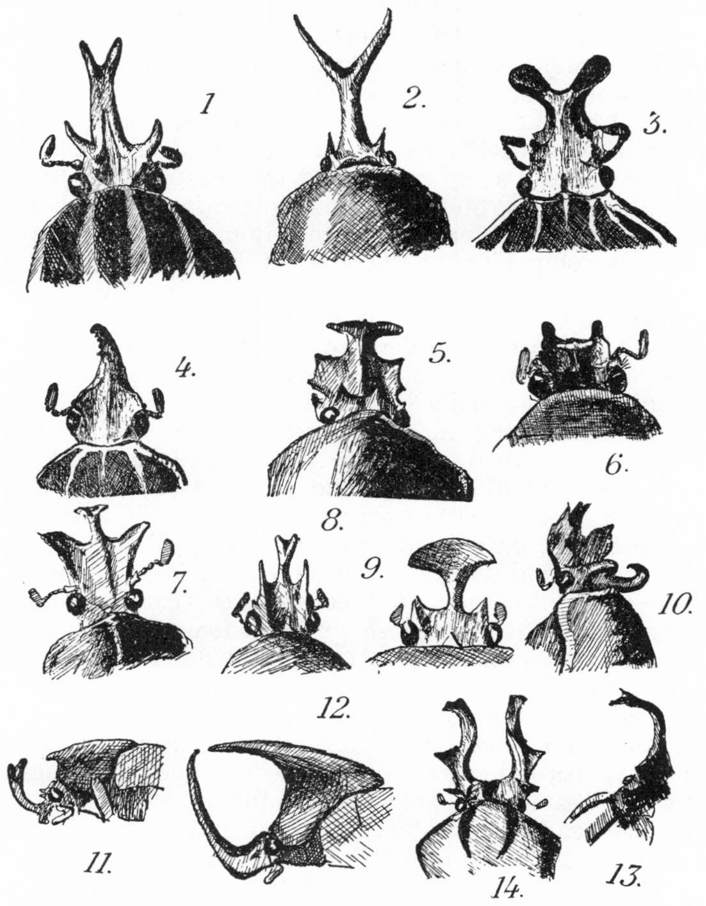

FIG. 95.—Specific variations in detail of heterogonic organ (cephalic horn) in different species of Goliath beetles (Goliathidae).

1, *Chelorina polyphemus*. 2, *Eudicella euthalia*. 3, *Goliathus giganteus*. 4, *Mecynorhina torquata*. 5, *Taurina nireus*. 6, *Stephanorhina guttata*. 7, *Neptunides polychrous*. 8, *Taurina longiceps*. 9, *Trigonophorus delesserti*. 10, *Rhanzania bertolinii*. 11, *Mycteristes rhinophyllus*. 12, *Theodosia magnifica*. (11 and 12 resemble Dynastids in possessing prothoracic horns). 13 and 14, *Dicranocephalus wallichii*, two views.

must further suppose that other independent genes are capable of modifying the size-relations (growth-ratio) of the organ in minor quantitative ways. (Fig. 95.)

Perhaps the most interesting evolutionary aspect of the study of relative growth is when we find a series of related forms possessing a markedly heterogonic organ, which not only differ in size, but also succeed each other in order of absolute size during evolutionary time. A classical example is afforded by the deer (Cervidae). It was long ago pointed out that in a broad way the antlers of deer increased in complexity from their first geological appearance up to the Pleistocene, much as do the antlers of an individual deer during its single lifetime, and this was adduced as an example of the laws of recapitulation. So in a sense it is, but it appears to owe its existence to the presence of a fundamental heterogony-mechanism for antler-growth, accompanied by increase of body-size both during ontogeny and phylogeny. Increase of absolute body-size automatically brings about disproportionate increase of antler-size, and increase of antler-size appears to be of necessity accompanied by greater complexity of branching (or palmation). And this applies equally to the individual and to the race.

Theoretically, the most interesting case is that of the Titanotheres, worked out by H. F. Osborn (1929, and earlier papers there cited). These, like so many other mammals, begin their evolutionary career as small organisms, and steadily increase in size until they become extinct. They also begin hornless, and end with a single bifurcated frontal horn.

But—and this is the salient feature of this example—Osborn has been able to distinguish at least four distinct lines of descent in the group, characterized by differences in skull-shape, dentition, leg-size, etc., and presumably mode of life; and in each of the groups we meet with the same phenomenon of small and hornless forms steadily increasing in size and eventually becoming horned. Thus we find the origin of horns of the same type, growing from the same location, taking place independently in four separate groups: and Osborn insists that this must be interpreted as true orthogenesis—in the strict sense of the term, as implying predetermined variation of the germ-plasm, not merely of directional evolution.

However, as pointed out earlier (Huxley, 1924; and see Sturtevant, 1924), our studies make it clear that this interpretation is not necessary. Granted (*a*) that there existed in the germ-plasm of the ancestor of the four lines of descent the hereditary basis of growth-mechanism for a frontal horn, and (*b*) that increase of size up to a certain limit was advantageous for Titanotheres in general, as would seem inherently probable, then the results follow without any need for invoking orthogenesis. Natural selection would account for the increase of absolute size, and increase of absolute size would evoke the latent potentialities of the horn's growth-mechanism in the history of the race just as surely as it does, for instance, in the individual deer to-day, or as it undoubtedly did in the individual Titanothere in the Oligocene. No alteration in the hereditary basis for horn-growth need be involved at all, whether determinate or otherwise: the visible horn-changes are automatic results of the effect of size-changes in the unaltered mechanism for horn-growth.[1]

The only remarkable thing about this case is that the mechanism for horn-growth must have been represented in the germ-plasm before it could ever operate—before any single Titanothere was large enough to have a horn at all. And this is undoubtedly surprising. Once more, it makes it extremely difficult to assign any adaptive value to the horns, at any rate in the early stages of their evolutionary development, when they were represented only by the merest incipient knobs, and still more when they merely existed potentially![2]

To take a case nearer home, Parsons (1927) maintains that the British race is showing evolutionary change, because the mean proportions of British skulls have slightly changed

[1] Doubtless some quantitative modifications in growth-ratio of horn took place, as evidenced by the difference in relative horn-size in the various groups; but this does not affect the main argument.

Champy (1924, p. 155) takes an almost identical view. He writes: 'Le phénomène d'orthogenèse ou rectigradation, lorsqu'il accompagne une augmentation de taille, n'est pas un phénomène évolutif, c'est un phenomène purement physiologique d'équilibre nutritif. La seule évolution qu'il y ait à expliquer est l'augmentation de taille.'

[2] The Asiatic Titanothere genus Embolotherium is distinguished not only by possessing a very large and peculiarly-shaped horn, but by this being formed entirely from the nasal bones, instead of from both nasals and frontals, as in all other genera. During both the ontogeny and the phylogeny of other late Titanotheres, there is a tendency for the horns to grow in an anterior direction as well as to increase in size. It may be suggested that the condition in Embolotherium represents a further step in this evolution—the actual shifting of the growth-centre from the fronto-nasal to the nasal region. This would be a shift of a whole subsidiary growth-region with a gradient of its own, along a main gradient. Somewhat similar shifts of quantitative growth-centres within a single growth-gradient have been already recorded. The most relevant case is the shift of the growth-centre of the female abdomen of Pinnotheres to the terminal segment in place of the sub-terminal segment where it is situated in most Brachyura (p. 95); cf. also p. 117 (Copepod body-regions).

during historic times. This may be so ; but caution is indicated. If it should prove to be the case, as is quite possible on the evidence, that the various dimensions of the skull alter at slightly different rates with increased absolute size, then the change observed by Parsons may be merely the effect of the increase of mean stature which appears undoubtedly to have occurred in Europe in the past few centuries. And since this is in all probability phenotypic in origin, the change in skull-proportions, though of course in a sense an evolutionary change, has far less significance than we should at first sight be inclined to assign to it. (See Hooton, 1931.)

Parsons himself states that the mean male stature of the upper and middle classes has risen rapidly by perhaps three inches to about 5 ft. 9 in., whereas the mean for the whole population is only about 5 ft. 5 in., and that for the stunted slum and factory populations perhaps an inch or so less. The high stature of the best-nourished classes is a modern phenomenon, due to good food, exercise, etc., and was not realized among the early Saxons (mean 5 ft. 6 in.).

The change in skull-proportions to which Parsons refers is one in increased relative height (taken as percentage-ratio of (auricular) height in relation to total head-size—measured as [length + breadth + height] as a standard). As a matter of fact, when we make a correlation-table between relative skull-height and absolute skull-size as measured by Parsons's figure for [length + breadth + height], we obtain a definite correlation (Table XIII).

Naturally it would be much more satisfactory to make a correlation table for all the individual measurements available, but this must be a task for the physical anthropologists.

We should also expect to find a correlation between total stature and total head-size : it is noteworthy that the lowest values are for poor neighbourhoods in London and for soldiers in the eighteenth century ; and the educated twentieth-century type, which we know to have a mean stature much above that of previous centuries, has values both for absolute skull-size and relative skull-height far beyond any previous limit.

As regards relative breadth (cephalic index), it seems probable that different skull-types may show a change with age (size) in either of two opposite directions, but here the evidence is conflicting. In any case Parsons's data are most simply, though not necessarily, to be interpreted as a further ex-

TABLE XIII

Relative skull-height, per cent	460·1–465	465·1–470	470·1–475	475·1–480	480·1–485
272·6–277·5					G
267·6–272·5			F		
262·6–267·5					
257·6–262·5	B, E		A		
252·6–257·5	C	D			

Absolute skull-size (l. + br. + h.), mm.

A Mean of the three main racial stocks of Britain, early Mediterranean (Neolithic), Alpine (Beaker), and Nordic (Anglo-Saxons).
B Fourteenth and fifteenth century villagers, Northants.
C London plague skulls, seventeenth century, probably poor stock.
D Eighteenth century, London, poor neighbourhood.
E Soldiers, eighteenth century.
F Twentieth century, poor classes.
G Twentieth century, educated and well-to-do classes.

ample of change of relative proportions with change of absolute size.

As suggested by Hooton (1931, p. 408), a similar effect may be responsible for the changes in skull-form of the descendents of immigrants into the U.S.A., made familiar by the work of Boas.

All the cases we have so far considered have another aspect of interest. They are all examples of what Darwin called correlated variation, where change in one character automatically brings about change in another character (unless counter-selection operates against the correlated change). Thus they relieve us of the necessity for seeking utilitarian explanations for the correlated changes; natural selection need not be invoked to account for the details of such characters.

The burden on natural selection is also appreciably lightened by the demonstration of the existence of growth-gradients. As pointed out in Chapter IV, the method of Cartesian transformation of related forms employed by D'Arcy Thompson shows that the great majority of the striking changes in shape and relative size of parts necessary for the evolution of a sunfish (Orthagoriscus) from a Diodon-like ancestor not only need not but cannot have been separately evolved ; they can only be interpreted as due primarily to a single main change in the form of the growth-gradient along the animal's axis. And the factual evidence we possess, as well as *a priori* reasoning, indicates that a mutation can act on a growth-gradient as a whole, thus simultaneously altering the proportions of a large number of parts.

An even more fully worked-out example is that of the conformation of domestic breeds of sheep analysed by Hammond (l. c.). As we have already seen in Chapter III, Hammond has been able to show that the changes in proportions which take place in foetal and post-natal life in improved breeds of 'domestic' sheep represent merely an accentuation of qualitatively similar changes which occur in unimproved breeds and in wild species (Fig. 96). In other words, there is a greater intensity of the relevant growth-changes, which means primarily a steepening of the growth-gradients concerned. Here again, the characters actually affected by selection are quite few, namely the growth-gradients ; whereas the number of what are usually called 'characters' which are altered during the process is very large.

In general, we may say that the existence of growth-gradients gives opportunity for mutation and selection to affect a number of parts in a correlated way, thus greatly simplifying the picture of the genetic and selective processes at work.

The cases previously considered concern organs with specific heterogony mechanisms. D'Arcy Thompson (l. c.) has drawn attention at some length to the *functional* changes necessitated by increase in absolute size. These are for the most part well known, such as the need for increased relative weight of skeleton in large land animals, the need for increase of absorptive surface of the gut with increase of bulk to be nourished (on which topic a recent essay of R. Hesse (1927) may be consulted for further details), and so forth. In any case, they do not so much concern us here, since we are chiefly dealing with the automatic and primarily non-adaptive effects of inherent growth-mechanisms.

An exhaustive botanical study of the relation of size to internal morphology has recently been published by Wardlaw (1924–8), but the details are of a technical nature and I cannot enter into them here. The conclusion which chiefly interests us is that certain characteristic types of internal morphology in vascular plants (e.g. polystely) need have no phyletic value, since they appear to be modifications directly dependent upon increase in size. These and other results have been sum-

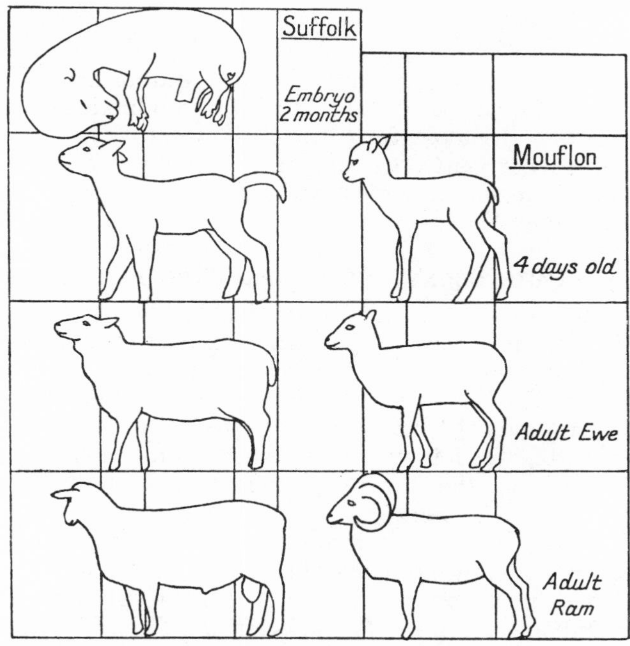

FIG. 96.—Changes in proportions during growth in wild sheep and improved breeds of domestic sheep (see text).

marized in a book by Bower (1930). The general conclusion to be drawn from these botanical studies would seem to be that in this case apparent orthogenesis is to be explained by parallel adaptive evolution due to size-increase.

We may take one example of the individual changes correlated with size increase (Bower, l. c., p. 16). As we pass up from rhizome to stem in *Psilotum triquetrum*, the size of the organ increases. The size of the tracheidal tract also increases, and its form also changes, exposing more surface than it would have if it remained approximately cylindrical. This is summarized in the following table, abridged from Bower's Table I.

Diameter of stele, mm. . . .	0·17	0·34	0·29	0·53	0·61	0·75
Surface-volume ratio of tracheidal tract	9·26	5·90	5·40	3·64	3·20	3·00
Surface-volume ratio of equivalent cylinder	6·82	4·15	1·88	1·08	0·91	0·81

i.e. the actual surface : volume ratio diminishes to about 30 per cent of its initial value : but if the conducting tissue had remained cylindrical in form the ratio would have decreased to about 13 per cent. Changes precisely similar in principle occur in phylogenetic series with increase of size : see e.g. Bower's Table X.

§ 4. Heterogony and Comparative Physiology

Disproportionate change of relative size with change of absolute size may have important results in studies in comparative physiology. (See also Murr, p. 260.)

This is well brought out by Klatt in an interesting paper (1919). He finds the relation of heart-weight to body-weight in vertebrates to follow our familiar formula for constant growth-partition, the (mean) coefficient, or as he calls it, the 'heart-exponent', being 0·83: i.e. heart-weight = constant × body-weight $^{0·83}$. It is interesting to find that the value appears to be very similar whether the size-differences considered are intra- or inter-specific; this result differs from that of Dubois, etc. (l. c.), on the brain, a difference which may presumably be correlated with the prompt functional regulation of the size of the heart to the work it is called upon to do.

In passing, it may be noted that the heart as a whole appears to have a different growth-coefficient from the right ventricle and the auricles. These latter, according to Hasebrock (1927), have a coefficient close to 2/3, which would indicate their functional dependence upon surface-area, or upon general metabolism, which in warm-blooded forms is nearly proportional to surface-area; while the left ventricle has a much higher coefficient, nearly in direct proportion to body-weight.

But from the point of view of comparative physiology,

the interest lies in the following considerations. It has usually been customary, as was done by Parrot (1894) for birds, to compare the cardiac efficiency of different species according to their *percentage* heart-weights. From what has been said, this is entirely erroneous; variations in cardiac efficiency can only be determined in relation to the mean value to be expected at a given body-size for the exponential growth-partition formula, which will be given by the fractional coefficient b in the formula $y = bx^k$. Clearly, the percentage method will put large birds at a disadvantage.

TABLE XIV

A	Per cent. heart-weight	B	True relative heart-size = fractional constant (b)
Capercaillie	7·81	Magpie	0·0228
Magpie	9·34	Chaffinch	0·0240
Wild Swan	11·78	Capercaillie	0·0322
Chaffinch	14·16	Hobby	0·0402
Peregrine Falcon	14·91	Peregrine Falcon	0·0467
Hobby	16·98	Song Thrush	0·0528
Song Thrush	25·64	Wild Swan	0·0550

The table shows for a few birds from Parrot's list the order of cardiac efficiency (in ascending relative heart-size) as determined (A) by percentage methods and (B) by the correct method. It will be seen how different the two are.

Dubois (l. c.) and Lapicque (l. c.) have given us a similar method for judging of the true degree of 'cephalization' (true relative brain-size) in vertebrates. Thus the true relative brain-size (fractional constant b in the growth-partition formula) for the mouse family is 0·07, for the cat family 0·31–0·34, for tailed monkeys 0·4–0·5, for anthropoid apes 0·7, for man 2·8, whereas percentage values give no information of any value at all.

A similar method of approach can be utilized for many organs. Huxley (1927c) has used it for relative egg-weight in birds. Here it is found that the exponent (partition-coefficient of material between egg and body) is not constant over the size-range covered by birds, but diminishes steadily with absolute size, from 0·9 or 0·95 for the smallest birds to about 0·7 for the largest. It is suggested that it would be most advantageous to have a linear relation (exponent = 1·0), but that physiological difficulties connected with surface-area (which would give a limiting value of 0·67) interfere with this, and that they interfere increasingly with increased absolute size. It is interesting to find the large Ratites lie very close to the theoretical curve, in spite of their relatively huge bulk and their flightless habits. (Fig. 97; see also p. 264.)

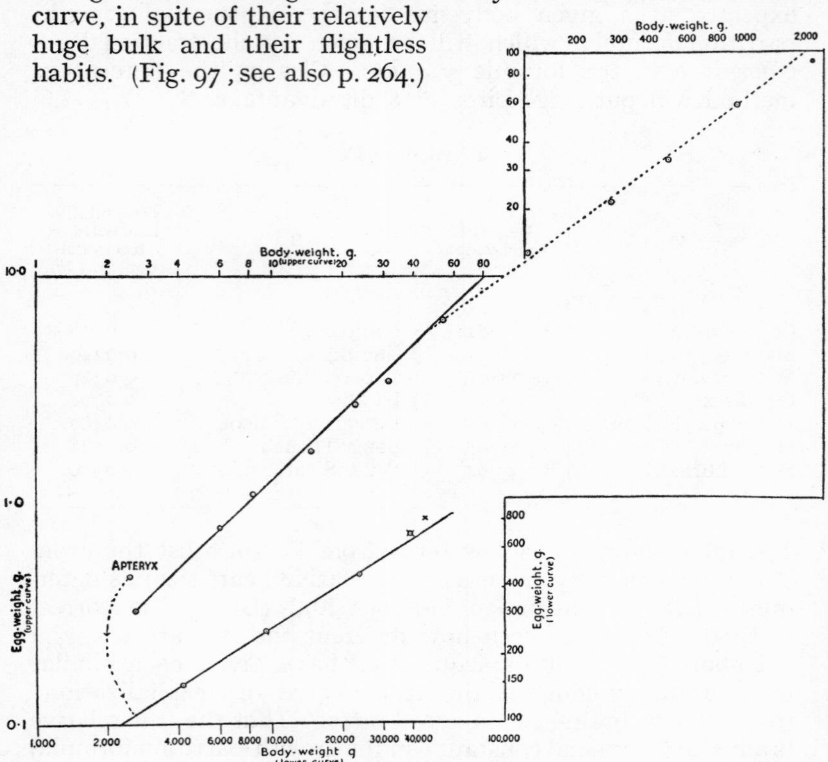

FIG. 97.—Graph showing change of egg-weight with body-weight in 432 species of birds: Means by weight-classes, logarithmic plotting.

o, Carinatae; ×, Ratitae (excluding Apteryx, which is plotted separately); ¤ mean for last class of Carinatae combined with Ratitae (except Apteryx).
The graph begins with the upper curve; $k = 1·0$. The value of k then sinks (dotted line), finally approximating to 0·67 (solid line, below).

Each large group has its own curve, differing chiefly in the fractional constant b. The modifiability of relative egg-size in relation to adaptive needs is clearly seen in the abnormally low position of the value for species which are reproductive parasites upon smaller birds, like *Cuculus canorus*, and in the high values of *b* found for most forms with precocial young.

In regard to the physiological side of the problem, it is important to distinguish two very different ways in which change in relative size of a part may be brought about. We have already discussed certain implications of this distinction, but have not set it forth in precise form. (See also pp. 257–8.)

(1) An organ may be capable of functional adjustment in size. The best example of this is the heart.

(2) An organ may possess a differential coefficient of growth-partition, which operates wholly or largely irrespective of functional demands, e.g. the antlers of deer, the large chela of male Uca, the limb of *Amblystoma tigrinum* grafted on to *A. punctatum*, the pituitary of rabbits (Robb, 1929), etc. Increase of absolute size of an animal will in both cases bring about an alteration of relative size of the part, but in the first case owing to functional hypertrophy, in the second owing to the specific growth-intensity of the organ, which in its turn is presumably due to a specific growth-promoting substance. And doubtless in many cases, both mechanisms will operate simultaneously.

Some cases of so-called functional hypertrophy may turn out to belong to the second group, e.g. the enlargement of the remaining testis after unilateral castration is probably not functional in the usual sense at all, but depends upon the existence of a specific partition-coefficient for testis-tissue relative to total size. (See Domm and Juhn, 1927, and also p. 257.)

The distinction, as I have suggested, is usually not an absolute one. If one prevents a limb from exerting its normal function, e.g. by tying it up at birth, its growth will be markedly subnormal. With regard to heart-weight, etc., it is *a priori* almost certain that the heart, in addition to its capacities for marked functional size and regulation, possesses a primary partition-coefficient which will differ from species to species. What we can say, however, is that in some cases the differences observed are due solely or mainly to functional regulation, in others solely or mainly to differences in inherent growth-capacity.

But there are cases in which organs attain their definitive relative size entirely owing to one or other of the two methods. The size of the organs of holometabolous insects must depend entirely on their growth-partition coefficient; while it is

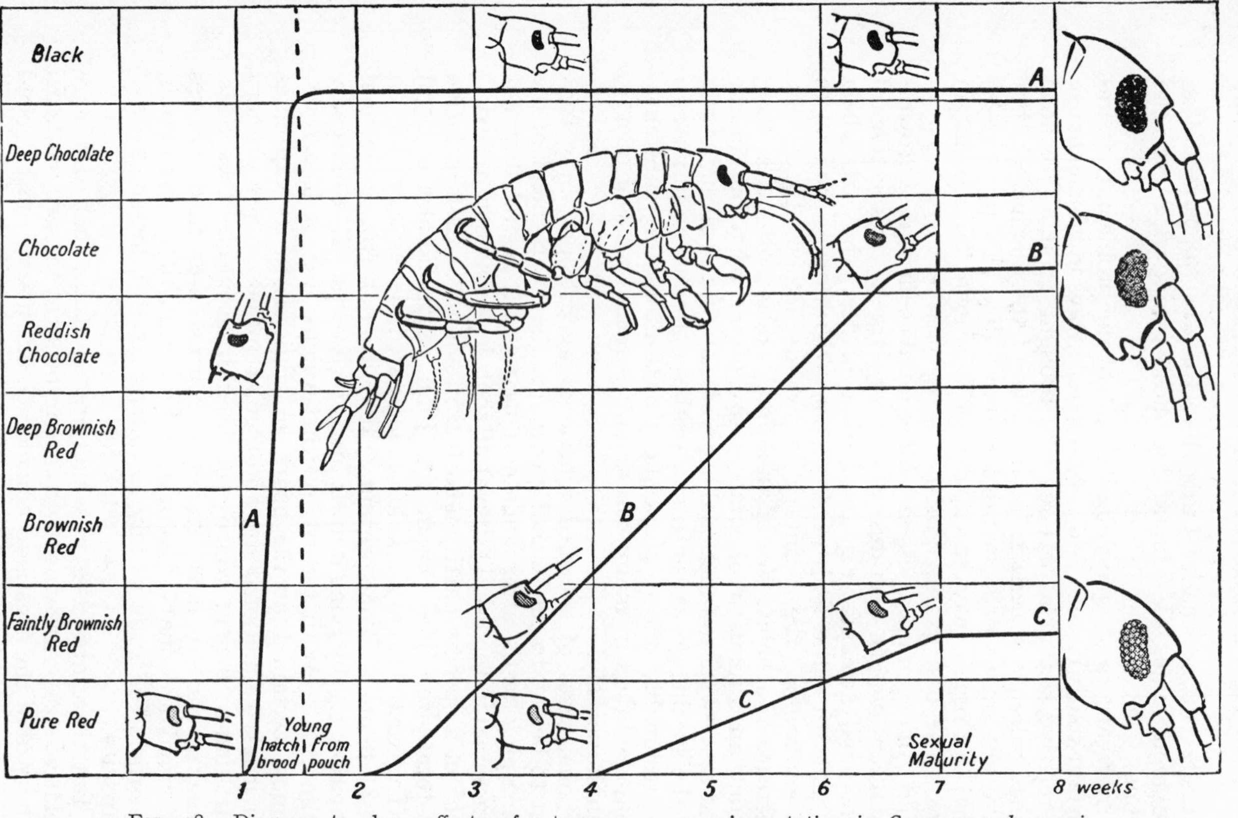

FIG. 98.—Diagram to show effects of rate-genes on eye-pigmentation in *Gammarus chevreuxi*.

Centre, an adult with black no-white eye. The scale on the left gives degrees of darkening of the eye from scarlet to black. Below, time taken in weeks from fertilization to reach various stages (at 23° C.).

A, curve of darkening of black no-white eye; B—B, C—C, curve of darkening, of rapid-darkening, and slow-darkening red no-white eyes respectively.

Both are of genetic composition *bbCCww*, as against BBCC*ww* for black no-white, but rapid-darkening possess a dominant rate-gene S, whereas the slow-darkening possess the recessive *ss*.

highly probable that the size of tendons in the mammalian body is entirely or almost entirely due to functional response.

Thus, when we compare organs of different relative size in two individuals, races, or species of different absolute size, our analysis should be devoted to establishing (1) whether the difference is due to the organ's capacity for functional size-regulation; (2) whether it is a mere consequential effect of increase of absolute size, as is the case with the relatively large antlers of Red Deer imported into New Zealand from Scotland, or of the relatively large mandibles of absolutely large male Lucanidae; or (3) whether it is due also to genetic alteration in the growth-partition coefficient; or (4), of course, to a combination of these in differing degrees.

§ 5. Relative Growth and Genetics

We next come to the subject of genetics. It is clear that the chief genetic factors controlling any organs which show a constant differential growth-ratio or a constant growth-partition coefficient must be *rate-genes*, or genes which determine the rate of a developmental process. Such factors have been deduced or demonstrated for numerous characters, notably in the sex-determining genes of moths (Goldschmidt, 1923, 1927) and the eye-colour genes of the crustacean Gammarus (Ford and Huxley, 1929) (for other cases see the references in the works cited, and also Ford, 1929, Bělehrádek and Huxley, 1930); see Figs. 98, 99.

There are two general points which are worth stressing. The only case where the curve for the development of such characters has been quantitatively obtained by direct measurement is that of the eye-colour of Gammarus. Here it appears clear that the actual form of the curve is determined by two factors—an inherent greater or less velocity of pigment-deposition, and some regulating mechanism which relates this to the growth of the body as a whole. This is clearly brought out by two facts—first, the relations in slow-growing mutants; secondly, the slight lightening of colour which occurs in the eyes of certain strains of Gammarus after sexual maturity. 'Slow growth' is a recessive mutation, which at first shows abnormally dark eyes. This appears to be due to the fact that the eggs are of normal size, that a pigment precursor is formed in them in some simple relation to egg-size, and then, when the time comes for the deposition of pigment, the eye-area, over which it is to be spread, is abnor-

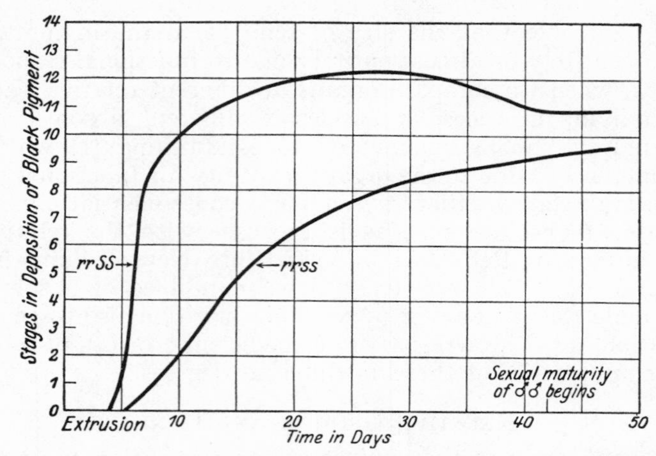

FIG. 99.—Actual rate of darkening (smoothed curves) in 1,000 rapid-darkening (*rr*SS) and 1,000 slow-darkening (*rrss*) specimens of *Gammarus chevreuxi* at 23° C. (compare Fig. 98).

mally small owing to the slow growth. Later, however, the eye-colour is regulated towards what is 'normal' for the particular rate-genes involved; this appears due to the fact that further formation of pigment occurs in relation to body-size (Ford, 1928); Fig. 100.

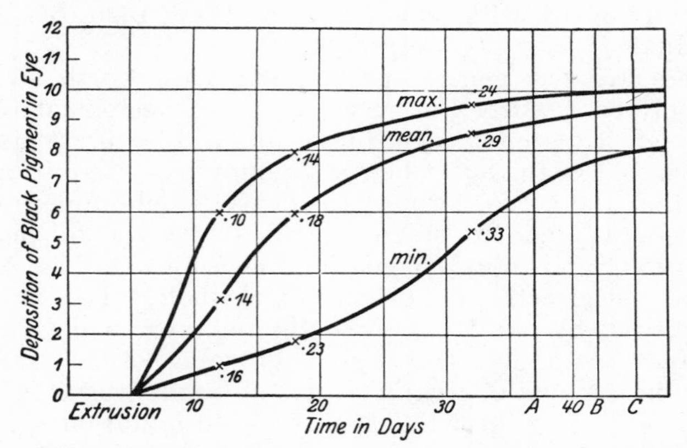

FIG. 100.—Effect of size (rate of body-growth) on rate of eye-darkening in a single genetic strain of *Gammarus chevreuxi*.

All specimens were males of constitution *rrss*, kept at 23° C. The figures on the curves give eye-length in mm.

Max., curve of darkening in slow-growing specimens, which attain sexual maturity at time represented by vertical line C; min., curve for fast-growing specimens (maturity attained at A); mean, curve for mean of population (maturity attained at B).

The second point, the slight lightening of eye-colour after maturity, is due to the expansion in size of the individual eye-facets, causing the same amount of pigment to be spread over a greater area (Ford and Huxley, 1927).

This complex relationship is precisely similar to what we have found to obtain for the relative growth of an organ such as a limb (see Chapter VI), where two factors are also at work, the inherent growth-potential of the organ, and some mechanism regulating this in relation to the growth of the body as a whole. Since the relative factor occurs in both cases, it is not to be expected that the curves for characters controlled by rate-genes will shed light directly on the nature of the biochemical processes involved, even if plotted against time. It would be theoretically better to plot them against some quantitative character of the animal, e.g. the amount of some important substance in the body, or even the total body-bulk. In this way we should obtain a growth-partition coefficient for the chemical substance responsible for the character, just as we obtain a similar coefficient for the growth of heterogonic organs when their size is plotted, not against time, but against body-size; and it is these coefficients which are of the greater biological importance, and are quantitatively comparable one with another. That this method may be valuable is seen in the already cited work of Teissier (1931).

The second general point is this. Just as the proportions of a fiddler-crab are continuously changing, so may the eye-colour of a Gammarus change throughout life, or at least until long after sexual maturity. But a holometabolous insect has its adult proportions fixed once for all at the attainment of the imaginal instar; and similarly even characters that are determined by rate-factors will in holometabolous insects become fixed at the same period—the final moult will take a cross-section across the developmental curves whose results will only be made visible at this instant. Thus a static is substituted for a dynamic picture.[1]

We may thus suppose that such a gradation as that presented by the eye-colour of various multiple allelomorphic series (e.g. of the white-eye locus in Drosophila) represents merely a

[1] This is not rigidly true: there are, for instance, some eye-colours in Drosophila which change with the age of the imago, but such changes play a minor part as compared with those in, e.g., Gammarus, and it is quite possible that they are not of the same nature; in any case they are not known to be related to size.

cross-section across a series of curves for a number of similar processes occurring at different rates, such as are actually visible in Gammarus, but could in Drosophila only be detected by chemical means or by deduction.

The Gammarus eye-colour curves also show the phenomenon of an *equilibrium-position*—i.e. the degree of pigmentation finally ceases to change. This merely means that pigment and eye-area are now increasing at the same rate, instead of the pigment-formation proceeding at a more rapid rate as at first—in other words, it corresponds to a change from heterogony to isogony, such as we have seen to occur in regard to the size-relations of many organs (Chapter I, etc.). In the case of rate-genes, when this change occurs before birth, we get an apparently fixed character ; but we may either observe (Gammarus wild-type eye) or deduce (human black eyes) that rate-factors have been operative in the prenatal period.

With all these similarities, we may with some confidence assume that the proportions of parts, at least in so far as they are determined during the period of auxano-differentiation and not during that of histo-differentiation, will be found to be determined by rate-factors operating according to the same rules as have been found to hold for the rate-factors controlling pigmentation in the eye of Gammarus.

For an interesting corollary of this, we may look to the work of Davenport (1923 ; his p. 151) on human body-build. He has plotted the frequency for relative chest-girth (chest-girth : stature) in man at various ages in a three-dimensional form. If the mean values alone were taken, the age-change would consist in a marked downward slope of the curve from birth to late adolescence, followed by a slight upward tendency, and then often a stable equilibrium-position. When the frequency model is examined, however, it is seen not only that there is very great variation at all ages, but that there is a well-marked multimodality. The general shape of the curve indicates that we are dealing with interrelated rate-factors ; the multimodality shows that there exist a few main genetic types, each corresponding to one sort of body-build and determined by a particular gene or set of genes influencing relative growth of chest-girth and/or stature (Fig. 101).

An important corollary of all this is that we should expect numerous genetically-determined characters concerning relative size of parts to alter with absolute size of body, without there being any change in their genetic basis. The unusual proportions of the limbs, e.g. in human pygmies, may probably be accounted for, wholly or in large part, along these lines.

Similarly, the general sexual differences in limb-proportions in man, women usually having relatively short arms, and relatively shorter legs, with in either case the distal portion of the limb short relatively to the proximal portion (see e.g. Hooton, 1931, p. 255) may also be due to altered time-relations of relative growth. The curves given by Davenport (1926)

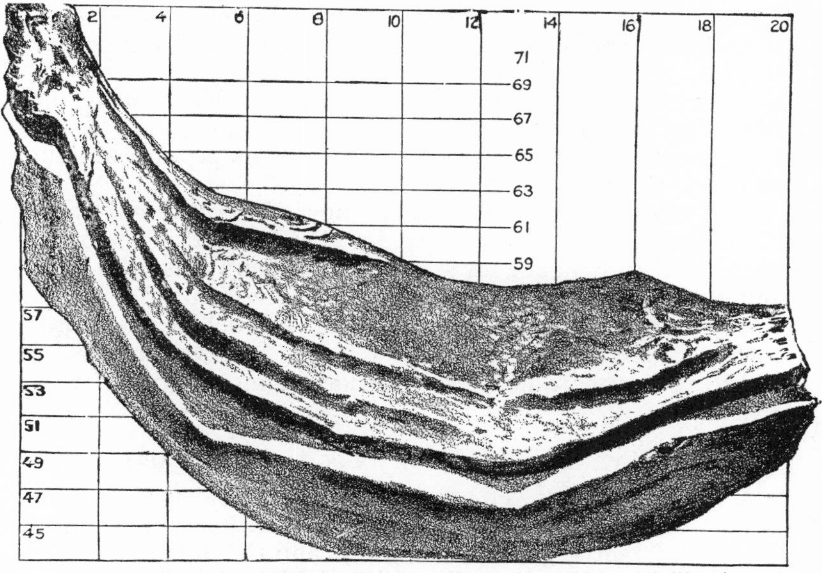

FIG. 101.—Body-build in Human Beings. View of a solid model, illuminated from the top of the page, showing distribution of relative chest-girth in man from birth to 20 years.

Ordinates, relative chest-girth (chest girth as a percentage of stature). Abscissae, age in years. The mean relative chest-girth diminishes to about 15 years, then rises again. There are several well-marked modes for relative chest-girth which persist throughout, indicating the presence of distinct genetic types.

give strong confirmation to this view. See also Wallis (1932) who finds, as is to be expected on the basis of the ideas set forth in this volume, that relative limb-size in boys and girls is more strongly correlated with absolute height than with age.

If the main principles involved would appear to be comparatively simple, the detailed complexity which may occur in certain cases can be realized by reference to the problem of the size of the eye and its two lobes in various members of the bar-eye series of Drosophila (for details see Goldschmidt, 1927; and notably Hersch, 1928).

An interesting case where genes independently control growth-intensity in length and in breadth is afforded by the studies of Sinnott and his associates (see Sinnott and Hammond, 1930), who find that fruit-shape in the gourd Cucurbita depends on the interaction of various separate factors of this nature. Sinnott and Durham (1929) find that most of the form-differences depend primarily on the growth of the carpellary tissue, the wall of the ovary and fruit merely following the form-changes induced by the central region. The results of Sinnott (1930) may be interpreted to mean that there exist various genes controlling size and shape independently, but that the shape-factors generally control *growth-ratios*, and therefore there is a progressive change of shape with absolute size in individual development.

An instance of a single gene having a differential but apparently graded effect on a localized region is found in the house-mouse, where the gene for short ears produces a marked shortening of the skull, especially in the anterior-palate region; a very marked diminution in the height just behind the incisors with a slighter diminution posteriorly, and an increase in width, rather more marked posteriorly than anteriorly (Snell, 1931).

§ 6. Relative Growth, Embryology, and Recapitulation

A rather different set of evolutionary problems from those we considered previously has light shed on it by a consideration of differential growth-coefficients and the rate-genes which we must postulate to regulate them. These are the problems of recapitulation and of vestigial organs.

There are numerous cases of recapitulation which cannot, in my opinion, be accounted for along these lines, e.g. the presence of notochord and gill-clefts in the embryos of the highest vertebrates, but there are many others which do receive some explanation from this source. E.g. we find that the relatively long-armed Gibbons have a fetus which, though longer-armed than that of other anthropoids, is relatively shorter-armed than the adult (Schultz, 1926). Presumably the line of least biological resistance in altering proportions of limbs, etc., is to modify relative growth-rates, rather than to modify the original partition of material in the early embryo between organ and rest-of-body. And if this is so, the recapitulatory phenomenon follows automatically. Similar reasoning will apply to such cases as the negative heterogony of the tail of fetal man or positive heterogony of the toes of the Jacana after hatching (Beebe; see p. 262).

As regards vestigial organs, the arm-chair critic often demands of the evolutionist how the last stages in their reduction could occur through selection, and why, if reduction has gone as far as it has, it could not go on to total disappearance. In the light of our knowledge of relative growth, we may retort that we would *expect* the organ to be formed of normal or only slightly reduced relative size at its first origin, but then to be rendered vestigial in the adult by being endowed with negative heterogony. If rate-genes are as common as they appear to be, then what we have called the line of biological least resistance would be to produce adult vestigiality of an organ by reducing its growth-coefficient. So long as it is reduced to the requisite degree of insignificance at birth (or at whatever period a larger bulk would be deleterious), there is no need for reduction of its growth-rate to be pressed further. But the negative heterogony with which it is endowed will continue to operate, and it will therefore continue to grow relatively smaller with increase of absolute size. This last fact may account for the apparently useless degree of reduction seen in some vestigial organs, e.g. that of the whale's hind-limb. The degree of reduction may be useless considered in relation to the adult, but the relative size in the adult may be merely a secondary result of the degree of negative heterogony needed to get the organ out of the way, so to speak, before birth. In addition threshold mechanisms will possibly be at work, so that the organ, after progressive reduction, eventually disappears entirely.

In such cases quite small differences in growth-ratio, if the range of absolute size over which they operate is considerable, will make quite large differences in final relative size, a fact which indubitably will help to account for the high variability of vestigial organs. Even when the organ itself never grows, as in the imaginal structures of insects with a metamorphosis, a similar degree of variability may be brought about by relatively small variations in the rate-genes responsible. As Goldschmidt has shown by his classical researches on sex in Lymantria, the end-result in such cases depends on the amount of some effective substance produced by the gene at the critical period for differentiation of the organ affected.

Since both the rate of production of the substance and the precise time of differentiation can be varied both by environmental and genetic agencies, there is room for considerable elasticity in the end-result provided that this is between the normal upper and lower thresholds for the development of the character in question. E.g., in normal male Lymantria an adequate excess of male-determining over female-determining substance is present well before the period of differentiation; but in female intersexes this excess of male-determining substance is transformed into an excess of female-determining substance during the differentiation phase. Thus whereas normal males are not affected as regards their sex-characters by environmental agencies such as temperature, similar agencies applied during the development of an intersex can exert a marked influence on the degree of intersexuality.

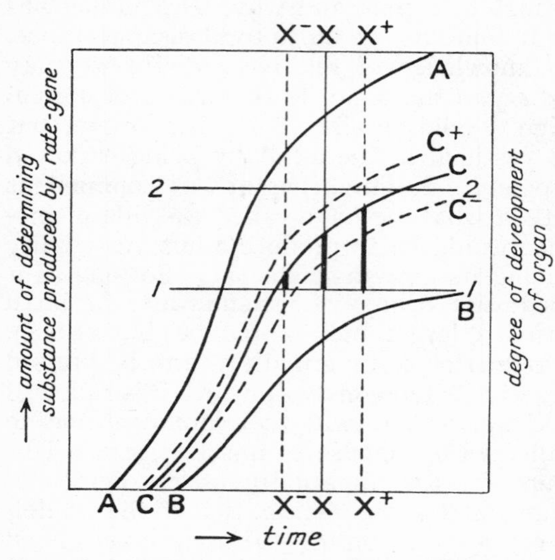

FIG. 102.—Hypothetical curves to illustrate the relation of rate-genes to vestigial organs.

1—1, lower threshold of amount of determining substance which must be produced by the rate-gene before differentiation of the organ can begin. 2—2, upper threshold; if this is reached during the period of differentiation, complete development of the organ occurs. X—X, effective time of differentiation. This may be varied so as to occur earlier (X^-) or later (X^+). A—A, action of rate-gene promoting normal complete development of organ; the full effective amount of determining substance is always produced before the period of differentiation. B—B, action of rate-gene whose intensity has been reduced to produce the total evolutionary disappearance of the organ. The lower threshold of determining substance is never reached before the period of differentiation. C—C, action of rate-gene normally promoting a vestigial development of the organ. Slight variations in its intensity or in the time of differentiation will produce large differences in the degree of development of the organ.

So with vestigial organs (Fig. 102). If the production of a limb-determining substance, for instance, be normally above the necessary threshold well before the onset of the phase of differentiation during metamorphosis, the limbs will be stable organs, little affected in their proportions or characters by environmental variations. Equally, if the rate-genes concerned are so much reduced in intensity that the limb-determining substances are well below the threshold during the whole differentiation period, the absence of limbs will be a stable character. But if the intensity of the rate-genes has only been reduced to a pitch at which a certain small amount of limb development can normally proceed, environmental agencies will quite certainly be capable of affecting the precise degree of this development within fairly wide limits. A good example of variability of this sort in vestigial organs is provided by the vestigial limbs of the stump-legged mayfly—*Campsurus segnis*, described and figured by A. H. Morgan (1929).

A somewhat similar fact, due to somewhat similar causes, is found in Gammarus (op. cit.). Normal black eyes, which are due to genes whose influence is to take melanin formation very rapidly up to saturation-point, are scarcely affected by temperature. Other genes slow down the process so much that the threshold for visible melanin-production is normally never reached. The permanently pure red eyes thus produced also are little affected by temperature. Genes of intermediate strength, however, which at room temperature produce a moderate darkening to red-brown, are markedly influenced in their effects by temperature-changes: at high temperatures the eyes of animals containing such genes are dark chocolate, at low temperatures almost pure scarlet (Fig. 103).

Finally, since rate-genes can obviously mutate both in the plus and the minus direction, so as to accelerate or slow down the processes which they control, it is clear that changes in rate-genes could as easily lead to the opposite of recapitulation as to recapitulation. Many examples of neoteny would fall under this head. See e.g. Mjöberg's discussion (1925) on the larviform females of certain Lycid beetles. De Beer (1930) has dealt fully with this point, and has proposed a useful terminology.

Finally, it is clear that numerous examples of von Baer's Law, that related forms tend to resemble each other more in the early stages of their development than when adult, need have nothing whatever to do with Haeckel's reformulation of von Baer's Law, or with recapitulation, but are simply consequences of the laws of relative growth and of physiological genetics. For instance, the proportions of the forelimbs of man and all anthropoids are more similar in the fetus than in the adult (Schultz, 1926). But in every case the relative length of the fore-limb increases between fetus and adult. This happens to be 'recapitulatory' in excessively long-armed forms like the gibbon, but is the reverse of recapitulatory in man, who is undoubtedly descended from brachiating ancestors with relatively longer arms. The great similarity of the fetuses is due to the fact that in all cases the arms begin relatively small, and that final differences of proportion are due to the different degrees of positive heterogony which they exhibit. The 'recapitulation' by the new-born lamb in improved breeds of sheep of the approximate proportions of adult wild sheep (Chapters III, VII) is another

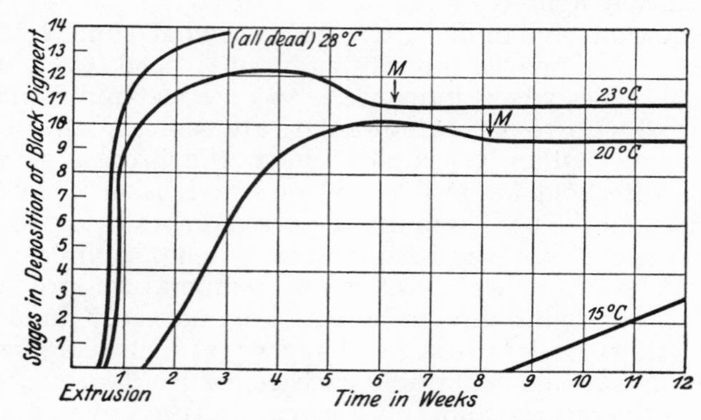

FIG. 103.—Effect of temperature on eye-darkening in a pure genetic strain (*rr*SS) of *Gammarus chevreuxi*.

At 10° C. no melanin is deposited. At 13°, deposition begins only after 5–6 months. The figure shows the facts for temperatures from 15° to 28°.

example of a change of this sort which happens to be recapitulatory.

All coloured Gammarus eyes so far investigated begin as pure red (Ford and Huxley, op. cit.). But we have in this no grounds whatever for supposing that Gammarus are descended from a red-eyed ancestor. It would appear to be a direct consequence of the empirical fact that melanin deposition can only begin after the red (lipochrome) pigment has been formed.

Again, the fact that both chelae of male Uca begin of female type, so that males and females are more alike when young, does not in the least prove that the males of any ancestral Uca ever possessed two chelae of existing or any other female type. In fact, all the probabilities are against this. It is due, we may suggest, rather to the fact that a large adult chela must begin its existence of small relative size, and that the form of the small chela has been moulded by natural selection into the female-type or feeding chela. A similar reasoning applies to the male type of juvenile female abdomen in all Brachyura: this affords no presumption whatever that ancestral female crabs when adult ever possessed an abdomen as narrow as that of the existing male. The presumption is quite different—that during Brachyuran evolution the male abdomen has been narrowed, the female broadened, and that it has proved an ontogenetic convenience to produce the abdomen at its first appearance, when it is relatively small, in a form similar to that of the male, from which the female type is derived by differential heterogony.

Numerous other examples could be given of this distinction between von Baer's and Haeckel's laws; but enough has been said to show the importance of the distinction, and the reason for the greater universality of von Baer's generalization.

Further, it is in general clear that rate-genes may mutate in either a plus or a minus direction, either accelerating or retarding the rates of the processes they affect. In the former case the effect will be in certain respects at least recapitulatory, since a condition which used to occur in the adult is now run through at an earlier stage. In the latter case, the effect will be anti-recapitulatory, since a condition which once characterized an earlier phase of development is now shifted to the adult phase. Neoteny is the most striking example of this effect; but we may also get single characters behaving in this way. When this occurs, previous adult characters, though still in a real sense potentially present, never appear because their formation is too long delayed: they are lost to the species by being driven off the time-scale of its development. For examples of this, notably in Helix, see the discussion in Ford and Huxley, 1927. (Fig. 104.)

De Beer (1930), in his résumé of this and kindred subjects, has given the useful term *paedomorphosis* to this type of effect. Bolk (1926) has drawn attention to the importance of such effects in the evolution of man, summing up the result as a process of 'fetalization', since in many respects post-natal or adult man resembles the fetal stages of apes. See also Kieslinger (1924).

Such phenomena are unintelligible on the Haeckelian doctrine. But they immediately fall into place when it is grasped that most examples of recapitulation constitute simply one side of a more general problem—the problem of altering the relative rates of growth and of other processes within the body.

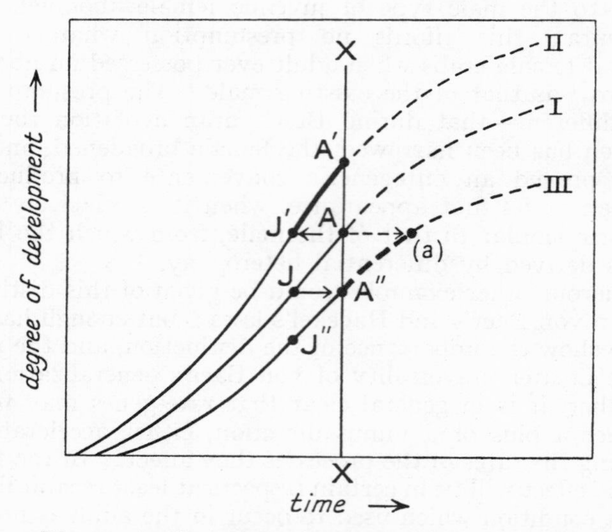

FIG. 104.—Diagram to illustrate positive and negative mutations in rate-factors, leading to recapitulation and paedomorphosis respectively.

X—X, period after which no further differentiation of the character is possible (adult phase in forms with fixed adult size; maximum size in forms with continuous growth). The curves represent the development of a character controlled by a rate-gene. The portions of the curves to the right of X—X (dotted) cannot be realized. I, condition in ancestral form; II, result of a mutation accelerating the process. The old adult condition (A) now becomes the juvenile condition (J′). A portion of the curve hitherto unrealized is now incorporated in the life-history (heavy line) leading to a new adult condition (A′). III, result of a mutation slowing down the process. The old juvenile condition (J) now becomes the adult condition (A″). A portion of the curve (heavy dotted line), including the old adult condition (*a*), is now extruded from the life-cycle by retardation.

§ 7. GENERAL APPROACH TO THE PROBLEM OF QUALITATIVE FORM-CHANGE

In general, it is clear that the effect of our knowledge of rate-factors and of the rules regulating relative growth will make it possible to analyse the problems involved in changes in proportion with a new insight.

In studying the growth of any part, it is by no means sufficient to determine its percentage change in relative size, as has often been the case in the past. A percentage change is the easiest to visualize by graphic methods, and is often very important in comparing related forms. But from the point of view of the underlying mechanism, it tells us little. A percentage change in relative size, as the work of Scammon has made clear (p. 132), may be merely an effect of the law of developmental direction and the time-handicap thus given to some organs as against others; the organ, once it begins its growth, may grow in linear proportions to the body as a whole, and yet show a change in relative size. Or a percentage change may be due solely to the heterogony of the organ, to its possessing a growth-coefficient above or below the body as a whole. Or thirdly, it may of course be and doubtless often is due to a combination of the two causes. In the first case, its relative growth will be according to the formula

$$y = bx + a \quad . \quad . \quad . \quad . \quad . \quad . \quad (1)$$

In the second case, according to the formula

$$y = bx^k \quad . \quad . \quad . \quad . \quad . \quad . \quad . \quad . \quad (2)$$

In the third case, according to the formula

$$y = bx^k + a \quad . \quad . \quad . \quad . \quad . \quad . \quad (3)$$

When $k = 1$, then formula (1) becomes equivalent to formula (3). And when in addition $a = 0$, it becomes equivalent to formula (2). Thus for those numerous cases where the growth-coefficient is close to unity, and the onset of development of the organ is not far from that of the standard part taken as representative of the rest of body (or from the mean value for all parts, when the body as a whole is taken as standard), the three formulae will serve equally well.

Formula (3) is the most inclusive, and should be taken as the theoretical basis for analysis; but a will often be negligible where heterogony is marked; and where heterogony is not marked, k will be so close to unity that Scammon's formula (1) will approximately apply. (See also Chap. IV, § 7.)

In addition to a precise analysis of growth after the organ has appeared, it is also of theoretical importance to know the time of its first appearance. From the work on Gammarus, we know that, as regards the straightforward rate-genes discussed in § 5, the slower the rate of the process which they control, the later is its visible onset.

This is presumably due to there being a maximum threshold which must be attained by the substance whose production they control before visible effects are produced. It is on general grounds probable that this is a widespread rule. In addition, genes have been discovered whose primary effect is on the time of onset of a process, and not on its rate (Ford and Huxley, op. cit.).

It should be possible by comparing related forms to discover whether time-relations of this second type are involved in addition to those of the first. We must further remember that growth-coefficients may change during ontogeny; they may change from strong to less strong positive heterogony, as in Uca large chela, or from isogony to positive heterogony, as in Maia large chela, etc. These facts must also have their genetic basis.

With an analysis such as this, we may hope for a fuller understanding of the processes involved in changes of proportion. One alteration in a single rate-gene may delay the first formation of an organ and also decrease the growth-coefficient once it is formed. Further, although the processes of histo-differentiation do not seem to follow the same laws of relative growth as those of auxano-differentiation, the quantitative intensity of the two kinds of growth-processes may well be controlled by the same genes.

In considering evolutionary changes in relative size, we must accordingly try to distinguish the various agencies which may be at work. It appears that these may be (*a*) mutations affecting the primary gradient of the early embryo, on which the time-relations of antero-posterior differentiation depend; (*b*) mutations affecting specific rate-genes; (*c*) mutations affecting specific 'time-genes'—genes controlling time of onset and not rate of processes. The processes controlled by the rate-genes and time-genes will be processes concerned with growth-gradients, whether of a major or minor nature: they will therefore always affect a number of parts in a correlated way.

Finally, it is at least possible, as we have seen in an earlier section, that the primary 'axial gradient' of the developing egg and early embryo, on which, we must suppose, depend the facts subsumed under the law of antero-posterior differentiation, itself continues to operate later as a growth-influencing gradient, as well as influencing the time-relations of differentiation. This would mean that mutations primarily selected because of their effect upon early development would have an effect upon proportional size in later life—an interesting example, if substantiated, of what Darwin called correlated variation.

§ 8. Conclusion

We have now completed our brief survey. Starting from the fact of obviously ' dysharmonic ' or heterogonic growth, we have discovered our first new empirical law—the law of constant differential growth-ratio. We have then recognized that it is only a special case of the law of differential growth-partition, which is the prime quantitative basis of relative growth. Passing on from that, we have found a further and quite unexpected empirical law—that the existence of a differential growth-ratio in an organ or region seems always to be associated with a growth-gradient culminating in a growth-centre ; or in other words, that the distribution of growth-potential is not marked by discontinuities or by frequent oscillations, but occurs in an orderly and continuously graded way. And we then showed that these localized growth-gradients were but special cases of growth-gradients permeating the whole body. These laws, however, only appear to apply to the stages of growth occurring after histological differentiation has been completed. Very rapid growth, obeying quite other laws, occurs during the earlier period. For these two phases of development, the terms histo-differentiation and auxano-differentiation are proposed.

After demonstrating that these growth-gradients were operative both in multiplicative and accretionary growth, giving rise to structures as dissimilar as a crustacean chela or a fowl's comb on the one hand, and a Nautilus shell or a rhinoceros horn on the other, we made it probable that the growth-gradients were either directly or indirectly correlated with the morphogenetic gradients or fields of Child, Weiss and others, and in general with the various polarized and field effects in the animal body.

In a discussion of the obscure subject of the physiological basis of growth-gradients, we discovered that the existence of a single appendage with high growth-ratio is associated with a slight increase of growth in the regions immediately posterior to it, but a slight decrease in those immediately anterior. The meaning of this remains quite unknown, but it has certain parallels in the field of regeneration and of experimental embryology. Further, the study of relative growth confirms that of regeneration in making us believe that the relative growth-rate (differential growth-ratio) of a part is determined in some way as an equilibrium between the growth of the part and the growth of the rest of the body. The role of hormones and of mutation in differential growth has been discussed, and the extent of our ignorance on this subject emphasized.

Finally, the bearings of the study of differential growth on other branches of biology have been discussed, and it has been shown that light is thereby shed upon such diverse problems as orthogenesis, recapitulation, vestigial organs, the existence of non-adaptive characters, physiological genetics, comparative physiology, and systematics.

I may conclude as I began, by a quotation from D'Arcy Thompson, to whose classical work all students of relative growth owe so much (*Growth and Form*, p. 719):

"The study of form may be descriptive merely, or it may become analytical. We begin by describing the shape of an object in the simple words of common speech: we end by defining it in the precise language of mathematics; and the one method tends to follow the other in strict scientific order and historical continuity".

BIBLIOGRAPHY

ABELOOS, M. (1928) : Sur la dysharmonie de croissance chez Planaria gonocephala Duges et sa reversibilité au cours du jeune ; *C.R. Soc. Biol.*, **98**, 917.

—— (1930) : Recherches expérimentales sur la croissance et la régénération chez les Planaires ; *Bull. Biol. France et Belg.*, **64**, 1–140.

ADOLPH, E. F. (1930) : Body Size as a Factor in the Metamorphosis of Frogs ; *Anat. Rec.*, **47**, 304.

ALLEN, B. M. (1918) : The Results of Thyroid Removal in the Larvae of Rana pipiens ; *Journ. Exper. Zool.*, **24**, 489–519.

—— (1919) : The Development of the Thyroid Glands of Bufo and their normal relation to Metamorphosis ; *Journ. Morph.*, **32**, 489–507.

ALPATOV, W. W. (1930) : Phenotypical Variation in Body and Cell-size of *Drosophila melanogaster* ; *Biol. Bull.*, **58**, 85–103.

ANDERSON, B. G. (1931) : Relative Growth in *Daphnia Magna* ; *Anat. Rec.*, **51**, Suppl., p. 55.

ANTONIUS, H. O. (1920) : Bemerkung über einige Saugetierschädel von Sardinien ; *Kon. Akad. Wetens. Amsterdam*, **29**, 26, vi. 20, p. 254.

APPLETON, A. B. (1929) : The Relation between the Rate of Growth and the Rate of Ossification in the Fœtus of the Rabbit (preliminary communication) ; *Proc. Assoc. Anat.*, 1929, p. 23.

ATKINS, D. (1926) : The Moulting Stages of the Pea-crab (*Pinnotheres pisum*) ; *Journ. Mar. Biol. Assn.*, **14**, 475.

BALFOUR-BROWNE, F. (1909) : The Life-history of the Agrionid Dragonfly ; *Proc. Zool. Soc. Lond.*, 1909, 253–84.

BATESON, W. (1894) : Materials for the Study of Variation. London : Macmillan.

BATESON, W. and BRINDLEY, H. H. (1892) : On some Cases of Variation in Secondary Sexual Characters statistically examined ; *Proc. Zool. Soc.* (1892), 585.

BEAN, R. B. (1924) : The Pulse of Growth in Man. A preliminary report ; *Anat. Rec.*, **28**, 45–61.

BECHER, A. (1923) : (Sheep-dog Skull). *Arch. Naturgesch.* (A.) 1923, (9).

DE BEER, G. R. (1930) : Embryology and Evolution. Oxford : Clarendon Press.

BĚLEHRÁDEK, J. and HUXLEY, J. S. (1930) : The Rate of Eye-growth and its Variation in *Gammarus chevreuxi* ; *Journ. Exper. Biol.*, **7**, 37.

BENAZZI, M. (1929) : Manifestazioni quantitative della rigenerazione negli insetti ; *Rivista di Biologia*, **11**, fasc. 5–6, p. 9.

BENOIT, J. (1927A) : Obtention chez les coqs d'un état intermédiaire de la crête . . . ; *C.R. Soc. Biol.*, **97**, 275.

—— (1927B) : La croissance conditionnée de la crête . . . ; *C.R. Soc. Biol.*, **97**, 279.

BERKSON, J. (1929) : Growth Changes in Physical Correlation—height, weight and chest-circumference, males ; *Human Biol.*, **1**, 462–502.

BERRILL, N. J. (1931) : Regeneration in *Sabella pavonina* (Sav.) and other Sabellid Worms ; *Journ. Exper. Zool.*, **58**, 495–523.

BERTALANFFY, LUDWIG (1928) : Kritische Theorie der Formbildung. Berlin : Borntraeger.

BLYTH, J. S. S., DODDS, E. C. and GALLIMORE, E. J. (1931) : Observations on the Assay of the Comb-growth Promoting Hormone ; *Journ. Physiol.*, **73**, 136–40.

BOLK, L. (1926) : Das Problem der Menschwerdung. Jena, 1926.

BOVET, D. (1930) : Les territoires de régénération ; leurs propriétés étudiées par la méthode de déviation du nerf ; *Rev. Suisse Zool.*, **37**, 83–145.

BOWER, F. O. (1930) : Size and Form in Plants, with special reference to the Primary Conducting Tracts. London : Macmillan.

BRAY, G. (1931) : Recent Advances in Allergy. London : Churchill, 1931.

BRODY, S. (1928) : An Analysis of the Course of Growth and Senescence. In *Growth*, by Robbins and other authors. Yale University Press.

BUSH, S. F. (1930) : Asymmetry and Relative Growth of Parts in the two Sexes of the Hermit-crab *Eupagurus prideauxi* ; *Arch. Entw. Mech.*, **123**, 39.

BUSH, S. F. and HUXLEY, J. S. (1930) : Distribution of Growth-activity in Eupagurus ; *Nature*, **126**, 240–1.

CALMAN, W. T. (1909) : Treatise on Zoology, Vol. VII. Appendiculata, 3rd fasc. Crustacea. London : Black.

CHAMPY, C. (1922) : (Thyroid) ; *Arch. Morph. Gen. Exp.*, 1922.

—— (1924) : Sexualité et Hormones. Paris : G. Doin.

—— (1929) : La croissance dysharmonique des caractères sexuels accessoires ; *Ann. Sci. Nat. Zool.*, (10) **12**, 193.

CHILD, C. M. (1915) : Senescence and Rejuvenation. Chicago.

CHOI, M. H. (1931) : Synchronism of Development in Parabiotic Amphibian Larvae ; *Folia Anatomica Japonica*, **9**. No. 5.

CLARK, A. (1927) : Comparative Physiology of the Heart. Cambridge Univ. Press.

CLAUSEN, H. J. (1929) : Rate of Histolysis of Anuran Tail Skin and Muscle during Metamorphosis ; *Anat. Rec.*, **44**, 218.

COGHILL, G. E. (1928) : VIII. (Growth of Nervous System) ; *Journ. Comp. Neurol.*, **45**.

COLLIP, J. B. (1930) : (Placental Hormones) ; *Canad. Med. Assoc. Journ.*, **23**, 631.

COTT, H. B. (1929) : Observations on the Natural History of the Racing-crab *Ocypoda ceratophthalma*, from Beira ; *Proc. Zool. Soc.* (1929), 755.

CUSHING, H. (1912) : The Pituitary Body and its Disorders ; Lippincott.

DAVENPORT, C. B. (1923): Body-build and its Inheritance; *Carnegie Inst., Washington, Publ.*, No. 329, p. 176.

DAVENPORT, C. B. (1926): Human Metamorphosis; *Amer. Journ. Phys. Anthrop.*, **9**, 205–32.

DAVENPORT, C. B. and SWINGLE, W. W. (1927): Effects of Operations upon the Thyroid Glands of Female Mice on the Growth of their Offspring; *Journ. Exper. Zool.*, **48**, 395–440.

DETWILER, S. R. (1930): Some Observations upon the Growth, Innervation and Function of Heteroplastic Limbs; *Journ. Exper. Zool.*, **57**, 183.

DIAKONOV, D. M. (1925): Experimental and Biometrical Investigations on Dimorphic Variability of Forficula; *Journ. Genetics*, **15**, 261.

DOMBROWSKI, VON (1889): Über die Geweihbildung der Rothirsche der Gegenwart . . .; De Weidmann, **20** (Nos. 15–19, 42, 44).

—— (1892): Die Gehörnbildung des Rehbockes des Gegenwart, etc.; Der Weidmann (1892–4), **22**, **23**, **24**.

DOMM, L. V. and JUHN, MARY (1927): Compensatory Hypertrophy of the Testes in Brown Leghorns; *Biol. Bull.*, **52**, 458–473.

DONALDSON, H. H. (1924): The Rat; Philadelphia: Wistar Inst. Mem.

DRENNAN, M. R. (1932): Pedomorphism in the Pre-Bushman Skull; *Amer. Journ. Phys. Anthrop.* (cited from Wistar Inst. Advance Abstract-sheet No. 179).

DUBOIS, E. (1914): Die Gesetzmässige Beziehung von Gehirn- zur Körpergrösse bei den Wirbeltieren; *Zeits. Morph. u. Anthrop.*, **18**.

—— (1918): On the Relation between the quantities of the Brain, the Neurone and its Parts, and the Size of the Body; *Acad. Sci. Amsterdam, Proceedings*, **20**, 1328.

—— (1922): Phylogenetic and Ontogenetic Increase of the Volumes of the Brain in Vertebrata; *K. Akad. Wetens. Amsterdam*, **25**, 330.

DUDICH, E. (1923): Ueber die Variation des Cyclommatus tarandus Thunberg; *Arch. f. Naturgesch.*, (A) 1923, (2) 62.

EIGENBRODT, H. J. (1930): The Somatic Effects of Temperature on a Homozygous Race of Drosophila; *Physiol. Zool.*, **3**, 392–411.

EMERSON, A. E. (1926): Development of a Soldier of Nasutitermes (Constrictotermes) . . .; *Zoologica*, **7**, 69.

EMERY, C. (1921): Quels sont les facteurs du Polymorphisme du sexe féminin chez les Fourmis?; *Rev. Gén. Sci.*, **32**.

ENTZ, G. (1927): Beiträge zur Kenntnis der Peridineen. II. resp. VII. Studien an Süsswasser-Ceratien; *Arch. f. Protistenk.*, **58**, 344–440.

FAURÉ-FREMIET, E. (1930): Growth and Differentiation of the colonies of *Zoothamnium alternans*; *Biol. Bull.*, **58**, 28.

FORD, E. (1930): Herring Investigations at Plymouth. VIII. The transition from larva to adolescent; *Journ. Mar. Biol. Assn.*, **16**, 723.

—— (1931A): Growth in Length during the Transition from Larva to Adolescent in the Pilchard and Sprat: *Journ. Mar. Biol. Assn.*, **17**. 977–85.

FORD, E. (1931B) : Changes in Length during the Larval Life and Metamorphosis of the Freshwater Eel (*Anguilla vulgaris Turt.*) ; *Journ. Mar. Biol. Assn.*, **17**. 987–1000.

FORD, E. B. (1928) : The Inheritance of Dwarfing in *Gammarus chevreuxi* ; *Journ. Genet.*, **20**, 43.

—— (1929) : The Physiology of Genetics ; *Eugen. Rev.*, **21**, 114.

FORD, E. B. and HUXLEY, J. S. (1929) : Genetic rate-factors in Gammarus ; *Arch. Entw. Mech.*, **117**, 67.

GABRITSCHEVSKY, E. (1930) : Les réductions régulatrices et les compensations hypertrophiques pendant l'ontogenèse et la régénération de l'araignée *Thomisus onistus* ; *Bull. Biol.*, **64**, 155–88.

GAJEWSKI, N. (1922) : Ueber die Variabilität bei Artemia salina ; *Intern. Rev. Ges. Hydrobiol.*, **10**, 139, 299.

GAUSE, C. F. (1931) : Ueber den Einfluss verkürzter larvaler Ernährungszeit auf die Eiergrösse von Drosophila ; *Biol. Zentralbl.*, **51**, 209.

GIESBRECHT, W. (1910) : Fauna und Flora des Golfes von Neapel ; **33**, Stomatopoda (1)

GOLDSCHMIDT, R. (1923) : The Mechanism and Physiology of Sex Determination ; London : Methuen.

—— (1927) : Physiologische Theorie der Vererbung ; Berlin : Springer.

GRAY, J. (1929) : The Kinetics of Growth ; *Brit. Journ. Exper. Biol.*, **6**, 248.

GRIFFINI, A. (1912) : Strane variazioni individuali in alcune specie di Coleopteri ; *Boll. Mat. Sci. Fis. Nat.*, Lodi, **13**, (8).

GUYENOT, E. and PONSE, K. (1930) : Territoires de régénération et transplantations ; *Bull. Biol. Fr. et Belg.*, **64**, 251–84.

HAMMETT, F. S. (1929A) : Thyroid and Differential Development ; *Endokrinologie*, **5**, 81.

—— (1929B) : Thyroid and Growth ; *Quart. Rev. Biol.*, **4**, 353.

HAMMOND, J. (1921) : On the Relative Growth and Development of Various Breeds and Crosses of Sheep ; *Journ. Agric. Sci.*, **11**, 267–407.

—— (1928) : Selection for Meat Production ; *Verh. v. Internat. Kongr. Vererbungswiss, Berlin*, 1927, **2**, 769–95.

—— (1928) : How Feeding affects Conformation ; *Farmer and Stockbreeder and Agric. Gaz.*, 10.12.28, p. 4.

—— (1929) : Probleme der Fleischerzeugung ; *Zeits. Züchtungskunde*, **4**, 543–62.

—— (1930) : Animal Production : the Influence of Natural Conditions on Animal Production ; *Trop. Agric.*, **7**, 147–50.

—— (1931) : Growth of the Rabbit in Relation to Flesh Production ; *Year Book Nat. Rabbit Council* (1931).

HARDESTY, MARY (1931) : The Structural Basis for the Response Comb of the Brown Leghorn Fowl to the Sex Hormones ; *Amer. Journ. Anat.*, **47**, 277–323.

HARE, F. (1932) : Polymorphism among the Sub-genera of Nasutitermes ; *Journ. Morph. and Physiol.* (in press).

HARRIS, H. A. (1931A) : The Comparative Aspect of Growth in Children ; *Lancet*, 28 Mar., 1931, p. 691.

HARRIS, (1931B) : The Anatomical and Physiological Characteristics and Development of Children between the ages of 7 and 11 ; Report. Consultative Committee on the Primary School, pp. 222–54.

HARRISON, R. G. (1924) : Some unexpected Results of the Heteroplastic Transplantation of Limbs ; *Proc. Nat. Acad. Sci.*, **10**, 69–74.

—— R. (1929A) : Heterotransplantation in Amphibian Embryos ; *Proc. X. Internat. Zool. Congr. Budapest*, 1927 (1929).

—— (1929B) : Correlation in the Development and Growth of the Eye studied by means of Heteroplastic Transplantation ; *Arch. Entw. Mech.*, **120**, 1.

HASEBROCK, K. (1917) : Die Entwicklungsmechanik des Herzwachstum ; *Pfluger's Archiv*, **168**.

HASEMAN, J. D. (1907A) : The Direction of Differentiation in regenerating Crustacean appendages : *Arch. Entw. Mech.*, **24**, 617.

—— (1907B) : The Reversal of the Direction of Differentiation in the Chelipeds of the Hermit-crab ; *Arch. Entw. Mech.*, **24**, 663.

HEATH, H. (1927) : Caste-formation in the Termite genus Termopsis ; *Journ. Morph. and Phys.*, **43**, 387.

—— (1928) : Fertile Termite Soldiers ; *Biol. Bull.*, **54**, 324.

HECHT, S. (1916) : Form and Growth in Fishes ; *Journ. Morph.*, **37**, 379.

HENDERSON, J. R. and MATTHAI, G. (1910) : On certain species of Palaemon from South India ; *Records of the Indian Museum*, **5**, 277–306.

HERMS, W. B. (1928) : The Effect of Different Quantities of Food during the Larval period on the sex-ratio and size of *Lucilia sericata* Meigen and *Theobaldia incidens* (Thom) ; *Journ. Econ. Entom.*, **21**, 720.

HERRICK, F. H. (1911) : Natural History of the American Lobster ; *Bull. U.S. Bureau Fisheries*, **29**, 149, Document No. 747.

HERSH, A. H. (1928) : Organic Correlation and its Modification in the bar series of Drosophila ; *Journ. Exper. Zool.*, **50**, 239.

HESSE, R. (1927) : Über Grenzen des Wachtums ; Jena, G. Fischer, 1927.

HICKLING, C. F. (1930) : A Contribution towards the Life-history of the Spur-dog ; *J. Mar. Biol. Ass.*, **16**, 529.

HINTON, M. A. C. (1926) : Monograph of the Voles and Lemmings (Microtinae) ; London, British Museum (Natural History).

HIRSCH, G. C. (1931) : The Theory of Fields of Restitution with special Reference to the Phenomena of Secretion ; *Biol. Rev.*, **6**, 88.

HOOTON, E. A. (1931) : Up from the Ape. London : Allen & Unwin.

HUBBS, CARL L. (1925) : The Metamorphosis of the Californian Ribbon Fish, *Trachypterus rex-salmonorum* ; *Papers Michigan Acad. Sci., Arts and Letters*, **5**, 469–76.

HUTT, F. B. (1929) : Sex Dimorphism and Variability in the Appendicular Skeleton of the Leghorn Fowl ; *Poultry Science*, **8**, 202–18.

HUNTSMAN, A. G. (1921) : The Effect of Light on Growth in the Mussel : *Proc. Roy. Soc. Canada*, **15**, (5), 23–8.

HUXLEY, J. S. (1921) : Studies in Dedifferentiation. II. Dedifferentiation and Resorption in Perophora ; *Q. J. Micr. Soc.*, **65**, 643.

HUXLEY, J. S. (1924A) : The Variation in the Width of the Abdomen in Immature Fiddler-crabs considered in relation to its relative Growth-rate ; *Amer. Nat.*, **58**, 468.

—— (1924B) : Constant Differential Growth-ratios and their Significance ; *Nature*, **114**, 895.

—— (1924C) : Early Embryonic Differentiation ; *Nature*, **114**, 3.

—— (1926) : The Annual Increment of the Antlers of the Red Deer (*Cervus elaphus*) ; *Proc. Zool. Soc.* (1926), 1021.

—— (1927A) : Further Work on Heterogonic Growth ; *Biol. Zentralbl.*, **47**, 151.

—— (1927B) : Discontinuous Variation and Heterogony in Forficula ; *Journ. Genetics*, **17**, 309.

—— (1927C) : On the Relation between Egg-weight and Body-weight in Birds ; *Journ. Linn. Soc.* (Zool.), **36**, 457.

—— (1927D) : Studies on Heterogonic Growth (IV). The bimodal cephalic horn of *Xylotrupes gideon* ; *Journ. Genetics*, **17**, 45.

—— (1931A) : The Relative Size of Antlers in Deer ; *Proc. Zool. Soc.* (1931), (819–64).

—— (1931B) : Notes on Differential Growth ; *Amer. Nat.*, **65**, 289–315.

—— (1931C) : Relative Growth of Mandibles in Stag-beetles (*Lucanidae*) ; *Linn. Soc. Journ.* (Zool.), Vol. 37 : 675–703.

HUXLEY, J. S. and CALLOW, F. S. (unpublished) : Note on the Asymmetry of male Fiddler-crabs (*Uca pugilator*).

HUXLEY, J. S. and FULTON, J. F. (1924) : The Influence of Temperature on the Action of Insulin ; *Nature*, 16.2.24, 234.

HUXLEY, J. S. and RICHARDS, O. W. (1931) : Studies in Heterogonic Growth (8). Changes in proportions in the abdomen of the shore-crab *Carcinus maenas* ; *Journ. Mar. Biol. Assn.*, **17**, 1001.

IMMS, A. D. (1920) : On the Structure and Biology of Archotermopsis ; *Phil. Trans. Roy. Soc. B.*, **209**, 75.

JACKSON, C. M. (1925) : The Effects of Inanition and Malnutrition upon Growth and Structure ; London and Philadelphia, 1925.

JUCCI, C. (1924) : Il polimorfismo unisessuale di un Acaro plumicolo ; *Boll. Lab. Zool. Gen. Agr. Portici.*, **18**, 79.

JUHN, M., FAULKNER, O. H., and GUSTAVSEN, R. G. (1931) : The Correlation of Rates of Growth and Hormone Threshold in the Feathers of Fowls ; *Journ. Exper. Zool.*, **58**, 69.

KALSHOVEN, L. G. (1930) : De Biologie van de Djatitermiet Kalotermis teitenae. . . . (H. Veenman & Zonen, Wageningen).

KEITH, A. (1923) : The Adaptational Machinery concerned in the Evolution of Man's Body ; *Nature*, 18·8·1923, Suppl.

KEMP, S. (1913) : Crustacea Decapoda. *Records of the Indian Museum*, **8** (1912 to 1922), 289–310.

—— (1914) : *Records of the Indian Museum*, **10**, 81.

—— (1915) : Fauna of the Chilka Lake, *Crustacea Decapoda, Memoirs of the Indian Museum*, **5**, 201–325.

KEYS, ANCEL B. (1928) : The Weight-length Relation in Fishes ; *Proc. Nat. Acad. Sci.*, **14**, 922–5.

KIESLINGER, A. (1924) : Neotenie, Persistenz, Degeneration ; *Proc. Konink. Akad. Wetens. Amsterdam*, **27**, 761–71.

KLATT, B. (1913) : Ueber den Einfluss der Gesamtgrösse auf das Schädelbild, etc. ; *Arch. Entw. Mech.*, **36**, 387–471.

—— (1919) : Zur methodik vergleichender metrischer Untersuchungen, besonders des Herzgewichtes ; *Biol. Zentralbl.*, **39**, 406.

KOZELKA, A. W. (1930) : Integumental Grafting as a Means of Analysing the Factors determining the Secondary Sexual Characters of the Domestic Fowl ; *Anat. Rec.*, **47**, 302.

KUHL, W. (1928) : Die Variabilität der abdominal Körperanhänge bei Forficula auricularia ; *Zts. Morph. Oekol. d. Tiere*, **12**, 299.

KUNKEL, B. W. and ROBERTSON, J. A. (1928) : Contributions to the Study of Relative Growth of Parts in *Gammarus chevreuxi* ; *Journ. Mar. Biol. Assn.*, **15**, 655.

LAMEERE, A. (1904) : L'évolution des ornements sexuels ; *Bull. Acad. Belg.* (1904), 1337.

—— (1915) : Les caractères sexuels des Prionides ; *Bull. Sci. Fr. et Belg.*, **49**, 1915.

LANDAUER, W. (1927) : Untersuchungen über Chondrodystrophie. I. Allgemeine Erscheinungen und Skelett chrondrodystrophischer Hühnerembryonen ; *Arch. Entw. Mech.*, **110**, 195–278.

—— (1929) : Thyrogenous Dwarfism (*Myxœdema infantilis*) in the Domestic Fowl ; *Amer. Journ. Anat.*, **43**, 1–20.

LAPICQUE, L. (1907) : Le poids encéphalique en fonction du poids corporal entre individues d'une même espèce ; *Bull. Mem. Soc. Anthrop.*, Paris (5), **8**, 313.

—— (1922) : Le poids du cerveau et l'intelligence. Chap. 2 in ' Traité de Psychologie ', ed. Dumas, Paris.

LAPICQUE, L. and GIROUD, A. (1923) : Sur le nombre des fibres nerveuses périphériques en fonction de la grandeur du corps ; *C.R. Soc. Biol.*, **88**, 43.

LATIMER, HOMER B. and AIKMAN, J. M. (1931) : The Pre-natal Growth of the Cat. I. The growth in weight of the head, trunk, fore-limbs, and hind-limbs ; *Anat. Rec.*, **48**, 1–26.

LIGHT, S. F. (1927) : A new and more exact Method of expressing important specific Characters of Termites ; *Univ. Cal. Publ. Entomol.*, **4**, 75–88.

LILLIE, F. R. and JUHN, M. (1932) : The Physiology of Development of Feathers ; *Physiol. Zool.*, cited from Wistar Inst. Advance Abstract-sheet.

LUND, E. J. (1923) : Threshold Densities of the Electric Current for Inhibition and Orientation of Growth in Obelia ; *Proc. Soc. Exper. Biol. and Med.*, **21**, 127–8.

—— (1928) : Relation between Continuous Bioelectric Currents and Cell Respiration. II. (*a*) A theory of continuous bio-electric currents and electric polarity of cells. II. (*b*) Theory of cell correlation ; *Journ. Exper. Zool.*, **51**, 265–90.

MACKINTOSH, N. A. and WHEELER, J. F. G. (1929) : Southern Blue and Fin Whales ; *Discovery Reports*, **1**, 257–540.

MARTIN, R. (1928) : Lehrbuch der Anthropologie. 2nd Ed. Jena, 1928.

MATTHEW, W. D. (1926): The Evolution of the Horse; *Quart. Rev. Biol.*, **1**, 139.

MEAD, C. H. (1930): A Quantitative Study in Human Teratology; *Human Biol.*, **2**, 1.

MJÖBERG, ERIC (1925): The Mystery of the so-called 'Trilobite Larvae' definitely solved; *Psyche*, **32**, 119–54.

MORGAN, A. H. (1929): The Mating Flight and the Vestigial Structures of the Stump-legged Mayfly, *Campsurus segnis* Needham; *Ann. Entomol. Soc. Amer.*, **22**, 61–8.

MORGAN, T. H. (1923A): The Development of Asymmetry in the Fiddler-crab; *Amer. Nat.*, **57**, 269.

—— (1923B): Further Evidence on Variation in the Width of the Abdomen in Immature Fiddler-crabs; *Amer. Nat.*, **57**, 274.

—— (1924):

NAÑAGAS, J. C. (1925): A Comparison of the Growth of the Body Dimensions of Anencephalic Human Fetuses with normal Fetal Growth as determined by Graphic Analysis and Empirical Formulae; *Amer. Journ. Anat.*, **35**, 455.

NEEDHAM, J. (1931): Chemical Embryology. Cambridge. Univ. Press. 3 vols.

NEVALONNYI, M. and PODHRADSKY, J. (1930): Der Einfluss der Schilddrüse und des Thymus auf die Skelettbildung . . .; *Vestnik Ceskoslov. Akad. Zemedelske.*, **6**, 1.

NOMURA, E. (1928): On the Relation between Weight and Dimensions in the Bivalves, *Tapes philippinarum* and *Cytherea meretrix*; *Sci. Rep. Tohoku Imp. Univ.* (4th Ser., Biol.), **3**, 113–24.

NOMURA, E. and SASAKI, K. (1928): On the Relation between Weight and Dimensions in the Gastropods, *Haliotis gigantea var. discus* and *Littorina sitchana*; *Sci. Rep. Tohoku Imp. Univ.* (4th Ser., Biol.), **3**, 125.

OLMSTED, J. M. D. and BAUMBERGER, J. P. (1923): Form and Growth of Grapsoid Crabs; *J. Morph.*, **38**, 279.

OSBORN, H. F. (1929): The Titanopheres of Ancient Wyoming, Dakota, and Nebraska; U.S. Dept. of the Interior Geol. Survey, Monograph No. 55.

PALENITSCHKO, Z. G. (1927): Zur vergleichenden Variabilität der Arten und Kasten bei den Ameisen; *Zeits. Morph. u. Ökol.*, **9**, 410.

PARKES, A. S. (1929): The Internal Secretions of the Ovary. London: Longmans Green & Co.

PARROT, C. (1894): Über die Grössenverhaltnisse des Herzens bei Vögeln; *Zool. Jahrb.* (Syst.), **7**.

PARSONS, F. G. (1927): The Englishman of the Future; Rept. Brit. Assoc. Leeds, 1927, Pres. Address, Section H.

PEARSALL, W. H. (1926): Growth Studies. V. Factors affecting the development and form of leaves; *Ann. Bot.*, **40**, 85–103.

—— (1927): Growth Studies. VI. On the relative sizes of growing plant organs; *Ann. Bot.*, **41**, 549.

PEARSALL, W. H. and HANBY, A. M. (1925): The Variation of Leaf Form in *Potamogeton perfoliatus*; *New Phyt.*, **24**, 112–20.
PEARSON, J. (1908): L.M.B.C. Memoir No. 16; Cancer. London: Williams & Norgate.
PERKINS, M. (1929): Growth Gradients and the Axial Gradients of the Animal Body; *Nature*, **124**, 299.
PETERSEN, CHR. (1921): Das Quotientengesetz: eine biologisch-statistische Untersuchung; Kopenhagen, Bianco Lunos, 1921.
PÉZARD, A. (1918): Le conditionnement physiologiques des caractères sexuels secondaires chez les Oiseaux.
—— (1921): Numerical Law of Regression of certain Secondary Sex Characters; *Journ. Gen. Physiol.*, **3**, 271.
PRZIBRAM, H. (1902): Intraindividuelle Variabilität der Carapax-dimensionen bei brachyuren Crustaceen; *Arch. Entw. Mech.*, **13**, 588.
—— (1917): Wachstumsmessungen an Sphodromantis bioculata Burm. III. Länge regenerierender und normaler Schreitbeine; *Arch. Entw. Mech.*, **43**, 1.
—— (1925): Direkte Temperaturabhängigkeit der Schwanzlänge bei Ratten, Mus (Epimys) decumanus Pall. und M. (E.) rattus L.; *Arch. Entw. Mech.*, **104**, 434.
—— (1927): Diskontinuität des Wachstums als eine Ursache diskontinuierlicher Variation bei Forficula; *Arch. Entw. Mech.*, **112**, 142.
—— (1930): Connecting Laws in Animal Morphology. London: Univ. London Press.
PURDY, C. and SHEARD, C. (1931): Electric Potentials in the Skin; *Anat. Rec.*, **51**, 25.

RAW, FRANK (1927): The Ontogenies of Trilobites, and their Significance; *Amer. Journ. Sci.*, **14**, 1–35, 131–49.
RITCHIE, J. (1920): The Influence of Man on Animal Life in Scotland; Cambridge Univ. Press.
ROBB, R. CUMMING (1929): On the Nature of Hereditary Size-limitation. II. The growth of parts in relation to the whole; *Brit. Journ. Exper. Biol.*, **6**, 311–24.
ROBBINS, W. J., BRODY, S., HOGAN, A. G., JACKSON, C. M. and GREENE, W. (1928): Growth. New Haven: Yale Univ. Press, 1928.
RÖRIG, (1901): (Deer Antlers); *Arch. Entw. Mech.*, **11**, 65.
RUUD, GUDRUN (1929): Heteronom-orthotopische Transplantationen von Extremitätenanlagen bei Axolotlembryonen; *Arch. Entw. Mech.*, **118**, 308–51.
RŮŽIČKA, V. (1921): Über Protoplasmahysteresis und eine Methode zur direkten Bestimmung derselben; *Pfl. Arch.*, **194**, 135.

SANTOS, F. V. (1929): Studies on Transplantation in Planaria; *Biol. Bull.*, **57**, 188–97.
SASAKI, K. (1928): On the Growth of *Telmessus cheiragonus*; *Sci. Rep. Tohoku Imp. Univ.* (4th Ser., Biol.), **3**, 805.
SCAMMON, R. E. and CALKINS, L. A. (1929): The Development and Growth of the External Dimensions of the Human Body in the Fetal Period. Univ. Minnesota Press: Minneapolis, p. 367.

SCHMALHAUSEN, J. (1927A) : Beiträge zur quantitativen Analyse der Formbildung. I. Über die Gesetzmässigkeiten des Embryonalen Wachstums : *Arch. Entw. Mech.*, **109**, 455–512.

—— (1927B) : Beiträge zur quantitativen Analyse der Formbildung. II. Das Problem des proportionalen Wachstums ; *Arch. Entw. Mech.*, **110**, 33–62.

—— (1930) : Über Wachstumformeln und Wachstumstheorien ; *Biol. Zentralbl.*, **50**, 292–307.

SCHMALHAUSEN, I. and STEPANOVA, J. (1926) : Das embryonale Wachstum des Extremitättenskelettes des Hühnchens ; *Arch. Entw. Mech.*, **108**, 729.

SCHULTZ, A. H. (1926) : Fetal Growth of Man and other Primates ; *Quart. Rev. Biol.*, **1**, 465.

—— (1930) ; The Skeleton of the Trunk and Limbs of Higher Primates ; *Human Biol.*, **2**, 303.

SCHWIND, J. L. (1931) : Heteroplastic Experiments on the Limb and Shoulder Girdle of Amblystoma ; *Journ. Exper. Zool.*, **59**, 265.

SEVERINGHAUS, A. E. (1930) : Gill Development in *Amblystoma punctatum* ; *Journ. Exper. Zool.*, **56**, 1.

SEXTON, E. W. (1924) : The Moulting and Growth-stages of Gammarus, with descriptions of the Normals and Intersexes of *G. chevreuxi* ; *Journ. Mar. Biol. Assn.*, **13**, 340.

SEYMOUR SEWELL, R. B. (1929) : The Copepoda of Indian Seas : Calanoida ; *Mem. Indian Mus.*, **10**, 1–221.

SHAW, M. (1928) : A Contribution to the Study of Relative Growth of parts in *Inachus dorsettensis* ; *Brit. Journ. Exper. Biol.*, **6**, 145.

SILVESTER, (1926) : Descrizione di particolori individui, etc. ; *Boll. Lab. Zool. Gen. & Agr.*, **19**, 1.

SINNOTT, E. W. (1930) : The Morphogenetic Relationships between Cell and Organ in the Petiole of Acer ; *Bull. Torrey Bot. Club*, **57**, 1–20.

SINNOTT, E. W. and DURHAM, G. B. (1929) : Developmental History of the Fruit in lines of *Cucurbita pepo* differing in fruit shape ; *Bot. Gazette*, **87**, 411–21.

SINNOTT, E. W. and HAMMOND, D. (1930) : Factorial Balance in the Determination of Fruit Shape in Cucurbita ; *Amer. Nat.*, **64**, 509–24.

SJOSTEDT, Y. (1925) : Revision der Termiten Afrikas ; *K. Svensk. Vetensk. akad. Handl.* (3), **3**, 1.

SMIRNOV, E. and ZHELOCHOVTSEV, A. N. (1926) : Veränderung der Merkmale bei Calliphora erythrocephala Mg. unter dem Einfluss verkürzter Ernährungsperiode der Larve ; *Arch. Entw. Mech.*, **108**, 579-95.

SMIRNOV, E. and ZHELOCHOVTSEV, A. N. (1931) : Das Gesetz der Altersveränderungen der Blattform bei Tropaeolum majus L. unter verschiedenen Beleuchtungsbedingungen : *Planta Arch. f. wissens. Bot.*, **15**, 299–354.

SMITH, G. W. (1906A) : Fauna and Flora des Golfes v. Neapel, Vol. **29** (Rhizocephala).

—— (1906B) : High and Low Dimorphism ; *Mit. Zool. Stat. Neapel*, **17**, 312.

SNELL, G. D. (1931) : Inheritance in the House Mouse: the linkage relations of short-ear, hairless, and naked ; *Genetics,* **16,** 42–74.

SPORN, E. (1926) ; Über die Gesetzmässigkeiten im Bau der Muschelgehäuse ; *Arch. Entw. Mech.,* **108.**

—— (1926) : Einiges über Ideal-, Normal- und Individual-formen der Blätter ; *Arch. Entw. Mech.,* **107.**

STOCKARD, C. R. (1931) : The Physical Basis of Personality. New York and London.

STREETER, G. L. (1930) : Deficiencies in Fœtal Tissues. . . . Contributions to Embryology, No. 126 (Carnegie Institution), **32.**

STURTEVANT, A. H. (1924) : An Interpretation of Orthogenesis ; *Science,* **59,** 579.

TAYLOR, W. P. (1915) : Description of a new subgenus (Arborimus) of Phenacomys, with a contribution to knowledge of the habits and distribution of *Phenacomys longicaudus* True ; *Proc. Cal. Acad. Sci.,* 4th series, **5,** 111–61.

TAZELAAR, M. A. (1930) : The Relative Growth of parts in *Palaemon carcinus* ; *Brit. Journ. Exper. Biol.,* **7,** 165.

TEISSIER, G. (1927) : La croissance nucléaire en fonction de la croissance cellulaire au cours de l'ovogenèse, chez Hydractinia echinata (Flem.) ; *C.R. Soc. Biol.,* **97,** 1524.

—— (1929) : Dysharmonies chimiques dans la croissance larvaire de Tenebrio molitor ; *C.R. Soc. Biol.,* **100,** 1171.

—— (1931) : Recherches morphologiques et physiologiques sur la croissance des insectes ; *Trav. Stat. Biol. Roscoff,* **9,** 29–238.

THIEL, M. E. (1926) : Formwachstumversuche an Sphaerium corneum ; *Arch. Entw. Mech.,* **108,** 87–137.

THOMPSON, C. B. (1917) ; Origin of the Castes of the Common Termite, *Leucotermis flavipes. Journ. Morph.,* **31,** 83.

THOMPSON, D'ARCY W. (1917) : Growth and Form : Cambridge Univ. Press, p. 793.

THOMSON, G. M. (1922) : The Naturalization of Plants and Animals in New Zealand : Cambridge Univ. Press.

TODD, T. WINGATE (1926) : Skeletal Adjustment in Jaw Growth ; *Dental Cosmos,* Dec., 1926, p. 17.

TUCKER, B. W. (1930) : On the Effects of an Epicaridan parasite, *Gyge branchialis,* on *Upogebia littoralis* ; *Quart. Journ. Micros. Sci.,* **74,** 1–118.

TWITTY, V. C. (1930) : Regulation in the Growth of Transplanted Eyes ; *Journ. Exper. Zool.,* **55,** 43.

TWITTY, V. C. and SCHWIND, J. L. (1931) : The Growth of Eyes and Limbs transplanted heteroplastically between two species of Amblystoma ; *Journ. Exper. Zool.,* **59,** 61–86.

UBISCH, L. v. (1915) : Über den Einfluss von Gleichgewichtsstörung auf die Regenerationsgeschwindigkeit ; *Arch. Entw. Mech.,* **41,** 237–50.

UBISCH, L. VON (1922) : Über die Aktivierung regenerativer Potenzen ; *Arch. Entw. Mech.,* **51,** 33–58.

—— (1923) : Das Differenzierungsgefälle des Amphibienkörpers und seine Auswirkungen ; *Arch. Entw. Mech.,* **52,** 624–70.

VANDEL, A. (1930): La production d'intercastes chez la fourmi *Pheidole pallidula* sous l'action de parasites du genre Mermis; *Bull. Biol.*, **64**, 457–94.

WACHS, H. (1914): Neue Versuche zur Wolff'schen Linsenregeneration; *Arch. Entw. Mech.*, **39**.

WADDINGTON, C. H. (1929): Notes on graphical methods of recording the dimensions of Ammonites; *Geol. Mag.*, **66**, 180–6.

WALLIS, R. SAWTELL (1932): Relative Growth of the Extremities from Two to Eighteen Years of Age; *Amer. Journ. Phys. Anthrop.* (cited from Wistar Inst. Advance Abstract-sheet No. 179).

WARDLAW, C. W. (1924–1928): Size in Relation to Internal Morphology. 1, 2 and 3; *Trans. Roy. Soc. Edinb.*, **53** (1924), **54** (1925) and **56** (1928).

WATANABE, Y. (1931): On the Physiological Axial Gradients of Chaetopod Annelids. II. Axial Gradients of Oxidizable Substance in Earthworms as determined by the Manoilov Reaction. *Sci. Repts. Tohoku Imp. Univ. 4th Series* (*Biol.*), **6**, 437–73.

WEISS, P. (1926): Morphodynamik: ein Einblick in die Gesetze der organischen Gestaltung an Hand von experimentellen Ergebnissen; *Abhandl. theoret. Biol.*, Heft 23; pp. 43.

WERNER, F. (1927): Wachstum und Formentwicklung der Cladocere macrothrix rosea; *Arch. Entw. Mech.*, **109**, 241–52.

WHEELER, W. M. (1920): Ants. New York: Columbia Univ. Press.

—— (1928): Mermis Parasites and Intercastes among Ants; *Journ. Exper. Zool.*, **50**, 165.

WILDER, I. W. (1924): The Relation of Growth to Metamorphosis in *Eurycea bislineata* (Green); *Journ. Exper. Zool.*, **40**, 1–112.

YERKES, R. M. (1901): A Study of Variation in the Fiddler-crab Gelasimus; *Proc. Amer. Acad. Arts. Sci.*, **36**, 417.

ZELENY, C. (1905): Compensatory Regulation; *Journ. Exper. Zool.*, **2**, 1.

ZUCKERMAN, S. (1926): Growth-changes in the Skull of the Baboon, *Papio porcarius*; *Proc. Zool. Soc.* (1926), 843.

ADDENDA

HERE I have summarized a few papers which I came across too late to insert in the body of the book.

Owing to an oversight, the valuable work of Robb (1929) has not been adequately discussed in the text. It has very interesting bearings on the relation of heterogon to endocrine control (see Chap. VI, § 4). He investigated the growth of various organs in a large (Flemish) and small (Polish) breed of rabbits, reaching about 6 and 3 kg. adult weight respectively, and in their F_1 hybrids.

He first found that the pituitary (weight) shows simple negative heterogony relative to (clean) body-weight, with the same growth-coefficient ($k = 0{\cdot}55$) in all three types; the curves all have the same point of origin. Adrenal weight shows simple positive heterogony, but k is higher for the small breed (1·34 as against 1·19 for the Flemish). As a result the relative weight attained by the adrenal in the adult Polish is just double what it is in the adult Flemish (0·2 as against 0·1 per cent); the hybrids show an intermediate value. The difference is due almost exclusively to an enlargement of the cortex.

The growth of the thyroid falls into two phases, one of negative heterogony (k about 0·53) up to about 600 g. body-weight, and one of positive heterogony (k about 1·12) from then on.

The testis, like the adrenal, attains a higher relative weight in the dwarf than in the giant race (0·3 as against 0·12 per cent, with a value of 0·2 in the hybrids), and the heterogony curves against body-weight are, of course, very dissimilar in the two breeds. But when testis weight is plotted logarithmically against adrenal weight, the two breeds show almost identical curves, negatively heterogonic ($k = 0{\cdot}74$) up to 40 days of age, then with very high positive heterogony ($k = 2{\cdot}3$). Testis weight against pituitary weight, on the other hand, shows $k = 1{\cdot}4$ up to 40 days, but then $k = 5{\cdot}1$ for the giant, 5·8 for the dwarf race. This appears to indicate a more marked interdependence of testis and adrenal cortex than of testis and pituitary. The identical growth-coefficient of the pituitary in both races indicates that body-size here must primarily depend upon other factors than relative pituitary size.

Robb concludes with an interesting theoretical discussion, which cannot be summarized here. One point deserves mention: if an organ has a certain fraction in active heterogonic relation with the body, but also an inert fraction which is isogonic, then the formula for its growth will be $y = bx + c$. In some cases apparently irregular heterogony curves could be made to conform to the simple heterogony formula by correcting for such an inert fraction; but this for the moment remains speculative.

Werner (1927) gives an elaborate discussion of form-changes in the Cladoceran *Macrothrix rosea*. Certain parts appear to show positive, others negative heterogony. But the growth-changes are often complex, and would appear to indicate the existence of elaborate *gradient-fields* controlling growth.

Anderson (1931) finds that the simple heterogony formula applies to the growth of various parts in the Cladoceran *Daphnia magna*. Interestingly enough, while carapace length is positively heterogonic (relative to total length) until the time of maturity, after which it becomes isogonic or slightly negatively heterogonic, carapace height shows positive heterogony throughout life, although its growth-coefficient is lower after maturity.

Adolph (1930) has an interesting note on the interrelated effects of size and age upon metamorphosis in unoperated frog larvae. He finds that for a given brood, $(W - d)(A - e)$ is a constant, where W is body-weight at completion of metamorphosis, A the age in days at which forelimbs appeared, and d and e are constants. Thus no increase of size would permit metamorphosis to occur before e days, but metamorphosis would never occur if the animal never attained to body-weight d. We may conclude that the growth of the thyroid (or of the thyroid-controlling agency of the pituitary) normally shows a heterogonic relation to absolute size; but that time also, within limits, promotes its growth (see p. 39, n.).

Further evidence of change of proportions with change of size in termites (see p. 65) is afforded by the work of Light (1927). In this paper he confined himself to the soldier caste of Coprotermes. Thirteen species of different absolute size were measured. There was no correlation between absolute size and certain characters such as the length-width ratio of the head. But there was for other characters, e.g. the ratio of minimum to maximum breadth of head and of gula. The minimum head-breadth is anterior; the maximum gular-breadth is at a spot close to the maximum head-breadth, with maximum gular-breadth more anterior. Accordingly we find these ratios move in opposite directions with increasing absolute size, that for head increasing steadily from below 0·6 to over 0·65, that for gula decreasing steadily from over 0·7 almost to 0·5. This means that with increased absolute size, the lateral growth of the head (and its parts) is relatively greater anteriorly than in the region of maximum width. And this, we may presume, is correlated with a relative increase of jaw-size, though Light gives no data on this point.

With reference to the conclusions of Hecht (p. 38), the work of Keys (1928) also indicates that Hecht's assumption of form-constancy in teleost fish is not strictly true. He finds that in herrings, sardines, and Fundulus the weight increases faster than the cube of the length, which implies form-change. Somewhat similar results have been obtained by Hickling (1930) for the dogfish *Acanthias vulgaris*.[1]

In connexion with the law of antero-posterior development, and the graded changes it may induce (see p. 132), the following point, which has been brought to my notice by Professor R. J. S. McDowall, is of interest. Bray (1931) finds that the incidence of eczema on different regions of the body varies with age. The incidence for head and neck declines with age, that for extremities increases with age, while that for the trunk remains approximately constant. There is reason to believe that the incidence of ringworm behaves in a somewhat similar way.

Gajewski (1922) discusses the effect of salinity upon *Artemia salina*, whose form-changes continue long after sexual maturity. Increasing salinity diminishes final absolute size, and also changes bodily proportions, as seen in the following table (for females: males are similar).

Salinity (Baumez)	4°	7°	12°	18°	22° B*e*
Body-length, mm.	15·56	13·05	10·65	9·62	7·8
$\frac{\text{Post-abdomen}}{\text{abdomen}}$ ratio	0·92	1·02	1·20	1·30	1·42
Length-breadth ratios:					
Of 8th post-abdominal segment	2·3	2·7	3·5	4·5	5·0
Of 7th ditto	1·3	1·35	1·4	1·5	1·6
Of gill-sacs of 6th foot	2·0	1·9	1·8	—	1·7

[1] The same is indicated for the Bittering, Paracheilognathus, by the data of Shaw (1931), *Bull. Fan. Memor. Inst. Biol.*, **2**, 245.

The furca diminishes disproportionately with increasing salinity. It would be of great interest to investigate the growth-coefficients of various parts accurately in different salinities. This is an important addition to Chap. VI, § 7.

Murr (1929) [1] gives particulars concerning the relation between retinal cells and the body as a whole in various mammals, both as regards relative size, and relative developmental rate. The results are interesting both as regards individual ontogeny and comparative physiology.

Unpublished work which E. B. Ford kindly allows me to quote, on relative eye-size in males of the red no-white mutant of *Gammarus chevreuxi* at 23° C., shows that after a head-length of a little over 0·5 mm. has been attained, up to the maximum size (head-length 1·64 mm.) there is simple negative heterogony both of eye-length (dorso-ventral) and eye-breadth (antero-posterior) against head-length (antero-posterior), the growth-coefficients (k) being 0·89 and 0·71 respectively. Immediately after extrusion, however (head-length 0·25 — 0·3 mm.), both dimensions show positive heterogony (k nearly 2·5 for eye-length and over 1·5 for eye-breadth), subsequently diminishing regularly to reach their definitive values. As consequence eye-breadth relative to eye-length shows a steady negative heterogony with k a little below 0·8 during the 'definite' period and for a little before it, with a still lower k value (0·6 to 0·65) for the earliest stages. The early phase of positive eye-heterogony would seem to be due to the late development of the organ, which is not fully differentiated at the time of extrusion. (See Chap. IV, § 7.)

With reference to the correlation between growth-rate and sensitivity to female hormone in fowl feathers (p. 101), Lillie and Juhn (1931) find that this holds also in different parts of the single feather. This growth-gradient has important consequences for the development of certain feather-patterns.

With reference to bio-electric phenomena (p. 174), Purdy and Sheard (1931) find in human beings that there is a definite association of 'low metabolism with large differences of electrical potential as measured at the extremities of the body' and vice versa.

J. W. Buchanan (1930, *J. Exper. Zool.*, **57**, 307 and 455) establishes the existence of an antero-posterior osmotic

[1] Murr, E. (1929), Zur Entwicklungsphysiologie des Auges, II, *Biol. Zentralbl.*, **49**, 346.

gradient concerned with water-appropriation, analyses its action, and discusses it in relation to other 'axial gradients' of Planarians (see pp. 171–2).

Olmsted and Baumberger (1923) state that in the crabs *Hemigrapsus oregenensis*, *H. nudus*, and *Pachygrapsus crassipes*, carapace length increases in a linear relation with carapace width. Unfortunately they do not give their actual measurements, and their graphs, in which individual points are plotted, might equally well indicate slight heterogony. This is especially so with *P. crassipes*, where length seems to show slight positive heterogony relative to width. In *H. nudus* the points are too few for any conclusion, and in *H. oregonensis* the relation appears to show if anything a slight negative heterogony. There must be marked heterogony of the male chela, especially in *H. oregonensis*, where large males have chelae up to 30 per cent of total weight, while the value for large females is never over 7 per cent. In *Carcinus maenas*, Huxley and Richards (1931) find a slight heterogony of carapace length.

Mr. G. H. Locket kindly allows me to cite his results (unpublished) on the chelicerae of the spider *Theridion lineatum*. The jaw undergoes heterogonic growth and takes on its definitive appearance only at the last moult, at which sexual maturity is attained; prior to this it appears to be almost isogonic. (In the genus Linyphia, there are signs of heterogony at the penultimate moult.) The jaw (basal joint, paturon) of the adult male is much elongated, whereas that of the female remains more nearly similar to that of the juvenile stages. The adult jaw has a prominent tooth on its inner surface, and measurements can be made of the lengths proximal and distal to this. Sternal area was taken as standard body measurement, and the square root of this was used as a standard against which to plot linear jaw measurements. There is a moderate size-range in adult females, a considerable one in adult males (probably dependent mainly on differences in total moult-number). It is clear that in the formation of the male type (i) jaw-breadth and jaw-length are both positively heterogonic, but length much more so than breadth (k = about 1·9 as against about 1·3); (ii) in regard to length-growth, the basal region, proximal to the tooth, is roughly isogonic, while the distal region beyond the tooth is very highly heterogonic ($k = 2{\cdot}7$); (iii) the length of the falx (distal joint, unguis) is highly heterogonic ($k = 2{\cdot}6$), but less so than the distal region of the main jaw. Here we find differential growth operative definitely within a single segment of an appendage (compare pp. 81, 98).

In the female, the figures are more irregular, but jaw-area and falx-length clearly show negative heterogony, while the distal region of the jaw is distinctly positive. Jaw-length as a whole is approximately isogonic, which means that jaw-breadth must be negatively heterogonic.

In the spider *Dolomedes plantarius*, P. Bonnet (*La Mue, L'Autotomie et la Régénération chez les Araignées, Thèse*, Toulouse, 1930) in his Table 25 gives measurements of the lengths of legs at all instars from 2nd to 11th (adult) in two individuals. From these the mean increases per moult can be calculated ; and it is then found that all the legs are increasing at the same rate—i.e. for this region of the body there exists no growth-gradient comparable to that seen in hermit-crabs, etc. (p. 111).

W. Beebe (*Tropical Wild Life in British Guiana*, 1917, chaps. 18 and 19) gives some data as to the heterogony of the claws of the Jacana. That of the hind toe is greatest. In the embryo the claws are not dissimilar to those of other birds and the hind claw-length is about $\frac{1}{3}$ of the toe-length ; in half-grown chicks it is $\frac{1}{2}$, in adults about $\frac{2}{3}$.

The other claws grow more slowly, reaching about $\frac{2}{3}$ of the length of the hind-claw ; their relative breadth-growth is greater than that of the hind-claw. The heterogony of the claws begins only in the late embryo ; that of the toes begins much earlier.

He also refers to the bill of the aberrant cuckoo-like Ani, which is swollen in the adult. It, however does not begin its positive heterogony until after the young bird has left the nest. At hatching, it is typically cuckoo-like, though somewhat swollen. It would be interesting to obtain accurate measurements on these structures. The case of the Jacana is interesting, since the great length of toes and claws is quite definitely adaptive, allowing the bird to walk over floating leaves.

G. Duncker (1903 ; *Biometrika*, **2**, 307) attacks the problems of growth-correlation and asymmetry in male fiddler-crabs (see pp. 80 and 121) by means of standard biometrical methods. These, however, failed to give any very important biological information, e.g. as to growth-centres or growth-gradients, or at least nothing so clear-cut as is to be obtained by the simple methods of taking means for a number of size-classes. The growth-gradient in the large chela is indicated by the fact that the correlation between right and left side diminishes distally along the appendage. The correlations for lengths of merus, carpus and propus are, for right-handed males, 0·754, 0·698 and 0·473; for left-handed males, 0·789, 0·699, 0·549. He suggests that the chela asymmetry may be responsible for the asymmetry of other parts.

Marples (*Proc. Zool. Soc.*, 1931, p. 997) has given some interesting facts as to the percentage changes of different parts of birds' wings during development. For calculating growth-coefficients, he has kindly put at my disposal his original data on the Common Tern (the species on which the most numerous measurements were taken). For a proper analysis, considerably more measurements are needed, including measurements of some standard part of the body; but provisionally we can see that there are three distinct phases of growth, during each of which the relative growth-rates of different parts of the wing remain approximately constant. The first phase ends at hatching; the second, juvenile phase goes from hatching (wing-length below 40 mm.) to wing-length about 100 mm.; the third up to the largest adults (wing-length over 180 mm.).

The growth-coefficients (approximate only) of the lengths of ulna and radius, relative to humerus length, are as follows:

Phase.	1	2	3
Ulna	1·05	about 0·8	1·6
Radius	1·2	,, 0·8	1·45

The growth-gradient appears to centre in the radius in the 1st phase (though this may be due in part to the late differentiation of this terminal region); to be reversed, centering in the upper arm, in the 2nd phase; and again to change its form, centering in the fore-arm, in the final phase (cf. p. 34).

I have not discussed the enormous body of data given in Donaldson s *The Rat* (1924), since all the comprehensive tables there set forth do not give the actual values for the various organs, but calculated values. These values have been calculated in accordance with empirical formulae devised by Hatai to fit smooth curves to the data.

These formulae are of the following types:

$$y = a \log x + b$$
$$y = a (\log x + c) + b$$
$$y = ax + b \log x + c$$
$$y = ax + b (\log x + c) + d$$
$$y = (ax + b) + b (\log x + c) + d$$
$$y = ax^b$$

where y = organ-size, x = body-size, and a, b, c, d are constants. Most of these have no assignable biological significance.

If the measurements contained in the original papers were re-analysed, it is probable that a number of cases of simple heterogony would be revealed. I have done this for one or two organs. E.g. testis weight (S. Hatai, 1913, *Am. J. Anat.*, **15,** 87), after an early period of rather slow growth, where more data are needed, shows a good approximation to simple heterogony between body-weights 25 g. and 95 g., with growth-coefficient about 1·65. After this, it enters on a phase of negative heterogony, with k only about 0·4 to 0·45. The ovary shows a very similar set of three phases, but the points are more irregular.

The hypophysis shows an interesting sex-difference. From body-weights of 60 g. on, k for the male hypophysis is positive (k about 1·3), whereas for the female it is negative (k rather below 0·8). Below this size, the points are rather irregular, but those for both sexes appear to fall on a prolongation of the curve for large females.

For heart-weight (males) k is close to 0·8 from body-weight 140 g. on. Before that, the points are more scattered, but could be considered as fitting the same curve, though apparently with a temporary acceleration of relative growth from body-weight 60 to 120 g., later compensated for.

The kidneys (males) begin by being somewhat negatively heterogonic (k about 0·75), and then, after an irregular period, show definite positive heterogony from body-weight 180 g. on, with k close to 1·1. Lung-weight (males) is more irregular, but roughly approximates to a negative heterogony of growth-coefficient about 0·8 throughout. For further analyses along these lines the data should be re-grouped into larger size-classes.

T. C. Byerley (1932, *J. Exp. Biol.*, **9,** 15) has recently shown that in chick embryos allantois weight shows negative heterogony relative to egg-weight, being roughly proportional to the two-thirds power of egg-weight. It may be recalled that egg-weight, at least in large birds, is itself roughly proportional to the two-thirds power of body-weight (p. 226).

He further points out an important connexion between relative and absolute growth. In fowls, gut-weight shows negative heterogony (H. B. Latimer, 1924, *J. Agr. Res.*, **29,** 363). Byerley finds a linear relation between feed-consumption and gut-weight (up to sexual maturity), Accordingly there will be a steady decline in the amount of food ingested per unit of body-weight ; and this may be presumed to be responsible for the steady decline in percentage absolute growth-rate during this period.

SOME PROBLEMS IN THE STUDY OF ALLOMETRIC GROWTH

E. C. R. REEVE *and* JULIAN S. HUXLEY

Introduction

AS long ago as 1917, D'Arcy Thompson emphasized the fact that all but the simplest organisms reach their adult form by differential growth in different directions, and various authors have pointed out that certain organs show a marked progressive change of relative size with increase of absolute size, notably Pezard (1918) who described them as *heterogonic*, and Champy (1924) who spoke of *dysharmonic* growth. However, the first quantitative analysis of differential growth on a general basis was made twenty years ago, when Huxley (1924) used the formula $y = bx^k$ to describe the relation of the growth of a part or organ to that of the whole organism, and suggested that this equation, which implies a constant ratio between the specific growth-rates of the two dimensions compared, might express a general law of differential growth. It is of historical interest that, although Huxley was not aware of this at the time, the same formula had been used as early as 1891 by Snell, and later by Dubois (1898, 1914) to express brain-weight : body-weight relations in mammals, and thus to determine their 'cephalization coefficient'. Lapicque (1898) applied the same technique to birds, and Klatt (1919) used the formula to compare heart : body-weight relations in birds; but no general application of the formula seems to have been proposed by any of these authors.

The equation is now generally written $y = bx^\alpha$, and is termed the *simple allometry* formula. It may be given the form $\log y = \log b + \alpha \log x$, showing that, when the logarithms of two dimensions x and y obeying the law are plotted against one another, the points lie along a straight line. In practice the two dimensions are more frequently plotted on logarithmic (i.e. multiplicative) scales, illustrating the fact that the formula expresses a constant ratio between the specific or logarithmic growth-rates in time of the two dimensions.

Huxley and others found the formula adequate to describe a wide range of growth phenomena, including many cases of changes in proportions during growth in animals and plants, variations in proportions correlated with variations in absolute size in holometabolous insects, change of adult proportions with increasing absolute size in related species of animals and during the course of evolutionary change, and changes in the proportions of various chemical constituents of the growing organism (Huxley, 1932; Teissier, 1934; Hersh, 1934; J. Needham, 1934). A single straight line on the double logarithmic grid will not in every case describe the change of proportions throughout the growth period, and two or more successive lines have frequently been employed, with or without gaps indicating a sudden change in organ-size at a given body-size. This method of treatment is not always satisfactory, and there has been criticism of both the theoretical basis and the practical value of the allometry formula.

The very general occurrence of *growth gradients*, both along the primary axis of the body and within individual limbs which are not growing at the same rate as the body, has been extensively analysed by Huxley and received particular emphasis in his treatment of relative growth. The study of these gradients forms a connecting link between allometry and the original and highly illuminating method of Cartesian transformations devised by D'Arcy Thompson (1917, 1942) for comparing the shapes of related forms, since these transformations contain an implicit recognition of the general gradient pattern of growth.

Opinion is still divided on the question whether the allometry formula expresses a basic principle of differential growth or is merely an empirical formula, of limited value for analysing the underlying biological factors controlling proportions; and we propose here to examine the theoretical status of the formula and the related question of the types of deviation observed from simple allometry. Recent work on growth gradients will also be discussed, since these gradients, while independent of the nature of the differential growth curve, are most easily studied by the technique of simple allometry. The present discussion will be confined to problems connected with the allometry of proportions in animals, and we shall first examine some formal difficulties.

Terminology

Confusion has arisen from the variety of terms in use by earlier writers, but recent suggestions for a standard terminology have not yet been universally accepted. The original terms *dysharmonic* and *heterogonic* growth are unsatisfactory because of ambiguous or alternative meanings (Huxley and Teissier, 1936), and the general term *allometry* is now accepted to mean change of proportions (whether morphological or chemical) with increase of size, both within a single species (ontogenetic allometry) and between adults of related groups (absolute-size allometry). Ontogenetic and absolute-size allometry are distinguished by the terms *heterauxesis* and *allomorphosis* (J. Needham and Lerner, 1940; Huxley, J. Needham, and Lerner, 1941), and the latter term may be extended to include special cases such as the relation of egg size or hatching weight to adult size or weight in different species of birds (e.g. Amadon, 1943).

Growth is said to follow *simple allometry* when the formula $y = bx^{\alpha}$ holds for two organs x and y, the relation being described as *isometry* in the special case of $\alpha = 1$, which gives simple proportion between x and y with $y = 0$ when $x = 0$. For heterauxesis the terms *tachyauxesis*, *isauxesis*, and *bradyauxesis* have been used to define cases where the coefficient α is greater than, equal to, and less than unity (J. Needham and Lerner, 1940); or we may speak simply of *positive* and *negative* allometry (Huxley and Teissier, 1936), and define actual negative growth (α negative) by the term *enantiometry*. Of the two parameters of the equation, b is termed the *initial growth index*, α the *equilibrium constant*. For ontogenetic allometry, α has been termed the *growth constant*, *growth ratio* or *actual equilibrium constant*, as opposed to the *limiting equilibrium constant* in absolute-size comparisons (Huxley and Teissier, 1936).

A word of caution is perhaps necessary on the question of terms, since to define a concept too early by a new term may tend to obscure some of the basic problems under investigation by implying a certainty which does not exist. Thus the term *simple allometry* should not be taken to imply that the type of differential growth so defined is physiologically simpler than other types, although it is in a sense mathematically simpler. In the same way, the terms *actual* and *limiting equilibrium constant* carry a physiological implication which has yet to be justified, and it might be preferable to use more neutral terms.

Formal Difficulties

The logical status of the allometry equation has been questioned on the ground that the dimensions of the two sides are only equal when $\alpha = 1$ (Needham, 1934); but in fact they are always equal, since the dimensions of b vary with those of x (Lumer, 1939; Kavanagh and Richards, 1942). This may be shown by writing the differential equation of the allometry formula as $x^{-\alpha}\,dy - \alpha x^{-\alpha-1}y\,dx = 0$, which is evidently dimensionally valid and has dimensions $(u)^{-i\alpha}(v)^j$, where the units of measurement of x and y are respectively $(u)^i$ and $(v)^j$. The expression is an exact differential and integrates to $x^{-\alpha}y = b$. Integration is a process of summation which will not alter the dimensions, and the dimensions of b are therefore the same as those of $x^{-\alpha}y$, i.e. $(u)^{-i\alpha}(v)^j$. Thus if x and y are both measured in units of cm, then $u = v = cm$ and $i = j = 1$, so that b will be in units of $(cm)^{1-\alpha}$.

A more serious difficulty has been pointed out by Haldane (cf. Huxley, 1932, p. 81). If, as often occurs, the different parts of an organ (e.g. segments of a limb) show unequal constant differential growth ratios against body size, then the whole organ cannot exactly obey the allometry law against the same standard, and vice versa, since the sum of a number of expressions $b_r x^{\alpha_r}$, with different values of α_r, cannot be identical with a single expression bx^{α}. In the simple case of an organ of length L, consisting of two segments y and z, each of which grows with simple allometry against body size x, we have

$$L = y+z, \qquad y = b_1 x^{\alpha_1}, \quad \text{and} \quad z = b_2 x^{\alpha_2}.$$

Then
$$\frac{dL}{dx} = \frac{dy}{dx} + \frac{dz}{dx} = \frac{\alpha_1 y + \alpha_2 z}{x},$$

so that
$$\frac{dL}{L}\Big/\frac{dx}{x} = \frac{\alpha_1 y + \alpha_2 z}{y+z}.$$

The differential growth ratio of L against x is thus a weighted mean of α_1 and α_2 (the weights being the segment lengths y and z), and it will always lie between α_1 and α_2, tending towards whichever is higher as size increases. The case is easily generalized for a number of segments. It is clear that, when α_1 and α_2

do not differ much, the growth ratio of L against x will be very nearly constant; and Teissier (1934) has argued that this disposes in practice of the theoretical difficulty. But fairly sharp differences between the growth ratios of neighbouring segments of a limb have occasionally been found, as in *Palaemon carcinus* (see Huxley, 1932, p. 92), where α varies from 1·6 in the ischium to 2·9 in the propus; and it is doubtful whether in such cases the complete limb and the individual segments could both show a close approximation to simple allometry. In the examples so far examined, the small size of the individual segments and high variability due to genetic heterogeneity of the material make it impossible to decide whether the whole organ or the separate segments give the nearest approximation to simple allometry. Whatever its practical effect, the theoretical difficulty certainly becomes important when we wish to interpret the biological meaning of the allometry formula, and we shall return to this problem later.

Differential Growth and Growth in Time

One of the most important facts about the allometry formula is that it allows us to ignore the time relations of growth by relating the sizes of the different parts of the body to total size, regardless of age. In other words, it expresses the well-known fact that form during growth is a function of absolute size rather than of absolute age, in so far as these two measures of growth are independent. The elimination of time in studying change of form during growth is consistent with the fact that variations in nutrition and growth-rate do not in general affect proportions at a given absolute size (e.g. Moment, 1933), and is reflected in the frequently used concept of physiological age.

Nevertheless, it can be argued that under conditions of optimal growth both the individual organs and the whole organism have their own specific growth curves in time, and several attempts have been made to relate the allometry formula to these hypothetical curves, while the importance of bringing back time into the study of differential growth has more than once been emphasized (Needham, 1934; Bernstein, 1934).

An early attempt to relate allometric growth to time was made by Huxley (1932), in deriving the formula from simple postulates about growth: namely, that growth is a process of self-multiplication of living substance, that the rate of self-multiplication slows down with increasing age (or size) and is much affected by external conditions, and finally, that changes in the rate of self-multiplication affect all parts of the body equally. Thus the growth-rate of an organ is assumed to be proportional to its size (x) at any moment, to a specific constant (A) which varies with the organ, and to a general factor (G) which is the same for all organs; and we have for any two organs:

$$dx/dt = AxG, \quad dy/dt = ByG, \text{ giving } y = (x^{B/A}) \times \text{constant}.$$

Bernstein (1934) argued that G would not be the same for all organs, and supposed that each organ obeyed the logistic law of growth in time, $dx/dt = mx(X-x)$, where $X =$ final size of x. Eliminating time between two such equations, he concluded that a constant growth ratio would only obtain when the parts compared were small in comparison with their final sizes.

Lumer (1937) has developed this idea further by investigating the relation between organ and body size when both follow the same type of growth law in time, of the general form

$$\frac{1}{x}\frac{dx}{dt} = F(x, t).$$

He considered the simple and generalized autocatalytic and the Gompertz equations, each of which has been used to define growth in time. With any one of these time laws, the relation $y = bx^{\alpha}$ only occurs as an approximation at small sizes, or as a special case when the velocity constants for organ and body are equal. Generally with each equation there is a marked change in growth ratio as we approach final size, resulting from the fact that the equation assumes that the size of each part approaches asymptotically a final upper limit. Lumer concludes that in general with this type of growth the law of simple allometry cannot hold throughout development, and that the exponent α is not simply a ratio of velocity constants, but also involves other parameters, e.g. final size; and he suggests that the formula has been found widely applicable simply because growth curves are often approximately exponential during early stages, so that deviations are to be expected if measurement is carried near enough to final size. Such deviations have of course been found in some cases, and their significance will be discussed later.

Lumer's case, however, has been destroyed by Kavanagh and Richards (1942), who pointed out that sigmoid or determinate growth in time is by no means incompatible with simple allometry. In fact, if $dx/dt = f(x)$ is the sigmoid curve for one organ, and $y = bx^{\alpha}$, then the growth in time of the other organ will also be sigmoid, with the equation

$$dy/dt = \alpha b^{1/\alpha}.y^{1-1/\alpha}.f\{(y/b)^{1/\alpha}\}.$$

An important consequence is that two organs related by a constant differential growth ratio will not in general obey the same form of sigmoid growth law, since the integrated equations in time of x and y will not be of the same form unless $\alpha = 1$. Likewise the equations used by Lumer to express growth in time cannot apply both to the separate parts and to the whole organism, since the sum of several autocatalytic or several Gompertz equations with different velocity constants will not be exactly an autocatalytic or Gompertz equation (cf. Medawar, 1940)—in other words, the law of time growth for the whole organism will not be the same as that for its individual parts.

A variety of equations are in use for describing growth in time, the logistic equation being still in favour with Brody (1937), and it is interesting to find that not all are incompatible with the allometry formula when applied to both organ and body size. Thus Glaser (1938) proposes the equation

$$\log w = K \log(2t+1)+c$$

for which he has attempted to give a theoretical justification, while Zucker and Zucker (1942) have concluded that none of the curves usually employed to describe growth in time gives a satisfactory fit to their new data on growth of the rat under optimal nutritional conditions, even when several cycles of growth are allowed; but they obtain an excellent fit with the two equations $\log w = k \log t + c$ for growth up to weaning and $\log(W/A) = K/T$ from weaning onwards.

Glaser's equation is practically identical with the first one of

Zucker and Zucker,[1] and each of the three equations, when applied to two organs, gives a constant differential growth-ratio. Zucker, Hall, Young, and Zucker (1941) give a graph showing a very good fit of the allometry formula to dry fat-free femur weight against body weight in well-nourished female rats from 24 to 750 days of post-natal age, so that in this case at least both organ weight and total weight against age are closely represented by equations of the same form.

The difficulty of linking up differential growth with growth in time by reference to particular formulae is brought out by the recent verdict of Zucker *et al.* (1941), when reviewing the various equations that have been proposed for growth in time: 'No growth equation so far suggested is sufficiently well founded in theory to stand on its own feet—all must be judged by their relative success in fitting data which can be considered suitable.'

The same difficulty has been emphasized by Gray (1929); and Kostitzin (1939), discussing the very close fit of a logistic equation to a case of population growth, writes: 'It should be noted that nothing is as deceptive as the beautiful agreement between the observed and calculated values. A series of observed values is, in fact, equivalent to a narrow band rather than to a curve, and in this band can be traced a number of curves, corresponding very well with the conditions of the problem within the period of the observations, but showing real divergence beyond these limits.' A convincing example has been given by Kavanagh and Richards (1934), who obtained an excellent fit with an autocatalytic equation to a set of points actually obeying an entirely different formula, the probability integral

$$x = 1 - \frac{2}{\sqrt{\pi}} \int e^{-x^2}\, dx.$$

[1] The difference between the two equations is in fact due to an error in the derivation of Glaser's equation. Starting with the observation that percentage increase in fresh weight (w) of the chick during the t-th day is proportional to $1/[(t+1)^2 - t^2] = 1/[2t+1]$, Glaser writes:

$$(1/w)dw/dt = K/[2t+1].$$

But dw/dt is the average growth-rate during the interval $(t-\frac{1}{2})$ to $(t+\frac{1}{2})$, and not during the tth day, which is the interval t to $(t+1)$. Glaser's observation should therefore lead to the equation:

$$(1/w)dw/dt = K/[(t+\tfrac{1}{2})^2 - (t-\tfrac{1}{2})^2] = K/2t,$$

which is identical with the first equation of Zucker and Zucker.

We must conclude that, with the data at present available, no appeal to time-growth formulae can be accepted as either a criticism or a justification of the allometry formula. In fact, bearing in mind the many cases where the simple allometry formula is closely obeyed, incompatibility between this formula and any particular time-law must be interpreted as evidence against the general applicability of the latter.

Attempts to develop time-growth formulae have been largely confined to warm-blooded animals, in which a smooth sigmoid growth curve is obtained, and even here no formula yet proposed has a firm physiological basis. Further complications arise with growth in poikilothermic animals, discontinuous growth by moulting in arthropods, and discontinuous combined with determinate growth in holometabolous insects. No attempt appears to have been made to relate laws of time-growth to allometry in these forms, to which Lumer's arguments do not apply as they stand.

Failure to derive any law of allometry from laws of growth in time raises the more general question of the theoretical basis of the simple allometry formula, which will be examined in the next section.

The Theoretical Basis of the Allometry Formula

Huxley's derivation of the formula, as we have seen, is based on the assumptions that growth is essentially multiplicative and that changes in the rate of self-multiplication affect all parts of the body equally. Both these assumptions have been criticized, the first by Davenport (1934), who pointed out that growth is not always a process of self-multiplication, and the second by Medawar (1941), who argued that it cannot be accepted as an axiom of growth, but follows as an important consequence whenever a constant differential growth ratio has been established.

An alternative hypothesis has been put forward by Teissier (1934, 1937), who suggested that the key to the law lies in nutrition and the unequal 'appetites' of the different organs. He postulated that, as a first approximation, the quantity of food absorbed by an organ in a given time interval is proportional both to the mass of the organ and to the total quantity of food absorbed in this interval. Assuming that a definite fraction

(L_1) of the food consumed by the organ is used for growth, and that its inherent 'appetite' is represented by the constant factor (L_2), and putting x = mass of organ, m = mass of food it absorbs, q = quantity of food available, he writes: $dx = L_1\,dm$, $dm = L_2xq\,dt$, and therefore $dx = L_1L_2xq\,dt$. Similarly for another organ or the whole body, we have $dy = M_1M_2yq\,dt$, and eliminating $q\,dt$ we obtain

$$\frac{dy}{dx} = \frac{y}{x}\frac{M_1M_2}{L_1L_2}, \quad \text{or} \quad y = bx^{\alpha}, \quad \text{where} \quad \alpha = \frac{M_1M_2}{L_1L_2}.$$

Changes in the growth coefficients at critical points, of which Teissier has investigated a number of cases, are explained as due to changes in one or more of the four constants to which α is now related. In support of this hypothesis he quotes data on the growth of trout embryos, cultures of fibroblasts, and melon shoots, to show that the intensity of growth at any instant is proportional to the quantity of tissue capable of growth and to the total quantity of nutritive material available.

Teissier's hypothesis suffers from the practical disadvantage that it substitutes four 'constants' which cannot be measured for one which can. It seems most improbable that a constant proportion of the total food absorbed by an organ will be used for growth, since this would require a constant ratio between the metabolic and growth activities of an organ during development. Moreover, no reason is given why the various organs should use different proportions (L_1, M_1, &c.) of the food they absorb for growth. The evidence Teissier adduces for his theory does not in fact bear on these difficulties, and it seems unlikely that this method of approach will be profitable.

Robb (1929) postulated that the differences in relative growth-rates of organs are due to the different amounts of material available to each, and drew an analogy between the mechanism of this uneven distribution and the partition of a solute between two immiscible solvents, suggesting that the relative growth constant (α) might be a partition coefficient measuring the relative concentration of some growth-controlling substance in each tissue. The growth of each organ is thus assumed to be proportional to the quantity of growth-controlling substance available to it. No such substance has yet been discovered, and it is difficult to explain many of the less straightforward phenomena of differential growth on this basis.

Huxley (1932) has also used the term 'growth partition' in rather a different sense when describing certain complex cases, such as the return to normal proportions by regeneration after autotomy or removal of tissue, and the annual growth of antlers in deer, which cannot be discussed in terms of a constant differential growth-ratio. Here, growth of the regenerating organ or antler is at first very much faster than that of the body, but slows down as it approaches the limiting size for its particular body-size; and it is the final relation between organ and body dimensions which shows an allometric relation, and seems to be controlled in accordance with an equilibrium constant —or in this case a constant coefficient of growth limitation. No physiological mechanism was proposed for the control of such phenomena, though it is clear that there are factors maintaining an equilibrium of proportions, and acting in such a way that the greater the disturbance of the equilibrium the faster is the rate of return towards normal proportions (cf. Przibram, 1930).

An experimental attack on the problem of the control of proportions has been made by Twitty, on the basis of transplantation experiments between larvae of *Amblystoma tigrinum* and *A. punctatum*, and between young and old larvae of the same species. When organs, such as eyes or limbs, are grafted into hosts of the same species but of different age from the donor, size regulation is brought about by the transplanted organ growing faster or slower than the corresponding organ in the host, according as the host is older or younger than the donor. As a result, the new organ tends to acquire the normal size for that of the host. Even more significant is the fact that eyes grafted into hosts older and younger than the donor have their growth-rates accelerated above normal and retarded below normal, respectively, suggesting that the internal environment becomes more favourable to growth as the individual grows older. A complementary experiment shows that the growth response of the eye to a given internal environment declines as the eye grows older, since a younger eye is accelerated, an older one retarded in growth, when grafted to a host of intermediate age.

The theory of competition for the available food-supply by organs of different physiological ages (and therefore with different assimilative capacities) was considered inadequate to explain these facts, since not only would a young eye transplanted to an older host show significant growth under conditions of complete starvation, when the host eye could not grow at all and the host body actually shrank in size, but the transplant grew just as well when the tail was also made to regenerate, thus producing a large additional drain on the nutritive substances in the blood. If competitive restriction was not manifest under such extreme conditions as these, it could hardly be important under conditions of normal growth.

In order to explain these results, Twitty has suggested that size regulation is controlled during development as an equilibrium between the decreasing specific assimilative capacity of the organ and the increasing nutritive opportunities afforded by the internal environment. Thus a growing organ transferred to an older host would grow faster than normal as a result of the combination of its previous assimilative capacity with an environment (the host's blood-stream) more favourable to growth than that of the young donor. This would account for the young transplant catching up in size with the host's equivalent organ; but we must also suppose that the faster growth-rate induced in the transplant results in an increased rate of fall in its assimilative capacity, so that by the time it has reached the same size as the normal host organ it has lost its growth advantage over the latter.

Twitty and Wagtendonk (1940) measured the concentration of various constituents of the blood in *A. tigrinum* of different ages, in seeking evidence for changes in its nutritional level during development, and found that while the concentration of sugar and total nitrogen remain essentially constant during larval growth, that of non-protein nitrogen rises sharply with age, from which it is inferred that the concentration of protein nitrogen declines and that of amino-acid nitrogen (apparently the main constituent of non-protein nitrogen) increases with age. One similar determination for *A. punctatum* gave about the same concentrations as those for *A. tigrinum* of the same age—but nearly double the size—suggesting that the two species may offer equal nutritive opportunities at the same age, and in accord with the fact that organs transplanted between them maintain approximately their donor or genetic growth-rates. While these results are still tentative, they do suggest the probability of a rise of the nutritive level of the blood with age, at least to the extent that this level depends on the amino-acid concentration; and they thus bear out Twitty's hypothesis for the control of size regulation drawn from grafting experiments.

Since Twitty's key experiments were carried out with transplanted eyes, it is interesting that in rats Moment (1933) found the eyes to react differently from other organs to variations in nutritional level leading to variations in growth-rate. The size of the eyes depended primarily on age, and was unaffected by induced differences in growth-rate, while the size of most organs behaved as a function of body size rather than age. In pigs also the growth-rate of the eyes appears to be independent of the plane of nutrition, and they continue to grow under conditions of starvation, in contrast to most organs except the brain (cf. Pomeroy, 1941). Thus Twitty's hypothesis must be applied to other organs with caution.

Certain phenomena of differential growth, discussed by Huxley (1932), cannot be easily explained by this theory: for example, the very general occurrence of growth gradients both within each limb and along the body, and the phenomena of heterochely. The sizes of some organs, moreover, are known to be regulated within limits by functional or mechanical factors, e.g. heart, lungs, and ear; while further complications occur in the mutual adjustment of the various parts of organs to each other, as instanced by Harrison's well-known experiments with eyes compounded of eye cup of *A. tigrinum* and lens of larger *A. punctatum*. In the case of reversible heterochely in *Alpheus*, it appears that whichever chela is given a small start over the other will take on a higher growth-rate and develop into a crusher claw; and Darby (1934) has shown that when both claws are autotomized the one on the nipper side will become a crusher if given a start of about 40 hours over the other claw, although normal heterochely is reproduced if both claws are removed simultaneously (see also Dawes, 1934). Thus some process of competition or physiological dominance between the two claws does seem to be at work here.

No mathematical law of normal differential growth can be

deduced from Twitty's hypothesis, though it helps to explain the return towards normal proportions, at first very rapid but declining as the normal equilibrium is approached, which occurs in physiological and experimental regeneration. Twitty suggests that normal size-regulation depends partly on the declining assimilative capacities of the various organs; and since it may be inferred that there is a constant ratio between the rates of decline of these inherent capacities whenever two organs have been empirically shown to obey simple allometry, we may expect that attempts to influence experimentally the rate of assimilation of an organ will throw light on the mechanism of allometry.

One such attempt is due to Lerner and Gunns (1938), who compared the growth constants of the leg-bone lengths against wet weight in chick embryos incubated at different temperatures. Increased incubation temperature caused a marked increase in general growth-rate, but appeared to have no effect on the growth ratios of the bones when embryos were compared on a basis of body size. In other words, increased general growth-rate induced by high temperature was accompanied by proportional increments in the assimilative capacities of the various organs, so that body proportions at a given absolute size were the same as in normal animals.

From the above discussion it must be concluded that no satisfactory theoretical basis has yet been found for simple allometry. The 'axioms' of growth which were put forward to justify the general use of the formula are far from self-evident, and should perhaps be considered as no more than consequences of simple allometry where it has been found to occur. This draws attention to the importance of examining the constancy of the differential growth ratio and the nature of its variation during development. Technical and statistical difficulties arise here, which will be discussed later, but it may be noted that most studies of allometry have been made on small samples of non-homogeneous material, and the marked variations in organ size for a given body size make it usually impossible in such cases to decide whether the simple allometry formula gives an adequate description of the differential growth trend.

Several theoretical difficulties remain to be discussed in greater detail. Since growth is multiplicative in the general sense that what is produced by growth is itself normally capable of growing, it is natural to think in terms of the specific growth-rate $\frac{1}{x}\frac{dx}{dt}$ and to compare the growth of different organs on a logarithmic scale. The significant index then becomes the ratio of these specific growth-rates for two organs, and when this ratio is constant we obtain simple allometry. But the question how far the growth of different organs remains multiplicative in a strictly mathematical sense during development needs further study, since there is some evidence of an antagonism between mitosis and the characteristic chemical activity of differentiated cells (cf. J. Needham, 1933).

Certain types of growth clearly cannot be analysed in terms of self-multiplication. D'Arcy Thompson (1917, 1942) and Huxley (1932) have discussed in detail the accretionary type of growth instanced by shells and horns, where the rate of growth is not proportional to the amount of material already present. A more complex problem is the growth of bone. It has long been known that bone does not grow interstitially—i.e. by expansion of the whole mass—and Brash (1934) has concluded from his studies with madder-feeding that increase in the size of the skull is due primarily to external surface accretion and internal surface absorption rather than to growth at the ends of individual bones. According to this interpretation, the sutures play mainly a passive role in skull growth, since they are not centres of new bone-formation. However, there may be an appearance of growth at a suture, when one bone overlaps another to form a sutural plane which is not at right angles to the surface of the skull. In such cases, the overlapping bone will appear to creep gradually over its neighbour, as bone is added on the outer surface and removed from the inner surface of the skull, and the sutural line will gradually change its position on the skull as growth proceeds.

If this interpretation is correct, an individual skull-bone cannot be considered as a unit of either multiplicative or additive growth, since its rate of growth will depend on both the rates of surface accretion and the angles of the sutural planes at its ends. Further complications may occur, such as the gradual forward movement of the battery of cheek teeth in the horse's skull, while Todd and his associates (see Todd and Wharton, 1934) have described a complex bending of the face on the cranium during the later stages of development of some mammals.

It seems clear that no simple concept of multiplicative or additive growth can be applied to explain changes in skull proportions; nevertheless, the allometry formula has proved valuable in describing such changes during growth or evolutionary size increase in a number of cases, although simple allometry has not always been demonstrated—e.g. face length against cranium length in sheep-dogs, baboons (Huxley, 1932), ant-eaters (Reeve, 1940, 1941), and horses (Reeve and Murray, 1942); various dimensions in the skulls of certain fish (A. E. Needham, 1935), titanotheres (Hersh, 1934), dogs (Lumer, 1940), and mice (Green and Fekete, 1933). It is probable that a similar analysis of the growth process, applied to other organs or dimensions, would show equally clearly that some of them cannot be discussed in terms of multiplicative growth.

A more general difficulty, pointed out by Haldane, has already been mentioned, namely, that if the segments of a limb show simple allometry against body size, then the whole limb cannot show exact simple allometry against the same standard, since the sum of two self-multiplying systems is not a formally identical system. It follows that the shape of a growing organism cannot be fully analysed in terms of simple allometry. The theoretical difficulty is in no way lessened by the fact that deviations due to this cause are usually slight. Two alternatives which suggest themselves are that the organism may be divided at any stage into a unique set of 'unit' organs and dimensions between each of which the allometry law holds, or (as held by Huxley) that the allometry formula is a first approximation to a more general formula, applicable both to the whole organ or limb and its parts.

The former theory has not been explicitly put forward, but seems to be implied in Hersh's (1941) use of the phrases 'rational unitary parts given by the organism' and 'a unitary element in the sub-pattern of the skull'. This possibility, though attractive at first sight, presupposes too simplified a conception of the growth process, and is inconsistent with the very general presence of growth gradients, since there is evidence for believing that the gradient along a limb is expressed within each segment, making it impossible to take either the whole limb or the individual segment as the final unit of growth. Thus Huxley (unpublished) found a gradient in the relative growth-rates of the regions between consecutive spines on the lobster's chela, and other arthropod limbs have shown a similar continuity of the gradient (Huxley, 1932).

With regard to the second hypothesis, although the allometry formula has frequently been accepted as nothing more than a useful first approximation to the true law of differential growth, little attempt has been made to develop a more general formula. Robb (1929) suggested the addition of a constant to allow for the presence of dead material or cells no longer capable of growth, and wrote $y = bx^{\alpha}+c$, where the constant c might represent inert masses in one or both primordia, and could not therefore be given a direct physical significance. He showed that while the logarithmic plot of thyroid weight against cleaned body-weight of rabbits required two successive straight lines of notably different slope, a single straight line gave a good fit over the entire size-range after a suitable constant had been subtracted from one dimension. This fact, though striking, cannot be said to demonstrate the significance of the constant c, since the addition of an extra constant to a formula will always result in a better fit to a set of data, and without finding histological evidence to explain the extra constant, it is at least as reasonable to suppose that two stages of simple allometry intervene during the growth period.

Lumer (1937) has pointed out that Robb's formula can only be expected to take account of inert material in the organ y, and suggests that to allow for inert masses in both organs it would be necessary to write $y-c = b(x-d)^{\alpha}$, a formula which might well give a good fit to many complex cases of differential growth, in view of its two extra constants, but is theoretically doubtful and useless in practice, since the four constants could not be estimated even very roughly without quite prohibitive labour in computation.

These formulae suffer from the same defect as the allometry formula, in that they could not be applied to both the individual segments and the whole limb, and in fact they are only designed to take into account hypothetical constant masses of inert material, whereas it seems more reasonable to expect such masses, if they occur, to increase progressively in quantity during growth. Apart from the formulae derived from time functions, and already discussed, no other generalizations of the allometry formula appear to have been put forward. A major problem in any such generalization is that the introduction of further constants is likely to make the formula impossible to apply in practice, and at the same time no easier to justify theoretically than its simpler prototype. A necessary preliminary to such a development seems to be a much more detailed analysis of the nature of deviations from the simple allometry formula than has yet been attempted. We shall in the next section classify the most important examples of such deviations.

Deviations from Simple Allometry

One of the most serious difficulties, and often a neglected one, in studies of differential growth, is how to decide whether the growth trend of the data is adequately represented by a straight line on a double logarithmic grid. In many past studies the scatter of plotted means is such that any one of a variety of straight lines or curves could be employed with equal justification, and it has only been the conviction of the investigator in the universal significance of the simple allometry formula which has led him to choose it (and often to calculate the growth constants to several decimal places!). Such data will often seem to give just as close an approximation to linearity when plotted arithmetically, since the latter part of a parabolic curve will not usually deviate sufficiently from rectilinear growth, on an arithmetic grid, to give a visible curvature with the degree of scatter of points generally found. Thus Robb concluded that face length followed a single arithmetically linear relation against total skull-length in prehistoric and modern horses. A curvilinear trend was, however, discernible on his graphs, and became quite clear when face length was plotted against rest-of-skull length (Reeve and Murray, 1942). It should be noted, therefore, that to plot a part against another dimension which includes that part will always tend to obscure change of proportions, and should generally be avoided.

Simple statistical methods are not available for testing the linearity of the relationship between two variables when both are subject to error, as is the case with allometry data, and the investigator has generally been content to judge how the data should be fitted from the appearance of the double-logarithmic plot of means. The danger of this method soon becomes clear to anyone who replots almost any published allometry data, draws in the curves or lines he thinks most satisfactory, and compares them with those of the author.

A valuable graphical aid has been devised by Kavanagh and Richards (1942), based on the fact that a given rate of change of slope becomes progressively more obvious to the eye as the slope approaches the horizontal. They plot y/x against x on a double logarithmic grid, resulting in a curve which is a straight line when and only when the ordinary allometry plot is linear, and which has a slope of $(\alpha - 1)$ at the point where the latter curve has a slope of α. In consequence, with the usual range of values of α, the average slope of the new curve will be nearly horizontal, and deviations from linearity will be magnified. It will be worth the extra labour in many cases to plot $\log(y/x^{\alpha})$ against $\log x$, using an average value of α estimated graphically, since this will ensure that the average slope is always horizontal, and deviations from linearity will receive maximum expression. Unfortunately no method will bring out systematic deviations which are masked by large random variations in the data, and we shall therefore confine our discussion to those cases where extensive data make it possible to study the growth trend in some detail.

Many examples have been found which show a close obedience to the simple allometry law over long periods of growth, but among others three main types of deviation may be distinguished. These are (1) a definitely curvilinear trend on the double logarithmic plot, which may be termed *increasing* or *decreasing* allometry according to the behaviour of the growth ratio; (2) the presence of critical points marked by sudden changes of slope or gaps in the allometry line; and (3) more or less rhythmical variations about an average trend. Sometimes two of these types may appear as alternative interpretations between which it is not easy to decide.

Perhaps the most striking example of a curvilinear trend is that of rostrum length against body length in the spoonbill catfish, *Polyodon spathula* (Thompson, discussed by Hersh, 1941). The data cover the body-length range of 15·5 to 1,727 mm., and the log–log curve is strongly concave to the x-axis, its slope falling steadily from 4·0 to about 0·5; but Hersh points out that if the available data began at a body length of 200 mm., they could be very well interpreted as conforming to a single straight line with slope 0·7.

Very similar trends are suggested by the graphs of A. E. Needham (1935) for mandible length against body length in the bony fishes *Lepidosteus*, *Belone*, and *Hemirhamphus*, which also undergo a marked elongation of one or both jaws; but here the author has fitted two or even three successive straight lines to the allometry plot, and has attempted to justify the rather sudden changes of slope on histological grounds. This interpretation seems forced, since a single curve would certainly give a better fit to the graph of *H. far*, and for the other species the data are not adequate to show the growth trend in detail.

Such markedly curved trends have not often been demonstrated. Carapace length shows a gradually changing growth ratio against width in young stages of many Brachyura, but later growth approximates closely to simple allometry—e.g. *Cancer magister* (Weymouth and Mackay, 1936), *Cancer pagurus* (Mackay, 1941), *Carcinus maenas* (Day, 1935). In *Cancer magister* females the widths of the sixth and seventh abdominal segments show deviations from simple allometry against carapace width which might be interpreted as either a single curve or a sudden change of proportions at sexual maturity. In a very extensive series of measurements of *Carcinus maenas*, Day found it necessary to fit five successive allometry lines to growth of abdominal segment widths against carapace size in males and females. This must be accepted as a purely empirical description of the growth trends.

For chela dimensions against carapace length in the pistol crab, *Alpheus dentipes*, Dawes (1934) found a progressively diminishing growth ratio throughout the size range examined. Dawes and Huxley (1934) later suggested that this decline in growth ratio, particularly in the male crusher, was a modification of simple allometry, due to the fact that the claw begins with an excessively high growth-rate, which it proves physiologically impossible to maintain as the bulk of the crusher increases. Huxley (1932) has used a modification of this theory to explain the cases of diminishing allometry in holometabolous insects, suggesting that this is a consequence of growth in a closed system, the pupa, since large organs with high growth ratios will be unable to complete their growth before the rest of the body has appropriated the reserves of nutrient material.

The conclusions of Dawes and Huxley are open to criticism (Reeve, 1945) because internal evidence from Dawes's published data suggests that the apparent decline of growth ratio in the claws of *Alpheus dentipes* may be due to the presence of regenerating individuals in his larger size-groups. There is, moreover, little evidence of a steady decline of growth ratios in either of the single dimensions propus width and propus thickness, whose combination gives the measure of chela size used by Dawes, and this example cannot be accepted as a proved case of diminishing allometry.

There does not appear to be any general tendency for organs with an initially high growth ratio to show a progressive decline in the value of α during development. In fact, many organs previously growing faster than the body show a marked increase in growth ratio at the onset of sexual maturity, and most examples of diminishing allometry as yet established occur in special cases which cannot be taken as typical, such as growth of bone, or allomorphosis in holometabolous insects. Cases also occur where a large organ with a low growth ratio or approximate isometry during most of development must have had an early period with a high growth ratio. Examples are the growth of the limbs in sheep (Huxley, 1932) and cattle (Pontecorvo, 1939), and doubtless in many mammals where a relatively long limb at birth is necessary. A striking instance is the third leg of *Notonecta undulata* (Clark and Hersh, 1939), which is specialized for swimming and much enlarged, yet has a lower growth ratio than the other legs from the first instar onwards, and in fact grows more slowly than the body. But allometry of this type clearly has adaptive significance, and cannot be explained as a progressive failure to maintain an early high growth ratio.

Differential growth in the house-wren (Huggins, 1940) appears to show the type of deviation deduced by Lumer as a consequence of sigmoid growth. All dimensions measured follow simple allometry against total length from hatching to the eighth day, when the bird has reached about 80 per cent. of adult size. After this point the straight-line relationship breaks down for many dimensions, since there are both sharp changes and considerable individual variations in the growth ratios. The time growth of each dimension looks sigmoid, and Huggins suggests that the deviations from simple allometry as final size is approached may be due to the asymptotic nature of this time growth. The difficulties in Lumer's theory have already been discussed, and a possible alternative explanation of the breakdown in the allometry relationships is that the various organs reach the end of their growth periods at different times.

Lerner (1938) considered Lumer's theory inadequate to explain variations in differential growth-rate of leg length against body weight in the domestic fowl. Measurements made at 4, 8, 12, 16, and 20 weeks on each of a number of live birds showed a significant change in the growth ratio over this period, the average growth ratio being higher at 8–12 weeks than previously and afterwards falling again. But the four determinations of average growth ratio over the successive age intervals were not sufficient to show whether the allometry trend followed a smooth curve or two straight lines. Remarkably enough, when a double logarithmic plot was made of all points, regardless of age, they could be fitted quite well by a single straight line.

Critical points appear as either sudden changes of slope or actual gaps in the allometry line, and the most firmly established cases occur in Crustacea, in the growth of secondary sexual characters, where normally changes occur in α, b, or both. Their interpretation is complicated by the discontinuous nature of crustacean growth, since a change in growth ratio at a particular moult will appear either as a gap or simply as a bend in the allometry line, according to whether the specimens measured before and after the critical moult are separated or not. Thus Teissier (1933), following up the pioneer work of Geoffrey Smith on *Inachus* (see Huxley 1932), was able to separate male *Macropodia rostrata* into 'weak' and 'strong' forms by relative chela size. Large specimens were all of strong type and small ones of weak type, and in the intermediate sizes which contained both forms they could be separated by inspection with few exceptions. In a sample collected at a single time of year each form showed simple allometry of chela length against carapace length, with growth ratio 1·25; and a gap between the two parallel allometry lines indicated that strong males had chelae always about 35 per cent. longer than weak males with the same carapace length. But had it been impossible to separate the two forms, the allometry trend would have had the appearance of a straight line with two bends but no actual gaps, or possibly a sigmoid curve.

Teissier states that the two forms are separated by a single moult, which may occur at any size from 7 to 18 mm. carapace length, and he describes this as the 'puberty' or 'imaginal' moult. Analysis of a larger series, collected at three different times of year, showed a slight bend in the allometry line of each form, which occurred at about 12 mm. carapace length in weak and 14 mm. in strong males; but Teissier's frequency distributions for the different samples suggest the possibility that these further critical points may be due to seasonal variation in proportions. Female *Macropodia* could likewise be classified into 'adolescents' and 'adults' by relative abdomen width, and each group showed simple allometry with growth ratios 1·25 for adolescents and 1·0 for adults. Again, a wide gap between the two allometry lines indicates that relative abdomen-width is nearly doubled at the critical moult separating the two forms. In both sexes we can thus speak of adolescent and adult forms separated by a single moult, the so-called puberty moult, which introduces sexual maturity. Post-puberty moults can apparently follow in this species.

Much the same situation had previously been found by Shaw (1928) in *Inachus dorsettensis* (see Huxley, 1932, p. 68), but in *I. mauritanicus* (Smith, further analysed by Huxley, 1932, p. 53) the male chelae actually appear to decline in size and to change towards the female type in the interval between their first and second breeding-seasons.

Further complications are found in *Maia squinado* (Teissier, 1935), where it is possible to distinguish three stages of simple allometry in males, and possibly in females. Sharp increases in the growth ratios of the pereiopods, without a gap, separate the first two stages; while a further increase, accompanied by a definite gap in the allometry line, marks off the last from the second stage. The last group, which may again be described as adults, can be separated from the largest immature specimens by inspection, and one wonders whether the first two stages might not also be separated by a gap in the allometry line, if it were possible to separate all specimens measured before and after the critical moult ending the first stage. Teissier concludes that there are two critical moults in the growth of *Maia squinado*, the *pre-puberty* and *puberty* moults, between which probably two ordinary moults intervene. The puberty moult is in this case the final moult, so that variations in proportions during the last stage are due to allomorphosis. He finds evidence of two similar moults in females, but the changes in growth ratio shown by his graphs at the first critical point are very slight and not convincing. Allometry of abdomen width is very similar to that found in *Macropodia rostrata*.

Drach (1933) found three stages of simple allometry in growth of abdomen width in female *Portunus puber*, each stage showing the same growth ratio, and being distinguished from the previous one by a gap indicating a sudden increase in relative abdomen width. Specimens in the second stage could be separated from those in the first by minor anatomical changes making fertilization possible, and again indicating the occurrence of a definitive puberty moult. The discontinuity between the second and last stages is slight, and is stated to be due to the single moult following the puberty moult, so that variations in proportions in each of the last two stages are probably the result of allomorphosis. This interpretation is in marked contrast to the findings of Day (1935) on abdomen growth in female *Carcinus maenas*, but it seems likely that the five joined lines fitted by Day could have been reduced, and his analysis brought more into line with that of Drach, if a similar segregation of sexually mature and immature specimens had been attempted. This contrast in interpretations emphasizes the difficulties to be experienced in attempting to deduce the ontogenic curve of differential growth from mass statistics.

Apart from Crustacea, critical points have occasionally been described in the allometric growth of various vertebrates, but their validity as against smooth curves is sometimes in doubt. A. E. Needham's analysis of jaw growth in bony fishes has already been discussed, and rather similar breaks have been described in the differential growth of various dimensions in freshwater gar (*Lepidosteus*) by Huggins and Thompson (1942), but smooth curves would appear to give an equally realistic description of the allometry trends. Smooth curves would certainly seem to fit the graphs for lengths of tail, hind foot, ulna, radius, and scapula against body length in the mouse better than the pairs of straight lines drawn by Green and Fekete (1933), and it would be dangerous to attach any significance to the critical points indicated by their graphs.

Robb (1929) fitted two stages of simple allometry to weights of hypophysis and thyroid against body weight in two races of rabbits, but showed his lack of confidence in this interpretation by proposing the formula $y = bx^{\alpha}+c$ as an alternative. Teissier (1937) has also found breaks in the differential growth against total weight of various organs in the rat, using Donaldson's data, but some of his graphs strongly suggest a smooth curve. While there is no doubt that changes in growth equilibrium occur under hormonic control in vertebrates, further study is necessary to discover whether such changes can in general be realistically described graphically by critical points between straight allometry lines. A succession of allometry lines may nevertheless be valuable in giving an empirical description of a complex differential growth curve.

The third type of deviation to be considered is a rhythmical variation about an average trend of simple allometry, of which several cases have been discussed by Huxley (1932, p. 203), where he suggested the phenomenon might be general. The only case which appears to have received detailed study is growth of thoracic segments and abdomen in *Asellus aquaticus* (A. E. Needham 1937), in which the growth ratios of the widths of the eight thoracic segments and of abdomen width and length against total body length go through several cycles of declining amplitude before approximating to simple allometry. There is no doubt of the reality of these fluctuations, which are marked in all segments and all have roughly the same periodicity; and their presence suggests that the concentration of growth fluctuates in some way between the dimensions of length and breadth. It does not seem possible at present to suggest any explanation for this remarkable phenomenon, and few similar cases appear to have been described in other animals, though human growth appears definitely to show a succession of phases of predominantly length and breadth growth.

It may be noted that many allometry plots show variations

about the mean allometry line which could be interpreted as rhythmical fluctuations, since frequently several points above the line are followed by several below, and so on; but it is usually impossible to decide whether these deviations are random or systematic, and it would be unwise to assume that they have any biological significance.

In conclusion it may be said that, on the evidence at present available, simple allometry does not stand out as merely a special case of a more complex differential growth law, since marked deviations from simple allometry themselves tend to be special cases, attributable to special factors. The data available are still quite inadequate to settle this question, and more detailed study may yet show a fairly general occurrence of rhythmical or progressive deviations from simple allometry. Clearly it is of prime importance in future studies to find means of reducing the individual variations in proportions, and to examine the growth trend of the individual.

Growth Gradients

The pervasiveness of the growth-gradient pattern is very clearly demonstrated by the study of allometry, and in studying these general patterns of intensity of growth we return towards the more comprehensive type of form analysis envisaged by D'Arcy Thompson. In fact, a recognition of the gradient pattern is implicit in his Cartesian transformations (Huxley, 1932, chap. 4). Since the subject has been discussed in detail by Huxley, we shall confine ourselves here to more recent studies. The presence of a growth gradient can be demonstrated whether or not there is simple allometry, whenever we find a gradient in the growth ratios of a series of sections of the region considered, e.g. the segments of a crustacean limb or abdomen. These gradients in growth intensity should not be confused with gradients in absolute size of homologous organs at any particular moment.

The primary gradient along the body seems to be connected with the law of antero-posterior development. In Crustacea it is manifest in the growth ratios of successive bilateral appendages, and usually declines antero-posteriorly; but it may be complicated by the presence of a secondary sexual character with a high growth ratio, which appears to have the unexpected effect of slightly accelerating the growth of those appendages behind, and usually of depressing that of those in front (Huxley, 1932, chap. 4). An interesting modification has been found by Teissier (1935) in *Maia squinado*, in which, as we have already seen, differential growth of the pereiopods falls into three successive stages of simple allometry, of which the last is probably confined to a single moult. In males the merus lengths of the legs grow uniformly faster than carapace length, and also have a higher growth ratio than the chela merus except in the last stage of allometry. In this stage alone is a gradient found in the pereiopods. It thus seems to be a secondary phenomenon, associated with the late very high growth ratio of the chela. Growth ratios of merus length against carapace length for the three stages in males are shown in the accompanying table.

Growth ratios for merus : carapace length in male Maia

	First Stage	*Second Stage*	*Third Stage*
Chela	1·04	1·28	1·63
1st Walking leg . .	1·09	1·35	1·33
2nd ,, . .	1·09	1·35	1·30
3rd ,, . .	1·09	1·35	1·21
4th ,, . .	1·09	1·35	1·19

In females the chela merus always has a lower growth ratio than those of the walking-legs, and the latter show no growth gradient in any stage measured. It is interesting, therefore, to find in both sexes a definite antero-posterior gradient in absolute size of merus of the walking-legs throughout the size range covered. This would seem to have been determined by the law of antero-posterior development. No other segments of the walking-legs were measured.

Dawes (1934) found evidence of a slight antero-posterior gradient in the appendages of *Alpheus dentipes*. His values for the growth ratios against carapace length of lengths of third maxillipede and the five pereiopods are plotted in Text-fig. 1. The first pereiopod has been taken on the nipper side. Although the nipper has a higher growth ratio than the crusher, there is no bilateral asymmetry in the other appendages. Females show a regular antero-posterior gradient in the first four appendages, behind which the growth ratio appears to rise. In males the high growth ratio of the chela is associated with increased growth ratios both in front and behind, but the gradient in the walking-legs is rather flat, and of doubtful significance. This is in contrast with the depressive effect of a high growth ratio on appendages in front of it previously found. Here again there is a fairly marked antero-posterior gradient in absolute length

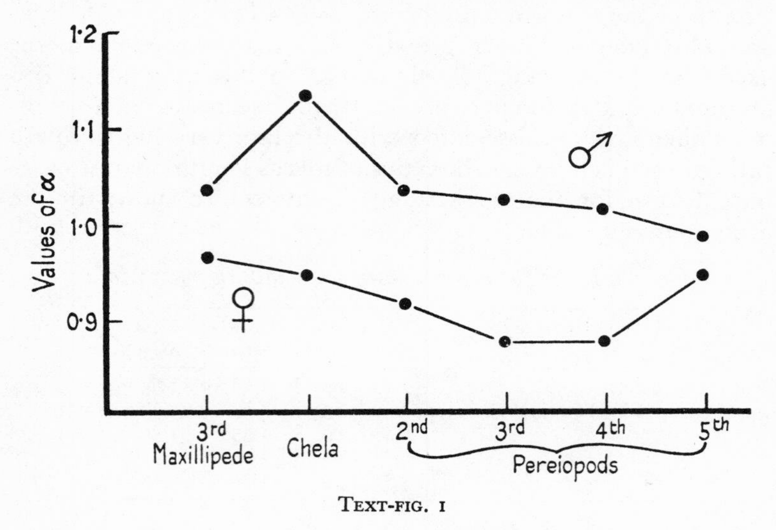

TEXT-FIG. 1

of walking-legs, which is correlated with the fact that the first legs (2nd pereiopods) are long and extremely slender chelate limbs, in contrast to the more posterior legs, which are stouter but shorter and hooked.

An unexpected discovery has been made in *Notonecta undulata*, where Clark was able to follow seventy-two individuals through development from egg to adult (Clark and Hersh, 1939). They report that, although the average measurements show a slight antero-posterior growth gradient in leg length in both sexes, when the growth ratios of individuals are examined separately, two types of gradient are found. About half the individuals of each sex showed the normal antero-posterior gradient, while half showed a gradient with high point in the middle leg. In addition, there were two aberrant females which did not fit into either group. The average growth coefficients of the three walking-legs in the two sexes are shown below, where the figures of Clark and Hersh have been rounded off to two decimal places. In each case the females show slightly higher growth ratios than the males.

		Growth ratios against body length of		
No. of individuals	*Growth centre*	*1st leg*	*2nd leg*	*3rd leg*
17 males . .	1st leg	1·05	1·00	0·92
20 females . .	1st leg	1·09	1·05	0·96
18 males . .	2nd leg	0·99	1·05	0·90
15 females . .	2nd leg	1·05	1·09	0·95

This is the first case in which it has been possible to compare the gradients of different individuals, and the variation in the order of growth intensity raises the general question whether other gradients of this slight degree of slope are common to all individuals. Slight growth gradients along each limb showed a central high point in the first leg and proximal high points in the other two, but it was not stated whether there was any individual variation in these gradients. It will be remembered that distal growth centres have usually only been found in limbs growing much faster than the body.

A rather similar example of individual variation in the shape of the gradient has been reported by Lerner (1941) in the hind limbs of poultry. Growth ratios of the long bones of the legs were compared in Leghorns, Bantams, and their reciprocal crosses, and while in each sex the average for all breeds showed a distal-proximal gradient, the same gradient was not found in all groups.

The most detailed study of growth of the abdomen in Decapoda yet undertaken is that of Day (1935) on *Carcinus maenas*, in which he estimated the growth ratios against carapace size of the breadths of the seven abdominal segments at different stages of growth. In the unsexable youngest specimens there is a gradient with growth centre in the third segment ($\alpha = 1{\cdot}34$) and a rather steep decline posteriorly to a value of 0·91 for the telson. During growth rather complex changes in the growth ratios take place, whose general result is a progressive decline in the growth ratios of the proximal segments and the appearance of a new growth centre in the sixth segment. This new centre is strongly marked in females, hardly noticeable in males.

Day points out that the change in position of the highest growth ratio might be interpreted as either a progressive migration distally of the original growth centre, or as due to the appearance during development of a secondary gradient, centred in the sixth segment, which swamps the primary growth centre in the third segment. The latter hypothesis seems to give the most plausible explanation of the fact that sex differences in segment width first appear in the region of the sixth segment, from which they seem to spread out over the abdomen, and become visible in the first three segments only at the time of sexual maturity. In males all segments except the telson show a progressive decline in growth ratio, and it would seem that the secondary growth centre makes its appearance in the sixth segment simply because the rate of decline of the growth ratio is highest in the first three segments and falls off progressively in a distal direction.

An interesting light is thrown on the factors underlying differential growth by the rare cases of abnormalities of proportions discovered in nature, of which two remarkable examples have been described. Williams and A. E. Needham (1937) found a young female *Carcinus maenas* which was precociously mature and carried developing berry. The carapace of this specimen measured only 12·1 by 15·2 mm., whereas minimum normal dimensions for females in berry are stated to be 22 by 28 mm. Marked precocious development was also shown by the abdomen, which had the relative proportions to be expected in a female twice the size; and the table below reproduces the comparison made by the authors between the precocious female and a normal female with the same carapace dimensions.

	Carapace (mm.)		*Width of abdominal segments (mm.)*						
	Length	*Breadth*	*1*	*2*	*3*	*4*	*5*	*6*	*Telson*
Precocious female .	12·1	15·2	4·71	5·22	5·76	5·36	4·86	3·95	1·91
Normal female with same carapace dimensions .	—	—	4·66	4·96	5·16	4·66	4·00	3·13	1·74
Per cent. excess over normal .	—	—	1	5	12	15	22	26	10

There is clear evidence of a gradient in percentage excess growth, with its highest point in segment 6, and the aberrant proportions of the abdomen are most simply explained as due to a precocious onset of the secondary growth-centre in the sixth segment.

The other abnormality referred to, which has been described by Young (1933), is a female *Cancer pagurus*, with its abdomen both bilaterally asymmetrical and abnormally long and wide. The pleopods are shorter on the right than on the left side, and on the right side the sixth abdominal segment (which possesses uropods in most decapods but is usually limbless in crabs) bears a supernumerary limb. This has the general plan of a uropod but resembles a walking-limb in the shape of its joints, the thickness of its cuticle, the hairiness and the possession of a hard black claw. A similar abnormality in a female *Carcinus maenas* was previously recorded in 1896 by Bethe, who could only conclude that it had been caused by some cells containing determinants appropriate for a walking-leg becoming lodged in the tail during early development. Young was able to connect up this phenomenon with the general theory of gradients by suggesting that the secondary gradient in the abdomen, centred in the sixth segment and running in the reverse direction to the primary antero-posterior gradient, had risen exceptionally steeply, with the result that the absolute rates of whatever processes are concerned approached the rates usually found in the thorax, and caused the supernumerary limb to develop the characteristics of a walking-leg. The steepness of the secondary gradient was shown by the excessive development of the posterior segments. This interesting speculation deserves to be tested out on other examples of serial heteromorphosis.

A few recent studies have been made on growth gradients in other groups of animals. Huggins and Huggins (1942) found no regular gradient along the body in the black-fish, *Centropristus striatus*, which they divided into nine body-regions for the purpose of the study. Further, the scales of each region showed little sign of a growth gradient and no correlation in growth ratio with the regions underlying them, and in fact the only gradient discovered was a postero-anterior one with α-values 0·94, 0·97, and 1·10 for the pectoral, pelvic, and caudal fins. The growth ratios of all regions examined in this species differed little from isometry, however.

Huggins and Thompson (1942) recorded more marked allometric growth in freshwater gar (*Lepidosteus* sp.), in which they found two or even three stages of simple allometry necessary to describe differential growth of the dimensions studied. Body length was divided into six sections by such points as the anterior border of the eye and the insertions of the various fins, and again no definite gradient was apparent. In both *L. platostomus* and *L. osseus* the regions with the highest growth ratios are at first from tip of snout to eye and the length of the longest caudal ray, but later the central regions of the body take on the highest growth-rate. It seems probable that factors of regional importance override any primary gradient which may exist.

The house-wren (Huggins 1940) presents a particularly interesting case of differential growth in birds. In the period of simple allometry from hatching to the eighth day, there is a postero-anterior gradient along the body axis, as shown by the following values of α: head 0·70, neck 0·96, body 1·02, tail 1·06. The bill, with growth ratio 1·32, is an exception to the general gradient; this is presumably of adaptive significance, since a long bill before hatching would be an obvious disadvantage. The postero-anterior gradient found after hatching is the reverse of that shown in the embryo, but it is possible that the latter depends on the time relations of growth (law of antero-posterior development), and not on the differential growth constants for the various regions. The wing shows a high point in the fore-arm, with lower values in both upper arm and manus, but the leg has a regular disto-proximal gradient, as indicated by the following α-values: thigh 1·29, tibio-tarsus 1·41, tarso-metatarsus 1·58, middle toe 1·57, middle nail 2·22. Values for hind toe and hind nail are rather lower (1·47 and 1·75), but show the same trend.

In man, Burt (1943) reports that he has found evidence of individual variation in differential growth-pattern, by applying the technique of multiple factor analysis to a series of successive measurements of the physical growth of children. The analysis gave two factors, the first a general size factor accounting for more than 60 per cent. of the total variance, and the second a bipolar type-factor, distinguishing persons with relatively excessive growth in 'vertical' directions (height, length of leg, arm, or trunk) from those with relatively excessive 'transverse' growth (iliac width, chest width, chest depth, weight). Burt points out an analogy between these two factors, influencing size and proportions respectively, and the distinction between isometric and allometric growth.

Summary and Conclusions

The present review has been concerned particularly with theoretical problems arising in the quantitative study of differential growth by the allometry technique, and our main conclusions are summarized below.

The simple allometry formula, $y = bx^{\alpha}$, has been widely used in the study of differential growth, since it often describes closely the change of proportions with size increase, and has the advantages that it is easy to apply, short-circuits the time factor (which is an erratic variable at least in poikilothermic forms), and takes into account the multiplicative nature of growth in general.

The theoretical implications of this formula are important in the interpretation of allometry data, and various attempts have been made to find a basis for it in the laws of time growth of organs and organisms, or in simple postulates about growth. These attempts are examined and found to be unconvincing, since no formula of time growth has yet been given a secure physiological basis, and most of those in use cannot from their nature apply to both the whole organism and its parts, while the various sets of postulates from which the allometry formula has been derived are themselves questionable.

A further theoretical difficulty arises from the fact that a complete limb and its individual segments cannot both show exact simple allometry against the same body-standard, so that the changes of proportions in an organism during development cannot be completely and exactly described in terms of constant differential growth ratios. Also, some dimensions to which the formula has been successfully applied, e.g. bone lengths and skull dimensions, certainly do not grow by intussusception or self-multiplication of tissue. Attempts to meet these difficulties by introducing a correction factor or deriving a significant and more general formula of differential growth have as yet met with no success. Statistical difficulties are likely to make it impossible to apply any complex formula in practice.

The uncertain theoretical basis of the allometry formula

suggests the need for a more careful study of the behaviour of differential growth ratios during development. Review of published data suggests that, while excellent conformity with the formula is often found, many of the data are not adequate to show whether the growth ratio remains substantially constant or undergoes moderate systematic changes during development; since too few specimens have been measured, and no attempt has been made to reduce the considerable variation in proportions found among specimens of the same size.

Deviations from simple allometry have been recorded in a number of cases. These usually take the form of either a progressive change in growth ratio, a sudden change to a new equilibrium at a particular growth stage, or a rhythmical fluctuation about an average trend. A few cases in vertebrates and holometabolous insects show clear evidence of a progressive change in growth ratio during development, and in others, while a deviation from simple allometry certainly occurs, the data are not sufficient to indicate whether the allometry trend is curvilinear or follows two consecutive straight lines. Crustacea show numerous examples of a change in growth equilibrium of secondary sexual characters at the moult introducing sexual maturity, but some more complex cases are discussed, in which there is evidence of two critical moults separating three periods of simple allometry. While similar changes of growth equilibrium must also occur under hormonic control in vertebrates, quantitative data are scarce, and it is not certain that such changes are in general best represented graphically as a break between two periods of simple allometry. Few definite cases of rhythmical fluctuations have been described, though it is possible that they are of widespread occurrence.

In general, while the data at present available are quite inadequate to show whether deviations from simple allometry are the rule, it is possible to regard those clear cases of deviations which have been recorded as special cases, attributable to special factors; and it cannot therefore be said that simple allometry stands out as merely a special case of a more complex differential growth law.

The technique of allometry makes possible a quantitative study of growth gradients, by comparison of the average or successive growth ratios of a series of organs, and some recent research along these lines is discussed. These gradients appear to be of general, though not perhaps of universal, occurrence. No general conclusion as to their physiological basis can as yet be drawn; but certain empirical rules have emerged—e.g. that high growth-rate of a region or organ is usually associated with a steep growth gradient having a distal or sub-distal high point, and that an appendage with an exceptionally high growth ratio appears to influence the growth of adjacent appendages.

REFERENCES

AMADON, D. (1943) *Auk*, **60**, 221.
BERNSTEIN, F. (1934) *Cold Spring Harbor Symp. Quant. Biol.* **2**, 209.
BRASH, J. C. (1934) *Edinburgh med. J.*, New Ser. **16**.
BRODY, S. (1937) *Growth*, **1**, 60.
BURT, C. (1943) *Nature*, **152**, 75.
CHAMPY, C. (1924) *Sexualité et Hormones*. Paris.
CLARK, L. B., and HERSH, A. H. (1939) *Growth*, **3**, 347.
DARBY, H. (1934) *Pub. Carnegie Inst.*, Washington, **435**.
DAVENPORT, C. B. (1934) *Cold Spring Harbor Symp. Quant. Biol.* **2**, 203.
DAWES, B. (1934) *Arch. EntwMech. Org.* **131**, 543.
—— and HUXLEY, J. S. (1934) *Nature*, **133**, 982.
DAY, J. H. (1935) *Dove Marine Lab. Reports*, 3rd Series, No. 3, 49.
DRACH, P. (1933) *C. R. Acad. Sci. Paris*, **197**, 93.
DUBOIS, E. (1898) *Arch. f. Anthropol.* **25**, 1, 423.
—— (1914) *Ztschr. Morph. u. Anthropol.* **18**.
GLASER, O. (1938) *Biol. Rev.* **13**, 20.
GRAY, J. (1929) *J. exp. Biol.* **6**, 248.
GREEN, C. V., and FEKETE, E. (1933) *J. exp. Zool.* **66**, 351.
HERSH, A. H. (1934) *Amer. Nat.* **68**, 537.
—— (1941) *Growth*, **5**, 113.
HUGGINS, R. A., and HUGGINS, S. E. (1942) Ibid. **6**, 135.
HUGGINS, S. E. (1940) Ibid. **4**, 225.
—— and THOMPSON, D. H. (1942) Ibid. **6**, 163.
HUXLEY, J. S. (1924) *Nature*, **114**, 895.
—— (1932) *Problems of Relative Growth*. London.
—— NEEDHAM, J., and LERNER, I. M. (1941) *Nature*, **148**, 225.
—— and TEISSIER, G. (1936) Ibid. **137**, 780.
KAVANAGH, A. J., and RICHARDS, O. W. (1934) *Amer. Nat.* **68**, 54.
—— —— (1942) *Proc. Rochester. Acad. Sci.* **8**, 150.
KLATT, B. (1919) *Biol. Zbl.* **39**, 406.
KOSTITZIN, V. A. (1939) *Mathematical Biology*, trans. T. H. Savory. London.
LAPICQUE, L. (1898) *C. R. Soc. Biol. Paris*, **50**, 62.
LERNER, I. M. (1938) *Growth*, **2**, 135.
—— (1941) Ibid. **5**, 1.

LERNER, I. M., and GUNNS, C. A. (1938) *Growth*, **2**, 261.
LUMER, H. (1937) *Growth*, **1**, 140.
—— (1939) *Amer. Nat.* **73**, 339.
—— (1940) Ibid. **74**, 439.
MACKAY, D. C. G. (1942) *Growth*, **6**, 251.
MEDAWAR, P. B. (1940) *Proc. roy. Soc., B*, **129**, 332.
—— (1941) *Nature*, **148**, 772.
MOMENT, G. B. (1933) *J. exp. Zool.* **65**, 659.
NEEDHAM, A. E. (1935) *Proc. zool. Soc. Lond.* 773.
—— (1937) Ibid. 289.
NEEDHAM, J. (1933) *Biol. Rev.* **8**, 180.
—— (1934) Ibid. **9**, 79.
—— and LERNER, I. M. (1940) *Nature*, **146**, 618.
PEZARD, A. (1918) *Bull. Biol. France. et Belg.* **52**, 1.
POMEROY, R. W. (1941) *J. agric. Soc.* **31**, 50.
PONTECORVO, G. (1939) Ibid. **29**, 111.
PRZIBRAM, H. (1930) *Connecting Laws in Animal Morphology.* London.
REEVE, E. C. R. (1940) *Proc. zool. Soc. Lond., A*, **110**, 47.
—— (1941) *Proc. zool. Soc. Lond., A*, **111**, 279.
—— (1945) *Nature* (in press).
—— and MURRAY, P. D. F. (1942) *Nature*, **150**, 402.
ROBB, R. C. (1929) *J. exp. Biol.* **6**, 311.
TEISSIER, G. (1928) *C. R. Soc. Biol. Paris*, **98**, 842.
—— (1933) *Bull. Biol. Fr. et Belg.* **67**, 401.
—— (1934) *Dysharmonies et Discontinuités dans la Croissance.* Paris.
—— (1935) *Trav. Stat. Biol. Roscoff*, **13**, 93.
—— (1937) *Les Lois Quantitatives de la Croissance.* Paris.
THOMPSON, D'ARCY W. (1917) *On Growth and Form.* Cambridge.
—— (1942) *On Growth and Form.* A new edition. Cambridge.
TODD, T. W., and WHARTON, R. E. (1934) *Amer. J. Anat.* **55**, 79.
TWITTY, V. C., and WAGTENDONK, W. J. (1940) *Growth*, **4**, 349.
WEYMOUTH, W., and MACKAY, D. C. G. (1936) *Proc. zool. Soc. Lond.* 17.
WILLIAMS, G., and NEEDHAM, A. E. (1937) Ibid. 161.
YOUNG, J. Z. (1933) *Nature*, **132**, 785.
ZUCKER, L., HALL, L., YOUNG, M., and ZUCKER, T. F. (1941) *Growth*, **5**, 415.
—— and ZUCKER, T. F. (1942) *J. gen. Physiol.* **25**, 445.

INDEX OF AUTHORS

SUBJECT INDEX

(*Figures in italics refer to pages with illustrations of the subject cited.*)

INDEX OF ORGANISMS

(*Figures in italics refer to pages with illustrations of the organisms cited.*)

A CATALOGUE OF SELECTED DOVER BOOKS
IN ALL FIELDS OF INTEREST

A CATALOGUE OF SELECTED DOVER BOOKS IN ALL FIELDS OF INTEREST

AMERICA'S OLD MASTERS, James T. Flexner. Four men emerged unexpectedly from provincial 18th century America to leadership in European art: Benjamin West, J. S. Copley, C. R. Peale, Gilbert Stuart. Brilliant coverage of lives and contributions. Revised, 1967 edition. 69 plates. 365pp. of text.
21806-6 Paperbound $3.00

FIRST FLOWERS OF OUR WILDERNESS: AMERICAN PAINTING, THE COLONIAL PERIOD, James T. Flexner. Painters, and regional painting traditions from earliest Colonial times up to the emergence of Copley, West and Peale Sr., Foster, Gustavus Hesselius, Feke, John Smibert and many anonymous painters in the primitive manner. Engaging presentation, with 162 illustrations. xxii + 368pp.
22180-6 Paperbound $3.50

THE LIGHT OF DISTANT SKIES: AMERICAN PAINTING, 1760-1835, James T. Flexner. The great generation of early American painters goes to Europe to learn and to teach: West, Copley, Gilbert Stuart and others. Allston, Trumbull, Morse; also contemporary American painters—primitives, derivatives, academics—who remained in America. 102 illustrations. xiii + 306pp. 22179-2 Paperbound $3.00

A HISTORY OF THE RISE AND PROGRESS OF THE ARTS OF DESIGN IN THE UNITED STATES, William Dunlap. Much the richest mine of information on early American painters, sculptors, architects, engravers, miniaturists, etc. The only source of information for scores of artists, the major primary source for many others. Unabridged reprint of rare original 1834 edition, with new introduction by James T. Flexner, and 394 new illustrations. Edited by Rita Weiss. 6⅝ x 9⅝.
21695-0, 21696-9, 21697-7 Three volumes, Paperbound $13.50

EPOCHS OF CHINESE AND JAPANESE ART, Ernest F. Fenollosa. From primitive Chinese art to the 20th century, thorough history, explanation of every important art period and form, including Japanese woodcuts; main stress on China and Japan, but Tibet, Korea also included. Still unexcelled for its detailed, rich coverage of cultural background, aesthetic elements, diffusion studies, particularly of the historical period. 2nd, 1913 edition. 242 illustrations. lii + 439pp. of text.
20364-6, 20365-4 Two volumes, Paperbound $6.00

THE GENTLE ART OF MAKING ENEMIES, James A. M. Whistler. Greatest wit of his day deflates Oscar Wilde, Ruskin, Swinburne; strikes back at inane critics, exhibitions, art journalism; aesthetics of impressionist revolution in most striking form. Highly readable classic by great painter. Reproduction of edition designed by Whistler. Introduction by Alfred Werner. xxxvi + 334pp.
21875-9 Paperbound $2.50

JIM WHITEWOLF: THE LIFE OF A KIOWA APACHE INDIAN, Charles S. Brant, editor. Spans transition between native life and acculturation period, 1880 on. Kiowa culture, personal life pattern, religion and the supernatural, the Ghost Dance, breakdown in the White Man's world, similar material. 1 map. xii + 144pp.
22015-X Paperbound $1.75

THE NATIVE TRIBES OF CENTRAL AUSTRALIA, Baldwin Spencer and F. J. Gillen. Basic book in anthropology, devoted to full coverage of the Arunta and Warramunga tribes; the source for knowledge about kinship systems, material and social culture, religion, etc. Still unsurpassed. 121 photographs, 89 drawings. xviii + 669pp. 21775-2 Paperbound $5.00

MALAY MAGIC, Walter W. Skeat. Classic (1900); still the definitive work on the folklore and popular religion of the Malay peninsula. Describes marriage rites, birth spirits and ceremonies, medicine, dances, games, war and weapons, etc. Extensive quotes from original sources, many magic charms translated into English. 35 illustrations. Preface by Charles Otto Blagden. xxiv + 685pp.
21760-4 Paperbound $4.00

HEAVENS ON EARTH: UTOPIAN COMMUNITIES IN AMERICA, 1680-1880, Mark Holloway. The finest nontechnical account of American utopias, from the early Woman in the Wilderness, Ephrata, Rappites to the enormous mid 19th-century efflorescence; Shakers, New Harmony, Equity Stores, Fourier's Phalanxes, Oneida, Amana, Fruitlands, etc. "Entertaining and very instructive." *Times Literary Supplement.* 15 illustrations. 246pp. 21593-8 Paperbound $2.00

LONDON LABOUR AND THE LONDON POOR, Henry Mayhew. Earliest (c. 1850) sociological study in English, describing myriad subcultures of London poor. Particularly remarkable for the thousands of pages of direct testimony taken from the lips of London prostitutes, thieves, beggars, street sellers, chimney-sweepers, street-musicians, "mudlarks," "pure-finders," rag-gatherers, "running-patterers," dock laborers, cab-men, and hundreds of others, quoted directly in this massive work. An extraordinarily vital picture of London emerges. 110 illustrations. Total of lxxvi + 1951pp. 6⅝ x 10.
21934-8, 21935-6, 21936-4, 21937-2 Four volumes, Paperbound $14.00

HISTORY OF THE LATER ROMAN EMPIRE, J. B. Bury. Eloquent, detailed reconstruction of Western and Byzantine Roman Empire by a major historian, from the death of Theodosius I (395 A.D.) to the death of Justinian (565). Extensive quotations from contemporary sources; full coverage of important Roman and foreign figures of the time. xxxiv + 965pp. 21829-5 Record, book, album. Monaural. $3.50

AN INTELLECTUAL AND CULTURAL HISTORY OF THE WESTERN WORLD, Harry Elmer Barnes. Monumental study, tracing the development of the accomplishments that make up human culture. Every aspect of man's achievement surveyed from its origins in the Paleolithic to the present day (1964); social structures, ideas, economic systems, art, literature, technology, mathematics, the sciences, medicine, religion, jurisprudence, etc. Evaluations of the contributions of scores of great men. 1964 edition, revised and edited by scholars in the many fields represented. Total of xxix + 1381pp. 21275-0, 21276-9, 21277-7 Three volumes, Paperbound $7.75

PLANETS, STARS AND GALAXIES: DESCRIPTIVE ASTRONOMY FOR BEGINNERS, A. E. Fanning. Comprehensive introductory survey of astronomy: the sun, solar system, stars, galaxies, universe, cosmology; up-to-date, including quasars, radio stars, etc. Preface by Prof. Donald Menzel. 24pp. of photographs. 189pp. 5¼ x 8¼.
21680-2 Paperbound $1.50

TEACH YOURSELF CALCULUS, P. Abbott. With a good background in algebra and trig, you can teach yourself calculus with this book. Simple, straightforward introduction to functions of all kinds, integration, differentiation, series, etc. "Students who are beginning to study calculus method will derive great help from this book." Faraday House Journal. 308pp. 20683-1 Clothbound $2.00

TEACH YOURSELF TRIGONOMETRY, P. Abbott. Geometrical foundations, indices and logarithms, ratios, angles, circular measure, etc. are presented in this sound, easy-to-use text. Excellent for the beginner or as a brush up, this text carries the student through the solution of triangles. 204pp. 20682-3 Clothbound $2.00

TEACH YOURSELF ANATOMY, David LeVay. Accurate, inclusive, profusely illustrated account of structure, skeleton, abdomen, muscles, nervous system, glands, brain, reproductive organs, evolution. "Quite the best and most readable account,' *Medical Officer.* 12 color plates. 164 figures. 311pp. 4¾ x 7.
21651-9 Clothbound $2.50

TEACH YOURSELF PHYSIOLOGY, David LeVay. Anatomical, biochemical bases; digestive, nervous, endocrine systems; metabolism; respiration; muscle; excretion; temperature control; reproduction. "Good elementary exposition," *The Lancet.* 6 color plates. 44 illustrations. 208pp. 4¼ x 7. 21658-6 Clothbound $2.50

THE FRIENDLY STARS, Martha Evans Martin. Classic has taught naked-eye observation of stars, planets to hundreds of thousands, still not surpassed for charm, lucidity, adequacy. Completely updated by Professor Donald H. Menzel, Harvard Observatory. 25 illustrations. 16 x 30 chart. x + 147pp. 21099-5 Paperbound $1.25

MUSIC OF THE SPHERES: THE MATERIAL UNIVERSE FROM ATOM TO QUASAR, SIMPLY EXPLAINED, Guy Murchie. Extremely broad, brilliantly written popular account begins with the solar system and reaches to dividing line between matter and nonmatter; latest understandings presented with exceptional clarity. Volume One: Planets, stars, galaxies, cosmology, geology, celestial mechanics, latest astronomical discoveries; Volume Two: Matter, atoms, waves, radiation, relativity, chemical action, heat, nuclear energy, quantum theory, music, light, color, probability, antimatter, antigravity, and similar topics. 319 figures. 1967 (second) edition. Total of xx + 644pp. 21809-0, 21810-4 Two volumes, Paperbound $5.00

OLD-TIME SCHOOLS AND SCHOOL BOOKS, Clifton Johnson. Illustrations and rhymes from early primers, abundant quotations from early textbooks, many anecdotes of school life enliven this study of elementary schools from Puritans to middle 19th century. Introduction by Carl Withers. 234 illustrations. xxiii + 381pp.
21031-6 Paperbound $2.50

AMERICAN FOOD AND GAME FISHES, David S. Jordan and Barton W. Evermann. Definitive source of information, detailed and accurate enough to enable the sportsman and nature lover to identify conclusively some 1,000 species and sub-species of North American fish, sought for food or sport. Coverage of range, physiology, habits, life history, food value. Best methods of capture, interest to the angler, advice on bait, fly-fishing, etc. 338 drawings and photographs. l + 574pp. 6⅝ x 9⅜.
22383-1 Paperbound $4.50

THE FROG BOOK, Mary C. Dickerson. Complete with extensive finding keys, over 300 photographs, and an introduction to the general biology of frogs and toads, this is the classic non-technical study of Northeastern and Central species. 58 species; 290 photographs and 16 color plates. xvii + 253pp.
21973-9 Paperbound $4.00

THE MOTH BOOK: A GUIDE TO THE MOTHS OF NORTH AMERICA, William J. Holland. Classical study, eagerly sought after and used for the past 60 years. Clear identification manual to more than 2,000 different moths, largest manual in existence. General information about moths, capturing, mounting, classifying, etc., followed by species by species descriptions. 263 illustrations plus 48 color plates show almost every species, full size. 1968 edition, preface, nomenclature changes by A. E. Brower. xxiv + 479pp. of text. 6½ x 9¼.
21948-8 Paperbound $5.00

THE SEA-BEACH AT EBB-TIDE, Augusta Foote Arnold. Interested amateur can identify hundreds of marine plants and animals on coasts of North America; marine algae; seaweeds; squids; hermit crabs; horse shoe crabs; shrimps; corals; sea anemones; etc. Species descriptions cover: structure; food; reproductive cycle; size; shape; color; habitat; etc. Over 600 drawings. 85 plates. xii + 490pp.
21949-6 Paperbound $3.50

COMMON BIRD SONGS, Donald J. Borror. 33⅓ 12-inch record presents songs of 60 important birds of the eastern United States. A thorough, serious record which provides several examples for each bird, showing different types of song, individual variations, etc. Inestimable identification aid for birdwatcher. 32-page booklet gives text about birds and songs, with illustration for each bird.
21829-5 Record, book, album. Monaural. $2.75

FADS AND FALLACIES IN THE NAME OF SCIENCE, Martin Gardner. Fair, witty appraisal of cranks and quacks of science: Atlantis, Lemuria, hollow earth, flat earth, Velikovsky, orgone energy, Dianetics, flying saucers, Bridey Murphy, food fads, medical fads, perpetual motion, etc. Formerly "In the Name of Science." x + 363pp.
20394-8 Paperbound $2.00

HOAXES, Curtis D. MacDougall. Exhaustive, unbelievably rich account of great hoaxes: Locke's moon hoax, Shakespearean forgeries, sea serpents, Loch Ness monster, Cardiff giant, John Wilkes Booth's mummy, Disumbrationist school of art, dozens more; also journalism, psychology of hoaxing. 54 illustrations. xi + 338pp.
20465-0 Paperbound $2.75

THE PHILOSOPHY OF THE UPANISHADS, Paul Deussen. Clear, detailed statement of upanishadic system of thought, generally considered among best available. History of these works, full exposition of system emergent from them, parallel concepts in the West. Translated by A. S. Geden. xiv + 429pp.
21616-0 Paperbound $3.00

LANGUAGE, TRUTH AND LOGIC, Alfred J. Ayer. Famous, remarkably clear introduction to the Vienna and Cambridge schools of Logical Positivism; function of philosophy, elimination of metaphysical thought, nature of analysis, similar topics. "Wish I had written it myself," Bertrand Russell. 2nd, 1946 edition. 160pp.
20010-8 Paperbound $1.35

THE GUIDE FOR THE PERPLEXED, Moses Maimonides. Great classic of medieval Judaism, major attempt to reconcile revealed religion (Pentateuch, commentaries) and Aristotelian philosophy. Enormously important in all Western thought. Unabridged Friedländer translation. 50-page introduction. lix + 414pp.
(USO) 20351-4 Paperbound $2.50

OCCULT AND SUPERNATURAL PHENOMENA, D. H. Rawcliffe. Full, serious study of the most persistent delusions of mankind: crystal gazing, mediumistic trance, stigmata, lycanthropy, fire walking, dowsing, telepathy, ghosts, ESP, etc., and their relation to common forms of abnormal psychology. Formerly *Illusions and Delusions of the Supernatural and the Occult.* iii + 551pp. 20503-7 Paperbound $3.50

THE EGYPTIAN BOOK OF THE DEAD: THE PAPYRUS OF ANI, E. A. Wallis Budge. Full hieroglyphic text, interlinear transliteration of sounds, word for word translation, then smooth, connected translation; Theban recension. Basic work in Ancient Egyptian civilization; now even more significant than ever for historical importance, dilation of consciousness, etc. clvi + 377pp. 6½ x 9¼.
21866-X Paperbound $3.95

PSYCHOLOGY OF MUSIC, Carl E. Seashore. Basic, thorough survey of everything known about psychology of music up to 1940's; essential reading for psychologists, musicologists. Physical acoustics; auditory apparatus; relationship of physical sound to perceived sound; role of the mind in sorting, altering, suppressing, creating sound sensations; musical learning, testing for ability, absolute pitch, other topics. Records of Caruso, Menuhin analyzed. 88 figures. xix + 408pp.
21851-1 Paperbound $2.75

THE I CHING (THE BOOK OF CHANGES), translated by James Legge. Complete translated text plus appendices by Confucius, of perhaps the most penetrating divination book ever compiled. Indispensable to all study of early Oriental civilizations. 3 plates. xxiii + 448pp. 21062-6 Paperbound $3.00

THE UPANISHADS, translated by Max Müller. Twelve classical upanishads: Chandogya, Kena, Aitareya, Kaushitaki, Isa, Katha, Mundaka, Taittiriyaka, Brhadaranyaka, Svetasvatara, Prasna, Maitriyana. 160-page introduction, analysis by Prof. Müller. Total of 826pp. 20398-0, 20399-9 Two volumes, Paperbound $5.00

POEMS OF ANNE BRADSTREET, edited with an introduction by Robert Hutchinson. A new selection of poems by America's first poet and perhaps the first significant woman poet in the English language. 48 poems display her development in works of considerable variety—love poems, domestic poems, religious meditations, formal elegies, "quaternions," etc. Notes, bibliography. viii + 222pp.
22160-1 Paperbound $2.00

THREE GOTHIC NOVELS: THE CASTLE OF OTRANTO BY HORACE WALPOLE; VATHEK BY WILLIAM BECKFORD; THE VAMPYRE BY JOHN POLIDORI, WITH FRAGMENT OF A NOVEL BY LORD BYRON, edited by E. F. Bleiler. The first Gothic novel, by Walpole; the finest Oriental tale in English, by Beckford; powerful Romantic supernatural story in versions by Polidori and Byron. All extremely important in history of literature; all still exciting, packed with supernatural thrills, ghosts, haunted castles, magic, etc. xl + 291pp.
21232-7 Paperbound $2.00

THE BEST TALES OF HOFFMANN, E. T. A. Hoffmann. 10 of Hoffmann's most important stories, in modern re-editings of standard translations: Nutcracker and the King of Mice, Signor Formica, Automata, The Sandman, Rath Krespel, The Golden Flowerpot, Master Martin the Cooper, The Mines of Falun, The King's Betrothed, A New Year's Eve Adventure. 7 illustrations by Hoffmann. Edited by E. F. Bleiler. xxxix + 419pp. 21793-0 Paperbound $2.50

GHOST AND HORROR STORIES OF AMBROSE BIERCE, Ambrose Bierce. 23 strikingly modern stories of the horrors latent in the human mind: The Eyes of the Panther, The Damned Thing, An Occurrence at Owl Creek Bridge, An Inhabitant of Carcosa, etc., plus the dream-essay, Visions of the Night. Edited by E. F. Bleiler. xxii + 199pp. 20767-6 Paperbound $1.50

BEST GHOST STORIES OF J. S. LEFANU, J. Sheridan LeFanu. Finest stories by Victorian master often considered greatest supernatural writer of all. Carmilla, Green Tea, The Haunted Baronet, The Familiar, and 12 others. Most never before available in the U. S. A. Edited by E. F. Bleiler. 8 illustrations from Victorian publications. xvii + 467pp. 20415-4 Paperbound $3.00

THE TIME STREAM, THE GREATEST ADVENTURE, AND THE PURPLE SAPPHIRE—THREE SCIENCE FICTION NOVELS, John Taine (Eric Temple Bell). Great American mathematician was also foremost science fiction novelist of the 1920's. *The Time Stream,* one of all-time classics, uses concepts of circular time; *The Greatest Adventure,* incredibly ancient biological experiments from Antarctica threaten to escape; The *Purple Sapphire,* superscience, lost races in Central Tibet, survivors of the Great Race. 4 illustrations by Frank R. Paul. v + 532pp.
21180-0 Paperbound $3.00

SEVEN SCIENCE FICTION NOVELS, H. G. Wells. The standard collection of the great novels. Complete, unabridged. *First Men in the Moon, Island of Dr. Moreau, War of the Worlds, Food of the Gods, Invisible Man, Time Machine, In the Days of the Comet.* Not only science fiction fans, but every educated person owes it to himself to read these novels. 1015pp. 20264-X Clothbound $5.00